PROGRESS IN BRAIN RESEARCH

VOLUME 100

NEUROSCIENCE: FROM THE MOLECULAR TO THE COGNITIVE

Other volumes in PROGRESS IN BRAIN RESEARCH

Volume 78: Transplantation into the Mammalian CNS, by D.M. Gash and J.R. Sladek, Jr. (Eds.) – 1988, ISBN 0-444-81012-9.

Volume 79: Nicotinic Receptors in the CNS, by A. Nordberg, K. Fuxe, B. Holmstedt and A. Sundwall (Eds.) – 1989, ISBN 0-444-81088-9.

Volume 80: Afferent Control of Posture and Locomotion, by J.H.J. Allum and M. Hulliger (Eds.) – 1989, ISBN 0-444-81225-3

Volume 81: The Central Neural Organization of Cardiovascular Control, by J. Ciriello, M.M. Caverson and C. Polosa (Eds.) – 1989, ISBN 0-444-81224-5

Volume 82: Neural Transplantation: From Molecular Basis to Clinical Applications, by S. Dunnett and S.-J. Richards (Eds.) – 1990, ISBN 0-444-81137-0

Volume 83: Understanding the Brain through the Hippocampus, by J. Storm–Mathison, J. Zimmer and O.P. Ottersen (Eds.) – 1990, ISBN 0-444-81149-4

Volume 84: Cholinergic Neurotransmission: Functional and Clinical Aspects, by S.-M. Aquilonius and P.-G. Gillberg (Eds.) – 1990, ISBN 0-444-81148-6

Volume 85: The Prefrontal Cortex: Its Structure, Function and Pathology, by H.B.M. Uylings, C.G. van Eden, J.P.C. de Bruin, M.A. Corner and M.G.P. Feenstra (Eds.) – 1991, ISBN 0-444-81124-9

Volume 86: Molecular and Cellular Mechanisms of Neuronal Plasticity in Normal Aging and Alzheimer's Disease, by P.D. Coleman, G.A. Higgins and C.H. Phelps (Eds.) – 1990, ISBN 0-444-81121-4

Volume 87: Role of the Forebrain in Sensation and Behavior, by G. Holstege (Ed.) – 1991, ISBN 0-444-81181-8

Volume 88: Neurology of the Locus Coeruleus, by C.D. Barnes and O. Pompeiano (Eds.) 1991, ISBN 0-444-81394-2

Volume 89: Protein Kinase C and its Substrates: Role in Neuronal Growth and Plasticity, by W.H. Gispen and A. Routtenberg (Eds.) – 1991, ISBN 0-444-81436-1

Volume 90: GABA in the Retina and Central Visual System, by R.R. Mize, R.E. Marc and A.M. Sillito (Eds.) – 1992, ISBN 0-444-81446-9

Volume 91: Circumventricular Organs and Brain Fluid Environment, by A. Ermisch, R. Landgraf and H.–J. Rühle (Eds.) – 1992, ISBN 0-444-81419-1

Volume 92: The Peptidergic Neuron, by J. Joosse, R.M. Buijs and F.J.H. Tilders (Eds.) – 1992, ISBN 0-444-81457-4

Volume 93: The Human Hypothalamus in Health and Disease, by D.F. Swaab, M.A. Hofman, M. Mirmiran, R. David and F.W. van Leeuwen (Eds.) – 1992, ISBN 0-444-89538-8

Volume 94: Neuronal-Astrocyte Interactions: Implications for normal and Pathological CNS Function, by A.C.H. Yu, L. Hertz, M.D. Norenberg, E. Syková and S. Waxman (Eds.) – 1992, ISBN 0-444-89537-x

Volume 95: The visually Responsive Neuron: from Basic Neurophysiology to Behavior, by T.P. Hicks, S. Molotchnikoff and T. Ono (Eds.) – 1993, ISBN 0-444-89492-6

Volume 96: Neurobiology of Ischamic Brain Damage, by K. Kogure, K.-A. Hossmann and B.K. Siesjö (Eds.) – 1993, ISBN 0-444-89603-1

Volume 97: Natural and Artificial Control of Hearing and Balance, by J.H.J. Allum, D.J. Allum-Mecklenburg, F.P. Harris and R. Probst (Eds.) – 1993, ISBN 0-444-81252-0

Volume 98: Cholinergic Function and Dysfunction, by A.C. Cuello (Ed.) – 1993, ISBN 0-444-89717-8

Volume 99: Chemical Signalling in the Basal Ganglia, by G.W. Arbuthnott and P.C. Emson (Eds.) – 1993, ISBN 0-444-81562-7

PROGRESS IN BRAIN RESEARCH

VOLUME 100

NEUROSCIENCE: FROM THE MOLECULAR TO THE COGNITIVE

EDITED BY

FLOYD E. BLOOM

Department of Neuropharmacology, The Scripps Clinic and Research Institute, La Jolla, CA 92307, USA

ELSEVIER
AMSTERDAM – LONDON – NEW YORK – TOKYO
1994

ISBN 0-444-81678-x (volume)
ISSN 0-444-80104-9 (series)

Elsevier Science B.V.
P.O. Box 211
1000 AE Amsterdam
The Netherlands

Library of Congress Cataloging-in-Publication Data

Neuroscience : from the molecular to the cognitive / edited by Floyd
E. Bloom.
p. cm. -- (Progress in brain research ; v. 100)
Includes bibliographical references and index.
ISBN 0-444-81678-X (alk. paper). -- ISBN 0-444-80104-9 (series : alk. paper)
1. Neurosciences. I. Bloom, Floyd E. II. Series.
[DNLM: 1. Central Nervous System--physiology. W1 PR667 v.100 1994 / WL 300 N4939 1994]
QP355.2.N49 1994
612.8'2--dc20
DNLM/DLC
for Library of Congress
94-12298
CIP

Printed on acid-free paper

Printed in The Netherlands

List of Contributors

H. Akil, Mental Health Research Institute, Department of Psychiatry, University of Michigan Medical School, Ann Arbor, MI 48105, USA
D. Bär, Department of Neurology, Rudolf Magnus Institute for Neurosciences, Utrecht University, Universiteitsweg 100, 3584 CG Utrecht, The Netherlands
F.E. Bloom, Department of Neuropharmacology, The Scripps Research Institute, La Jolla, CA, USA
H. Cameron, Laboratory of Neuroendocrinology, Rockefeller University, 1230 York Avenue, New York, NY 10021, USA
T.J. Carew, Department of Psychology, Yale University, POB 11-A, Yale Station, New Haven, CT 06520, USA
H.M. Chao, Laboratory of Neuroendocrinology, Rockefeller University, 1230 York Avenue, New York, NY 10021, USA
D.W. Choi, Department of Neurology and Center for the Study of Nervous System Injury, Box 8111, Washington University School of Medicine, 660 S. Euclid Avenue, St. Louis, MO 63110, USA
A.C. Cuello, McGill University, Department of Pharmacology and Therapeutics, McIntyre Medical Building, 3655 Drummond Street, Suite 1325, Montreal, Quebec, Canada H3G 1Y6
N.C. Danbolt, Anatomical Institute, University of Oslo, P.O.B. 1105 Blindern, N-0317 Oslo, Norway
I. Divac, Department of Medical Physiology, Panum Institute, University of Copenhagen, Blegdamsvej 3C, DK-2200 Copenhagen, Denmark
N.J. Emptage, Department of Psychology, Yale University, 2 Hillhouse Avenue, Box 11A Yale Station, New Haven, CT 06520, USA
S.Y. Felten, Department of Neurobiology and Anatomy, Box 603, University of Rochester School of Medicine and Dentistry, 601 Elmwood Avenue, Rochester, NY. 14642, USA
D.L. Felten, Department of Neurobiology and Anatomy, Box 603, University of Rochester School of Medicine and Dentistry, 601 Elmwood Avenue, Rochester, NY. 14642, USA
G. Fink, MRC Brain Metabolism Unit, University Department of Pharmacology, 1 George Square, Edinburgh EH8 9JZ, UK
L. Giovannelli, Department of Preclinical and Clinical Pharmacology, University of Florence, Viale Morgagni 65, 50134 Florence, Italy
W.H. Gispen, Department of Medical Pharmacology, Rudolf Magnus Institute for Neurosciences, Utrecht University, Universiteitweg 100, 3584 CG Utrecht, The Netherlands
E. Gould, Laboratory of Neuroendocrinology, Rockefeller University, 1230 York Avenue, New York, NY 10021, USA
W.T. Greenough, Beckman Institute, Department of Psychology, Neuroscience Program and Department of Cell and Structural Biology, University of Illinois, Urbana, IL, USA
M.E.R. Hallonet, Institut d'Embryologie Cellulaire et moléculaire, du CNRS et du Collège de France, 49bis avenue de la belle Gabrielle, 94736 Nogent-sur-Marne Cedex, France
R. Hari, Low Temperature Laboratory, Helsinki University of Technology, 02150 Espoo, Finland
M.A. Hofman, Graduate School of Neurosciences Amsterdam, Netherlands Institute for Brain Research, Meibergdreef 33, 1105 AZ Amsterdam, The Netherlands
B.T. Hyman, Neurology Service, Massachusetts General Hospital and Harvard Medical School, Boston, MA 02114, USA
G. Jaim-Etcheverry, Department of Cell Biology and Histology, School of Medicine, University of Buenos Aires, Paraguay 2155, 1121 Buenos Aires, Argentina
G.A.R. Johnston, Department of Pharmacology, The University of Sydney, NSW, 2006, Australia
J.D. Kocsis, Department of Neurology, Yale University School of Medicine, New Haven, CT 06510; and PVA/EPVA Neuroscience Research Center, VA Hospital, West Haven, CT 06516, USA
S.C. Landis, Department of Neurosciences, Case Western Reserve University, School of Medicine, Cleveland, OH 44106, USA
N.M. Le Douarin, Institut d'Embryologie Cellulaire et moléculaire, du CNRS et du Collège de France, 49bis avenue de la belle Gabrielle, 94736 Nogent-sur-Marne Cedex, France

V. Luine, Department of Psychology, Hunter College, New York 10021, NY, USA
A.M. Magarinos, Laboratory of Neuroendocrinology, Rockefeller University, 1230 York Avenue, New York, NY 10021, USA
P.J. Magistretti, Institut de Physiologie, Faculté de Médecine, Université de Lausanne, CH-1005 Lausanne, Switzerland
E.A. Marcus, Department of Biology, Yale University, 2 Hillhouse Avenue, Box 11A Yale Station, New Haven, CT 06520, USA
R. Marois, Interdepartmental Neuroscience Program, Yale University, 2 Hillhouse Avenue, Box 11A Yale Station, New Haven, CT 06520, USA
B.S. McEwen, Laboratory of Neuroendocrinology, Rockefeller University, 1230 York Avenue, New York, NY 10021, USA
J. Mendlewicz, Department of Psychiatry, Free University Clinics of Brussels, Erasme Hospital, route de Lennik 808, 1070 Brussels, Belgium
P.B. Molinoff, Department of Pharmacology, University of Pennsylvania School of Medicine, Philadelphia, PA 19104-6084, USA
O.P. Ottersen, Anatomical Institute, University of Oslo, P.O.B. 1105 Blindern, N-0317 Oslo, Norway
C. Pavlides, Laboratory of Neuroendocrinology, Rockefeller University, 1230 York Avenue, New York, NY 10021, USA
G. Pepeu, Department of Preclinical and Clinical Pharmacology, University of Florence, Viale Morgagni 65, 50134 Florence, Italy
O. Pompeiano, Dipartimento di Fisiologia e Biochimica, Via S. Zeno 31, 56127 Pisa, Italy
O. Pourquié, Institut d'Embryologie Cellulaire et moléculaire, du CNRS et du Collège de France, 49bis avenue de la belle Gabrielle, 94736 Nogent-sur-Marne Cedex, France
D.B. Pritchett, Department of Pharmacology, University of Pennsylvania School of Medicine, Philadelphia, PA 19104-6084, USA
G. Raisman, Norman and Sadie Lee Research Centre, Laboratory of Neurobiology, National Institute for Medical Research, The Ridgeway, Mill Hill, London NW7 1AA, UK
R. Ranney Mize, Department of Anatomy and the Neuroscience Center, Louisiana State University Medical Center, 1901 Perdido Street, New Orleans, LA 70112, USA
G.W. Rebeck, Neurology Service, Massachusetts General Hospital and Harvard Medical School, Boston, MA 02114, USA
L.P. Renaud, Neurosciences Unit, Loeb Research Institute, Ottawa Civic Hospital and University of Ottawa, Ottawa, Ontario, Canada K1Y 4E9
J.L. Roberts, Dr. Arthur M. Fishberg Research Center for Neurobiology, Mount Sinai School of Medicine, One Gustave Levy Place, New York, NY 10029, USA
P. Rudomin, Department of Physiology, Biophysics and Neurosciences, Centro de Investigación y de Estudios Avanzados, México D.F
M. Schachner, Department of Neurobiology, Swiss Federal Institute of Technology, 8093 Zurich, Switzerland
N.A. Simonian, Neurology Service, Massachusetts General Hospital and Harvard Medical School, Boston, MA 02114, USA
R.L. Spencer, Laboratory of Neuroendocrinology, Rockefeller University, 1230 York Avenue, New York, NY 10021, USA
C.N. Stefanis, Department of Psychiatry, Athens University Medical School, Eginition Hospital, 72–74 Vas Sophias Ave., 115 28 Athens, Greece
D.G. Stein, Institute of Animal Behavior, Rutgers, The State University of New Jersey, Newark, NJ 07102, USA
J. Storm-Mathisen, Anatomical Institute, University of Oslo, P.O.B. 1105 Blindern, N-0317 Oslo, Norway
D.F. Swaab, Graduate School of Neurosciences Amsterdam, Netherlands Institute for Brain Research, Meibergdreef 33, 1105 AZ Amsterdam, The Netherlands
D.A. Utzschneider, Department of Neurology, Yale University School of Medicine, New Haven, CT 06510; and PVA/EPVA Neuroscience Research Center, VA Hospital, West Haven, CT 06516, USA
J. Verhaagen, Department of Medical Pharmacology, Rudolf Magnus Institute for Neurosciences, Utrecht University, Universiteitweg 100, 3584 CG Utrecht, The Netherlands
X. Wang, Neuroscience Program, University of Illinois, Urbana, IL, USA
Y. Watanabe, Laboratory of Neuroendocrinology, Rockefeller University, 1230 York Avenue, New York, NY 10021, USA
S.J. Watson, Mental Health Research Institute, Department of Psychiatry, University of Michigan Medical School, Ann Arbor, MI 48105, USA

S.G. Waxman, Department of Neurology, LCI 708 Yale School of Medicine 333 Cedar Street New Haven, CT 06510, USA

I.J. Weiler, Beckman Institute and Department of Psychology, University of Illinois, Urbana, IL, USA

K. Williams, Department of Pharmacology, University of Pennsylvania School of Medicine, Philadelphia, PA 19104-6084, USA

C. Woolley, Laboratory of Neuroendocrinology, Rockefeller University, 1230 York Avenue, New York, NY 10021, USA

W.G. Young, Department of Neuropharmacology, The Scripps Research Institute, La Jolla, CA, USA

J. Zhong, Department of Pharmacology, University of Pennsylvania School of Medicine, Philadelphia, PA 19104-6084, USA

M.J. Zigmond, Department of Neuroscience, University of Pittsburgh, 570 Crawford Hall, Pittsburgh, PA 15260, USA

Preface

All editors like to regard their volumes as unique. Reflecting back through the series of the prior 99 volumes in this series, that premise is strongly supported. However, in all modesty, we would have to assert that the present volume carries at least two features that distinguish it from its predecessor volumes in the series: (1) it is a commemorative milestone issue; and (2) unlike most of the prior volumes, it does not provide an overview of a specific scientific meeting. Instead, for appropriate and creative reasons on the part of the Publisher, Elsevier Science, Amsterdam, it was determined to have a special number in the series to commemorate the one hundredth volume. The question then evolved as to exactly what form this commemorative effort would take and how it would be organized.

After due consideration of a variety of ideas by the editorial advisory board to the *Progress in Brain Research* series, the idea was put forward that the volume should reflect the views of the international body of active neuroscientists, to report on contemporary topics of their choice. Accordingly, authors were invited to write on any topic, given that their choice represented the topic most near and dear to their own efforts over a significant period of the recent past, and to which they would likely continue to be devoted in the future. The authors were also urged where possible to reflect on the evolution of their selected topics and their likely future course.

In that sense this volume reflects not "just" a single scientific meeting, but rather an overview sample of the problems and methodologies that epitomize brain research broadly at this special moment in the maturation of the field. The editorial advisory board were polled for their recommendations of the scientists to be invited, and the inevitably required selection of countries. The collection of essays in this volume therefore reflect these choices, and in addition, another inevitable reflection of the hectic environment of scientific publications, many of our invitees who accepted, were in the end, simply unable to provide their chapters within the deadline.

Given this background of its origination, it seems remarkable that the chapters comprising this volume assorted themselves so readily into five or six easily established categories of topics: developmental brain research, molecular brain research, integrative brain research, neuroplasticity, and neuro-psychiatric conditions. These topics do indeed seem to reflect well on the major streams of effort of our field when measured against the indices of the leading journals.

We are indebted to the authors for their diligent efforts to provide their chapters rapidly. We offer here a volume which reports on a sample of recognized leaders in the neuroscientific community at a significant instant in the history of this renowned series and in the evolution of the field. It is likely, given the high momentum of progress in this field, that many of these topics, and some wholly unexpected by our authors, will establish new directions for the next one hundred volumes.

F.E. Bloom
The Scripps Research Institute, La Jolla, CA

N. Spiteri
Elsevier Science, Amsterdam

Then and Now

Dominick P. Purpura

President, IBRO

Despite the harsh political climate that characterized the Cold War in 1961, scientists dedicated to the understanding of brain mechanisms overcame communication barriers to attend a "Colloquium on Brain Mechanisms" sponsored by the International Brain Research Organization. The central theme of the Colloquium was "Sensorimotor Integration". The formal papers and extensive discussions appeared as Volume One of *Progress in Brain Research* in 1963. Elsevier's prescience in recognizing the emergence of brain research as a major growth industry in the life sciences may have been as important as the reports of the Colloquium. Not surprisingly, much of what was discussed during the 1961 Pisa meeting was but a prologue to current problems. Alfred Fessard provided the Colloquium's last words: "In conclusion, it seems that in the general case of a multineuronal assembly engaged in a specific sensori-motor operation, the question of a special mechanism destined to confer upon it *in isolation* the quality of an 'entity' does not really exist. It can only be called 'integrated' in relation to all other congruent assemblies, at all levels of the neuraxis, which participate in the particular operation".

Three decades later "integration" remains the central goal of brain research, its Holy Grail – now defined as vertical integration from "the Molecular to Cognition", the theme of the present milestone volume of *Progress in Brain Research.* In the intervening years since the first volume, no aspect of the universe of discourse referred to now as "neuroscience" has escaped close pursuit in the hundred volumes of *Progress*. Examples abound here from considerations of the molecular subunit structure of transmembrane channels to complex cognitive processes. In keeping with the international spirit of the inaugural volume, the Editor has succeeded in encouraging sixty-six neuroscientists from fourteen countries to examine the canon of brain research and this they have accomplished with due respect for the constraints of space. Remarkable is it also that the Editor was able to elicit from many contributors their views and speculations and even some of their most prophetic visions. Milestones give pause for review to examine the present and map the future course of discovery.

It was not intended that this commemorative issue of *Progress in Brain Research* would be a comprehensive repository of the most recent advances in neuroscience. Rather it may be viewed as an affirmation of the coming of age of brain research as a scholarly endeavor that now permits objective inquiry into problems that have perplexed humankind since the dawn of human consciousness. To speak about "brain mechanisms" 30 years ago was to infer causality without process. Today even the most complex brain mechanisms are explicable in terms of cellular, molecular and genetic events: Reductionism as champion not challenger of Integration. None of the participants of the 1961 Colloquium ventured into the forbidden realm of pathobiology, unable as they were to grasp at fundamentals of molecular pathogenesis. Today, as this volume attests, neuroscience is at the threshold of comprehending the essential mechanisms that give rise to Alzheimer's disease, epilepsy, stroke, multiple sclerosis and serious mental disorders. Translating understanding to application, though arduous, augers well for the future of brain research and the human condition.

The Editor is to be further congratulated for alerting us in his own contribution to the ineluctable problem of data overload on the "infobahn". How neuroscience deals with the appalling mass of "facts" being disgorged daily from thousands of laboratories throughout the world will be as critical to the success of the neuroscience enterprise as any number of new conceptual advances. The world community of neuroscientists in IBRO now consists of more than 30 000 neuroscientists in over 70 countries. Considering the intellectual power of this workforce, it will surely not be another 30 years before the appearance of the 200th volume of *Progress in Brain Research.* Surely, the need for rapid communication will only intensify as new technologies drive new advances and vice versa. *Progress in Brain Research* will continue to inform us about the next revolution in neuroscience, the understanding of how the brain works in health and disease. But whether the new canon will be conveyed on paper or via the "infobahn" remains to be seen. Elsevier was midwife to the *Progress* series and hand maiden to brain research. Now we celebrate the marriage of neuroscience and informatics. The offspring of the consummation will further test Elsevier's prescience.

Contents

SECTION I

Developmental Brain Research

F. Bloom (Editor)
Progress in Brain Research, Vol. 100

CHAPTER 1

Cell migrations and establishment of neuronal connections in the developing brain: a study using the quail-chick chimera system

Nicole M. Le Douarin, Marc E.R. Hallonet and Olivier Pourquié

Institut d'Embryologie Cellulaire et moléculaire, du CNRS et du Collège de France, 49bis avenue de la belle Gabrielle, 94736 Nogent-sur-Marne Cedex, France

Introduction

In vertebrates, the first morphological sign of emergence of the nervous system is the appearance of the neural plate, the medio-dorsal line of which is in intimate contact with a mesodermal structure, the notochord. Laterally, the neural epithelium is separated from the presumptive superficial ectoderm by a transitional zone, the neural fold. The next important step in neurogenesis is the transformation of the initially flat neural anlage into a tubular structure. The dorsomedial part of the neural plate thus becomes medioventral and its lateral ridges reach the medio-dorsal line where they fuse to form the neural crest and the roof plate. During these morphogenetic events, the floor plate acquires important inductive properties under the influence of the notochord (Yamada et al., 1991) and the neural tube becomes divided into six compartments along the dorsoventral axis: ventrally the floor plate in contact with the notochord, the roof plate dorsally, and laterally the alar and basal plates corresponding respectively to the dorsal and ventral quarters of the neural tube. Although these territories have long been recognized and named, it is only recently that their developmental significance is being really investigated.

One of the characteristics of vertebrates among the chordates is the emergence of the neural crest and the development of the brain at the rostral end of the neural tube (Gans and Northcutt, 1983). The process of cephalization involves the activity of a number of genes (Simeone et al., 1992) as well as an intricate network of intercellular signalling. An important requirement in attempts to decipher how the brain is built is to be able to follow embryonic cells within the neural primordium while the complexity arises, i.e. from the early stages of the neural plate up to completion of neurogenesis. This involves the construction of fate maps and the tracing of neuroepithelial cells along their developmental history. By using the quail-chick chimera system we have undertaken a study in which movements and fate of the neuroepithelial territories were followed in the embryo in ovo during the entire period of development. We have thus been able to sort out what is the respective contribution of the alar and basal plates of the mes- and rhombencephalon to structures like the cerebellum and various brain stem nuclei (Hallonet et al., 1990; Tan and Le Douarin, 1991; Hallonet and Le Douarin, 1993). Moreover, we have isolated an immunoglobulin-like (Ig-like) cell surface glycoprotein (called BEN) whose expression is developmentally regulated in definite subsets of neurons (Pourquié et al., 1990, 1992a,b). On the basis of fate mapping results at the mid- and hindbrain levels and combinations of the quail-chick chimerism with anti-BEN immunoreactivity, we could show that establishment of connections between the inferior olivary nucleus and Purkinje cells of the cerebellar cortex involves BEN expression for the period of time corresponding to the growth of the climbing fibers and the onset of synaptogenesis (Pourquié et al., 1992b). The developmental significance of these results is further discussed.

Following the fate of neuroepithelial territories in embryonic chimeras

The quail-chick marker system was initially based on the nuclear structure of quail cells, characterized by the presence of a large mass of heterochromatin associated with the nucleolus in all embryonic and adult cell types of this species (*Coturnix coturnix japonica*). Thus, quail cells are easy to distinguish from chick cells in which the constitutive heterochromatin is evenly distributed in the nucleus as it is in most animal species (Le Douarin, 1969, 1973, 1982). Since the time this observation was made, other means have been devised to analyze the chimeric tissues in a more refined way. Species-specific and cell type specific antibodies have been produced and as indicated in Table I, some allow neuronal somas or neurites to be selectively labeled in one or the other of the two species. The antibodies that we have produced against the cell surface glycoprotein BEN (for bursal epithelium and neurons, Pourquié et al., 1990) have been largely used in the work reviewed in this article. The BEN protein is interesting because it is not only a neuronal marker but it displays developmentally regulated expression in several classes of neurons in both the central nervous system (CNS) and the peripheral nervous system (PNS). The activity of the encoding gene is likely to be related to important steps in neurogenesis. This is why we have particularly investigated the onset of BEN expression in several neuronal systems.

TABLE I

Species-specific antibodies recognizing either neurites or neurones of quail and chick

Species-specific antibodies	Quail	Chick
Chick anti-quail serum[a]	All cells	Nothing
Mouse Mabs		
39B11	Nothing	Neurites
37F5[b]	Nothing	Neuronal somas
CN	Nothing	Neurites
QN	Neurites	Nothing
CQN[c]	Neurites	Neurites
BEN1[d]	Peripherally projecting neurones, inferior olivary neurones	

[a]Lance-Jones and Lagenaur (1987); [b]Takagi et al. (1989); [c]Tanaka et al. (1990); [d]Pourquié et al. (1990).

The experimental design used to follow the fate of embryonic cells through the quail-chick marker system consists of substituting definite territories in the chick embryo with their quail counterpart from the same developmental stage. The taxonomic proximity of the two species and the fact that, at least during the first half of the developmental period, the two embryos have about the same size and chronology of development, make it possible to construct viable chimeras in ovo. Neural chimeras, in which parts of either the spinal cord (Kinutani et al., 1986) or the brain (Balaban et al., 1988; Hallonet et al., 1990; Teillet et al., 1991; Guy et al., 1992, 1993) have been implanted, can hatch. Chicken with a chimeric brain remain in a healthy condition for a significant period of time during which they can walk, fly and compete for food with other birds.

In quail to chick combinations, however, an immune rejection of the graft takes place several weeks after the host's immune system has become mature (Kinutani et al., 1986) and this constitutes a limitation to this technique, if the behavior of the chimeras is to be followed during long periods after birth. Interestingly, neural tissue grafts, performed before vascularization of the neuroepithelium, between animals of the same species albeit across major histocompatibility (MHC) barrier do not induce an immune response from the host and yield healthy birds. This was used to define which parts of the brain are responsible for the manifestation of a genetic form of epilepsy depending on an autosomal recessive gene in the chicken Fayoumi (Fepi) strain. Selective substitution of certain areas of encephalic vesicles in E2 (embryonic day 2) normal chicken by their counterpart from a Fepi chicken results in the transfer of the disease which can

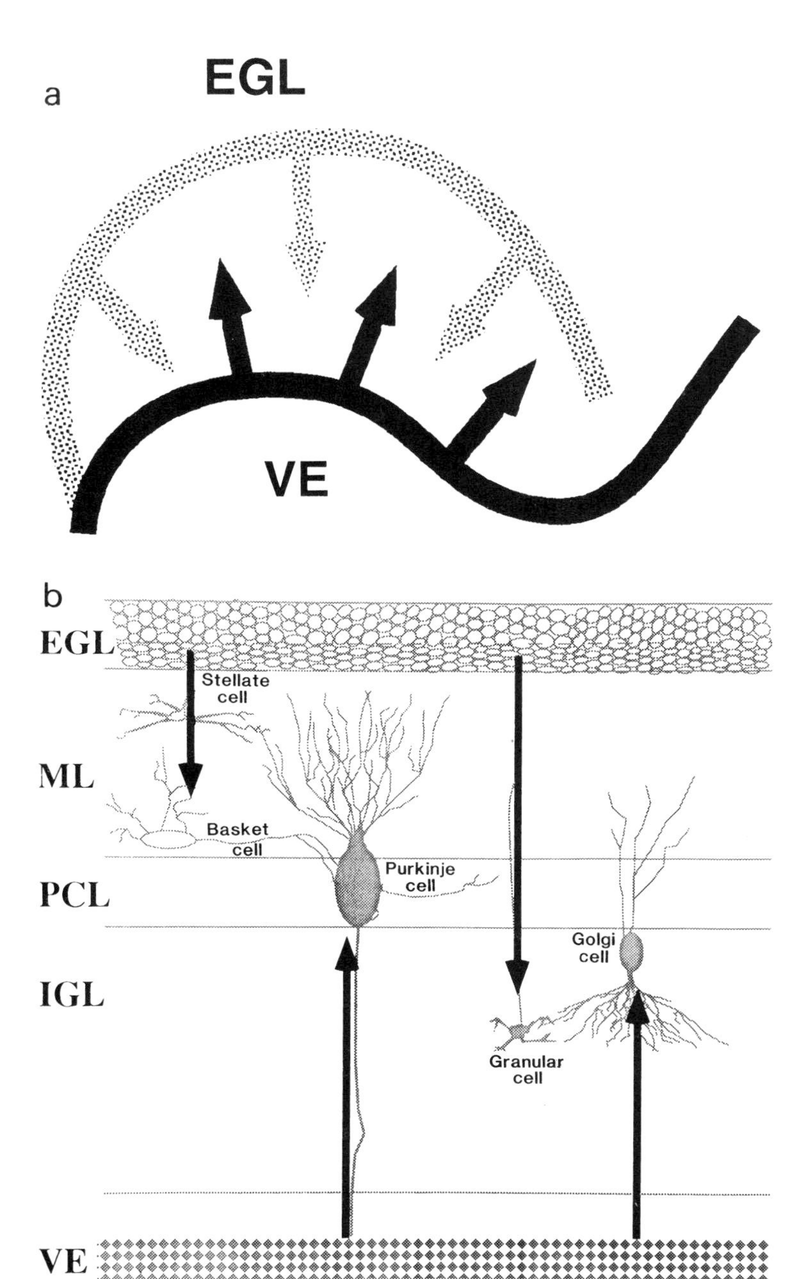

Fig. 1. (A) Schematic representation of the two proliferative layers of the cerebellum. The epithelium lining the fourth ventricule (VE) yields cells migrating centrifugally towards the periphery, whereas the external granular layer (EGL) produces cells migrating into the cerebellum. Arrows indicate the direction of these migrations. Rostral is to the right. (B) Summary of classical descriptions concerning the origin and migration of cells of the cerebellar cortex from their two germinative layers. Purkinje and Golgi cells are produced by the ventricular epithelium, whereas granular cells come from the EGL and follow a strictly radial migration along glial fibers. Interneurons of the molecular layer are classically described as deriving from the EGL (see Hallonet et al., 1990 and references therein). EGL, external granular layer; ML, molecular layer; PCL, Purkinje cell layer; IGL, internal granular layer; VE, ventricular epithelium.

be studied in the chimeras for prolonged periods of time (Teillet et al., 1991; Guy et al., 1992, 1993).

Mapping of the mesencephalon and rhombencephalon

When we started our work on the origin of the cerebellar anlage, it was well known that complex cell movements involving radial and tangential cell migrations take place during the ontogeny of this structure. Namely, radial outward migrations are responsible for positioning of Purkinje cells, whereas the cells originating from the so-called rhombic lip, and destined to form the external granular layer (EGL), undergo tangential migrations over the outer surface of the neuroepithelium. Later on, the cells of the EGL migrate radially inward, and it has been classically accepted that they form the molecular layer and the internal granular layer (IGL) (Fig. 1) (see Hallonet et al., 1990 and references therein). The problem we raised concerned (i) the limits of the cerebellar territory within the encephalic vesicles and (i) the origin of each cell type of the cerebellar cortex from one or the other of the two cell proliferative layers of the cerebellar anlage, i.e. the ventricular epithelium and the EGL. It was classically accepted that the cerebellum is derived from the metencephalic vesicle. Exchanges of the metencephalon (corresponding to prospective rhombomeres (r) 1 and 2) were performed between quail and chick at the 10–14 somites stages as represented in Fig. 2*A*. The results obtained were unexpected: a large part of the cerebellar cortex originated in fact from the mesencephalic vesicle. By doing unilateral grafts of the quail mes- and metencephalic vesicles in chick embryos, Martinez and Alvarado-Mallart (1989) reached a similar conclusion, i.e. that the presumptive cerebellar territory transgresses the primitive mes-metencephalic boundary.

A refined analysis of the contribution of r1 to the cerebellar cortex was done by grafting limited areas of the alar plate between quail and chick embryos over surfaces corresponding to 20° to 120° from the mediodorsal plan (Hallonet and Le Douarin, 1993) (Fig. 2*B*). It was found that the roof plate does not contribute to the cerebellum which arises exclusively from the alar plate. Interestingly, the metencephalic alar plate is able to yield all the cell types found in the cerebellar cortex (i.e. granular and Golgi cells as well as Purkinje neurons) including cells of the molecular layer (Fig. 3). In contrast, the mesencephalic contribution to the cerebellum does not concern granular neurons which are all derived from the metencephalon. In other words, the EGL is entirely derived from the anterior half of the metencephalon corresponding to r1. Moreover, the more dorsal the origin of the EGL cells the less rostral was the extent of their migration (see Hallonet and Le Douarin, 1993, for details).

The morphogenetic movements of the neural tube leading to the participation of the mesencephalic alar plate in cerebellar cortex could be visualized by following step by step the relative position of grafted and host's tissues. As represented in Fig. 4, a longitudinal morphogenetic distortion affects the neural tube from E5 onward and leads to the rostral displacement of its ventral aspect associated with the caudal displacement of its dorsal aspect. As a result, a large area of mesencephalic material is carried backward forming the rostromedial intrusion inserted into the metencephalic roof. The EGL, originating from the alar plate of r1, thus covers the whole cerebellar cortex through a movement of cells following a caudorostral and laterodorsal direction. The posterior half of the metencephalon (r2) gives rise to the choroid plexus and participates to the *medulla oblongata* with the myelencephalon.

Fig. 2. (*A*) Isotopic and isochronic graft of the metencephalic vesicle (in black) between quail and chick were performed at the 12-somite stage in ovo. (*B*) Three types of isotopic and isochronic reciprocal exchanges were performed at the 11- to 14-somite stage between quail and chick embryos. Graft types: (*A*) (I) bilateral rostral half of the metencephalic vesicle, (II) bilateral caudal half of the mesencephalic vesicle; (*C*) (III) unilateral caudal mesencephalon and rostral metencephalon. Extension of the grafts: (*B*,*D*) lateral extensions of bilateral and unilateral grafts, respectively. The lateral extensions of the grafted territories varied from the roof plate of the neural tube (R) to dorsal (d), lateroventral (lv) boundaries, limited at a maximum of 25°, 45°, 90° and 120° respectively from the sagittal plane (from Hallonet and Le Douarin, 1993).

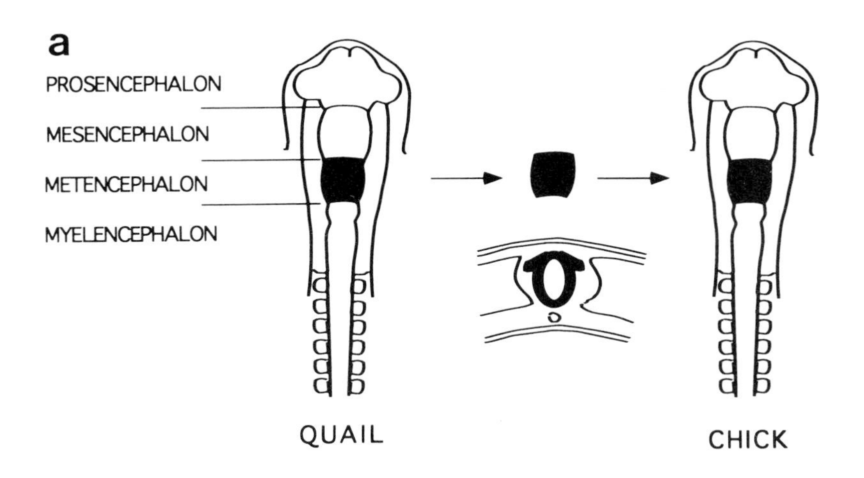
a
PROSENCEPHALON
MESENCEPHALON
METENCEPHALON
MYELENCEPHALON
QUAIL
CHICK

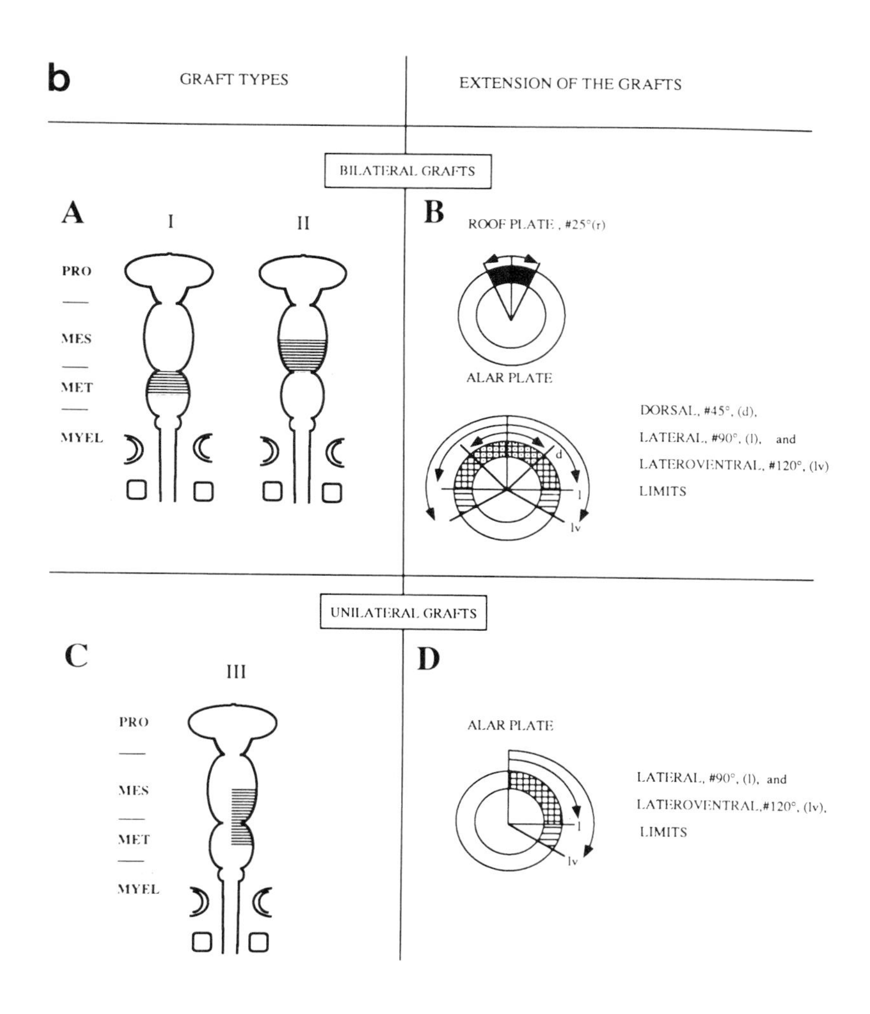
b
GRAFT TYPES
EXTENSION OF THE GRAFTS
BILATERAL GRAFTS
A
I
II
PRO
MES
MET
MYEL
B
ROOF PLATE, #25°(r)
ALAR PLATE
DORSAL, #45°, (d),
LATERAL, #90°, (l), and
LATEROVENTRAL, #120°, (lv)
LIMITS
d
l
lv
UNILATERAL GRAFTS
C
III
PRO
MES
MET
MYEL
D
ALAR PLATE
LATERAL, #90°, (l), and
LATEROVENTRAL, #120°, (lv),
LIMITS
l
lv

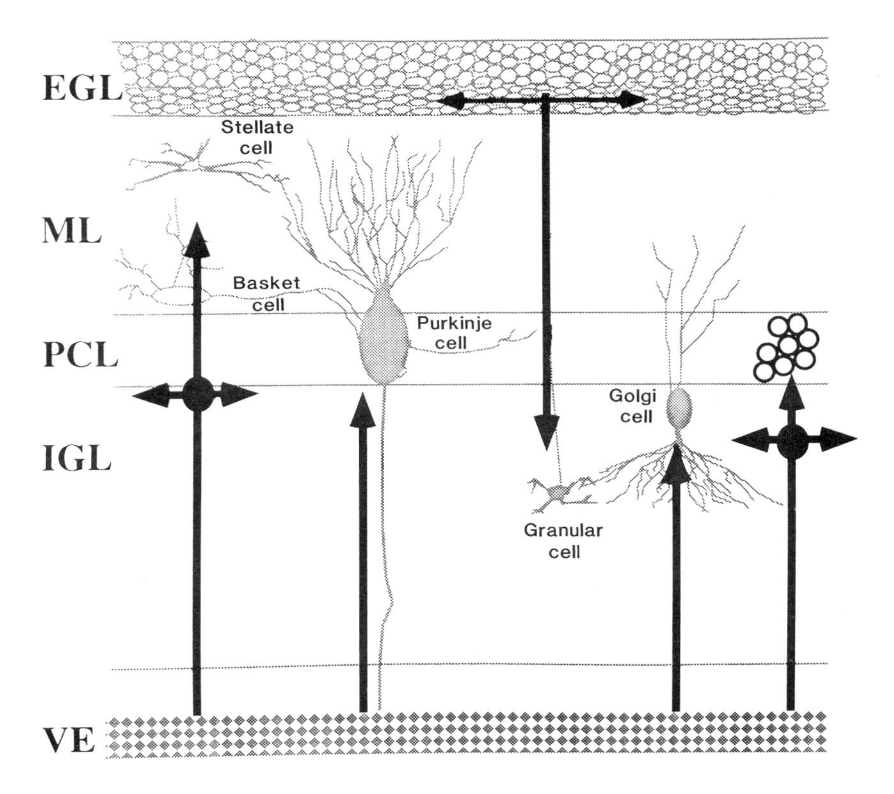

Fig. 3. Summary of the results of quail-chick chimeras analysis concerning the origin and migrations of cells of the cerebellar cortex from the two germinative layers of the cerebellum. Purkinje and Golgi cells come from the ventricular epithelium. Purkinje cells apparently follow a radial migration from the ventricular epithelium to their peripheral location. Granular cells come from the EGL and follow strictly radial migration along glial fibers but they can accomplish tangential migration in the deep levels of the EGL. The interneurons of the molecular layer do not originate from the EGL but from the ventricular epithelium as do the Purkinje and Golgi neurons. They seem to follow transversal migration. We also observe a population of small cells localized in the Purkinje cell layer. These cells follow radial migrations with a small tangential component.

Fate of the myelencephalon: respective contribution of basal and alar plate to the brain stem nuclei

The myelencephalon lies in a caudal position with respect to the metencephalon and gives rise to the *medulla oblongata*. It extends caudally from r3 to r8. Since its first description by His (1891), the dorsoventral migration of cells originating from the so-called "rhombic lip" forming the lateral ridges of the 4th ventricle roof has been well documented.

By using the quail-chick chimera system, we could analyze the respective contribution of the basal and alar plate to the nuclei deriving from the myelencephalon down to the level of somite 6 (Tan and Le Douarin, 1991). This was achieved by exchanging dorsal or ventral portions (i.e. halves or even quarters of the neural tube) of the myelencephalic vesicle between chick and quail embryos (Fig. 5). Chimerism analysis in these animals revealed that the motor nuclei of the abducens, facial, glossopharyngeal, vagal and of part of the trigeminal nerves are derived from the basal plate of the myelencephalon. In contrast, nuclei with essentially sensory components such as the *nuclei angularis, laminaris* or *magnocellularis* were shown to arise from the alar plate. Another category of associative nuclei, such as the precerebellar *inferior olivaris nucleus* and the *nuclei pontis lateralis* and

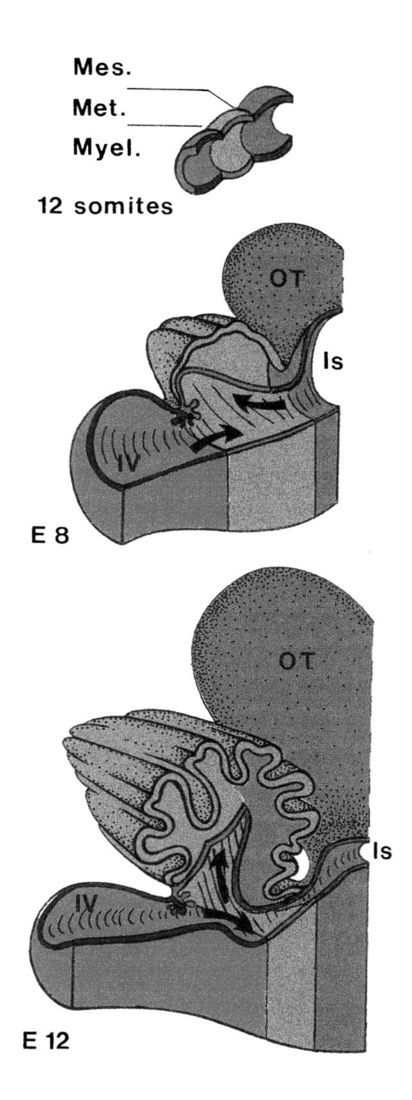

Fig. 4. Schematic three-dimensional reconstructions summarizing the morphogenetic movements modelling the mesencephalic, metencephalic and myelencephalic vesicles. During embryogenesis, mesencephalic material intrudes rostromedially into the dorsal extent of the metencephalic vesicle. Ventrally, a longitudinal movement rostrally displaces the floors of the myelencephalic, metencephalic and mesencephalic vesicles. Consequently, when the external granular layer, which is generated in the metencephalic vesicle, spreads over the cerebellar anlage, it covers the rostromedial part of the cerebellum, which is generated in the mesencephalic vesicle. Rostral lies to the right of the diagram. The reconstructions are viewed from the medial aspect (from Le Douarin, 1993). Mes, mesencephalic vesicle; Met, metencephalic vesicle; Myel, myelencephalic vesicle; Is, isthmus; IV, fourth ventricle; OT, optic tectum. Arrows indicate the direction of the morphogenetic movements modelling the neural tube.

medialis which provide the cerebellum with climbing and mossy fibers inputs respectively, originate from the alar plate. These nuclei undergo an extensive dorso-ventral migration, so that finally they lie ventrally to basal plate derivatives (Fig. 6). Nuclei such as the *reticularis gigantocellularis* or *subtrigeminalis* are of mixed origin, perhaps reflecting their origin from the intermediate zone between alar and basal plate.

Other types of grafts involving the replacement of the entire myelencephalon or of its lateral half allowed the extent of cell migration along the anteroposterior and dorsoventral axis to be evaluated by analyzing the chimerism at the ridges of the graft. It was thus possible to establish the myelencephalic contribution to certain important alar plate derived nuclei. In particular, the inferior olivary nucleus which provides the cerebellar cortex with climbing fibers was found to originate not only from the myelencephalon but also from the spinal cord down to the level of somite 6.

In contrast, the *nuclei pontis lateralis* and *medialis* which yield a contingent of mossy fibers, have a mixed myelencephalic and metencephalic origin. These experiments demonstrated in addition that nu-

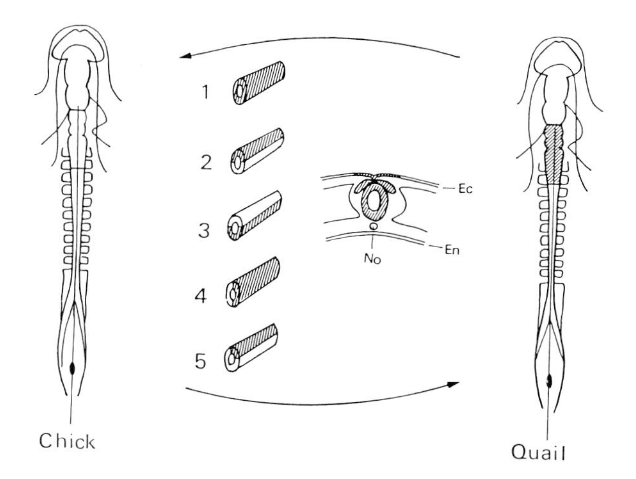

Fig. 5. Schematic representation of the five types of isotopic and isochronic transplantations of the myelencephalon between quail and chick embryos, performed at the 10- to 12- somite stage. 1, whole myelencephalon; 2, dorsal half of myelencephalon; 3, ventral half of myelencephalon; 4, right half of myelencephalon; 5, dorsal quarter of myelencephalon (from Tan and Le Douarin, 1991). Ec, ectoderm; En, endoderm; No, notochord.

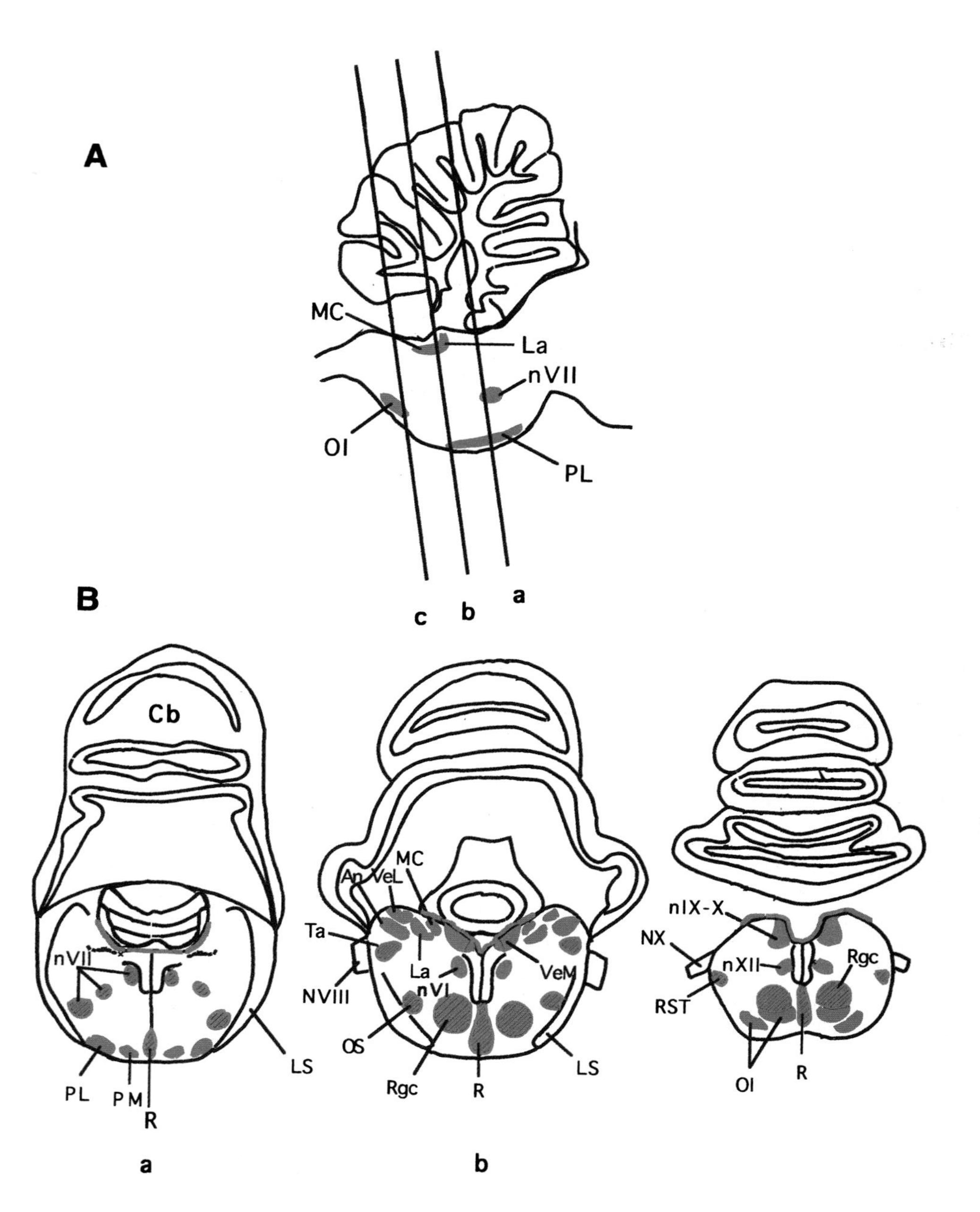

Fig. 6. Summary of the neuronal origin from the alar plate (in blue) and basal plate (in red) of the nuclei of the medulla oblongata. The hatched nuclei have a dual alar and basal plate origin. (A) Parasagittal section of post-hatched chick brain. Oblique lines indicate the positions of transverse sections shown in *B*. (B) Transverse sections of post-hatched chick hind brain. An, nucleus angularis; Cb, cerebellum; La, nucleus laminaris; LS, lemniscus spinalis; MC, nucleus magnocellularis; nVI, nucleus nervi abducentis; nVII, nucleus nervi facialis; NVIII, nerve VIII; nIX-X, nucleus nervi glossopharyngei and nucleus motorius dorsalis nervi vagi; NX, nervus vagus; nXII, nucleus nervi hypoglossi; OI, nucleus olivaris inferior; PL, nucleus pontis lateralis; PM, nucleus pontis medialis; R, nucleus raphe; Rgc, nucleus reticularis gigantocellularis; RST, nucleus reticularis subtrigeminalis; VeL, nucleus vestibularis lateralis; VeM, nucleus vestibularis medialis.

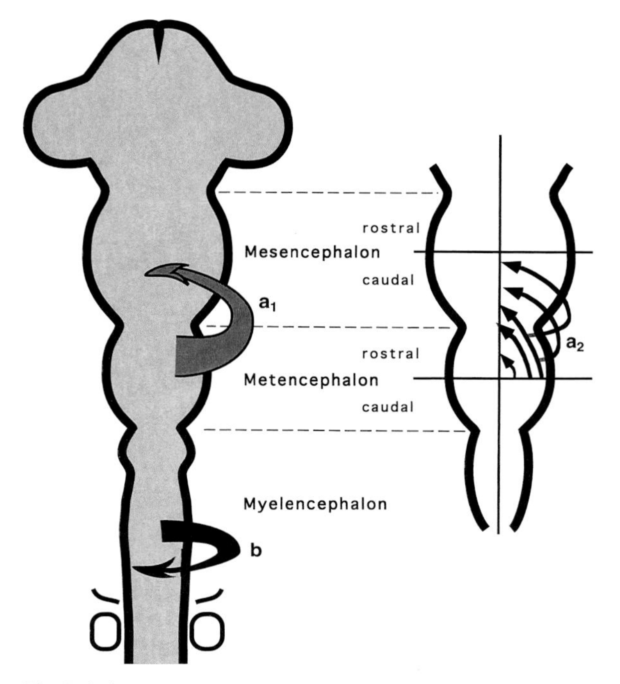

Fig. 7. Schematic representation of the migrations of cells from the anterior metencephalic alar plate leading to the formation of the EGL (top arrow labelled a_1). In a_2, it is shown that cells of the metencephalic wall located between 20° and 120° migrate anteriorly and towards the midline of the encephalic vesicles that they cover up to the mid-mesencephalic level. Cells located the most dorsally are those which undergo the less extended migration. In contrast the cells originating ventrally migrate further rostrally (see text and Hallonet and Le Douarin, 1993, for details). The arrow labelled b represents the dorsoventral migratory stream of the *medulla oblongata* which gives rise to the nuclei endowed of associative functions located ventrally in the brain stem.

clei derived from the alar plate such as the inferior olivary nucleus were made up of cells which accomplish extensive longitudinal as well as dorso-ventral migrations. In contrast, no significant longitudinal movements are detectable in the basal plate derived nuclei. This may be related to the original segmentation of the rhombencephalon into rhombomeres which seem to be separated by barriers to cell migration early in development (Fraser et al., 1990). The patterning of alar plate derivatives in contrast seems largely to escape from the rhombometameric constraints since alar plate derived cells navigate freely across these barriers.

In conclusion, the study of the fate map of the myelencephalon (Tan and Le Douarin, 1991) demonstrated that motor nuclei are derived from cells of basal plate origin migrating essentially along a radial direction; whereas cells forming most nuclei endowed with associative functions (such as the inferior olivary nucleus and most nuclei of the reticular formation) originate from the alar plate. They reach their ventral position after a dorso-ventral migration which could be clearly visualized by the quail-chick marker system.

The particular behavior of cells of the rhombic lip masks the original dorso-ventral polarity of the myelencephalon which is similar to that of the spinal cord, in which the alar plate gives rise to interneurons and the neural crest yields sensory ganglia (Hamburger, 1948; Wenger, 1950; Leber et al., 1990).

It is striking that at the level of the first rhombomere (r1) from which all the EGL of the cerebellum originates, cells of the alar plate migrate along laterodorsal and postero-anterior vectors as indicated in Fig. 7. Cells are endowed with a remarkably high proliferative capacity since they cover the entire cerebellar plate including its rostro-medial component which is derived from the mesencephalon (Fig. 4). These morphogenetic processes are under the control of developmental genes among which the Wnt1 gene has been identified (McMahon and Bradley, 1990; Thomas and Capecchi, 1990).

Cells of the alar plate originating from the myelencephalon are also endowed with a large proliferation potential and exhibit a unique migratory behavior which leads them to lie ventral to the progeny of the basal plate. The opposite movements of the metencephalic and myelencephalic cells are illustrated in Fig. 7.

One can say therefore that the initial dorsoventral polarity of the spinal cord in terms of fate and functions of cells originating from the basal and alar plates is conserved in the mid- and hindbrain.

In the next section, we consider the expression of BEN glycoprotein: its biochemical characteristics and its use to trace the climbing fibers from the inferior olivary nucleus to the Purkinje cells of the cerebellar cortex.

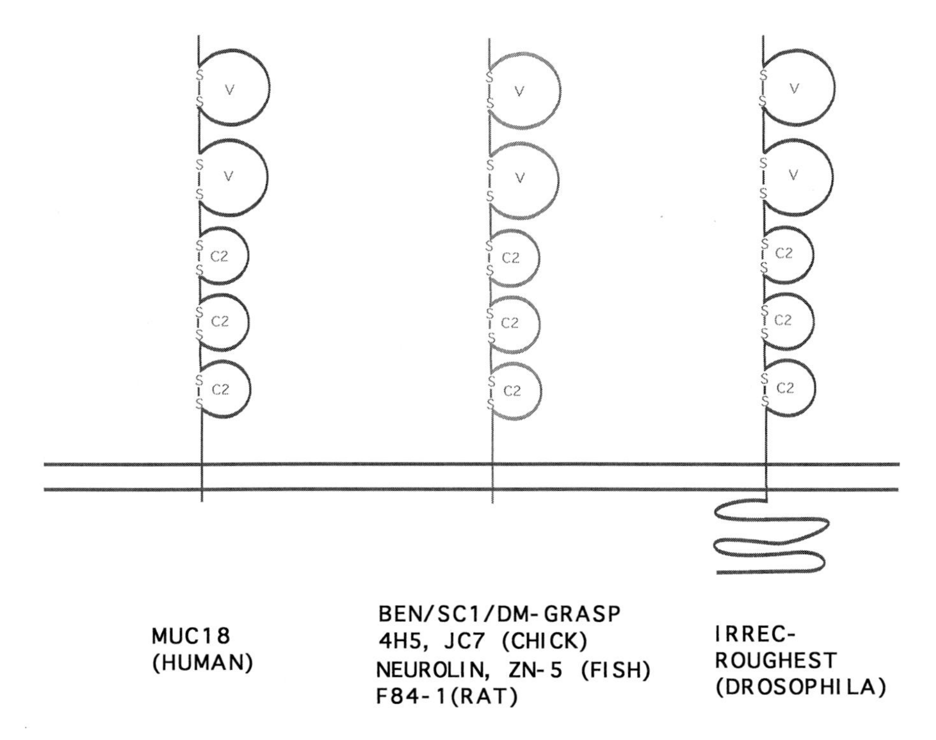

Fig. 8. Schematic representations of BEN, MUC 18 and IrreC molecules. These three molecules show a homology at the amino acid level of around 25% over the length of their extracellular domains. The first two Ig-domains have been classified as V-type for BEN and MUC 18 although this is not as clear for IrreC. The next three domains are of the C2 type.

Expression of BEN during spinal cord and brain stem development

BEN is an avian cell surface molecule of the Ig-superfamily which is widely expressed during chick embryonic development but is restricted to neuronal subsets in the nervous system. We have identified this molecule by means of a monoclonal antibody (Pourquié et al., 1990). SDS-PAGE analysis of the immuno-purification product obtained from the various cell types expressing BEN, i.e. the epithelium of the bursa of Fabricius, hemopoietic cells (thymocytes) and cells of the CNS, yielded molecules with slightly different M_r ranging from 110 000 for the epithelial form of BEN to 100 000 and 95 000 for its hemopoietic and neural forms, respectively.

Variations in *N*-glycosylation appeared to be mostly responsible for this molecular heterogeneity. This was further confirmed by the presence of the HNK-1 carbohydrate epitope only on the protein extracted from brain and thymus. We have microsequenced parts of the protein and used degenerate oligonucleotides to clone a cDNA encoding this molecule (Pourquié et al., 1992a). Its sequence turned out to be identical to that of the DM-GRASP (Burns et al., 1991) and SC1 (Tanaka et al., 1991) proteins also described in the

Fig. 9. (*A*) Transverse sections of a chick embryo immunostained with anti-BEN Mab, at E2.5. BEN-reactivity is conspicuous on motoneurons (MN) and in their axons forming the early spinal nerve (SN) and on the floor plate (FP) (Bar: 75 μm). (*B*) E4.5 embryo. BEN reactivity has appeared on DRG neurons and their axons contributing to the spinal nerve and forming in the spinal cord the dorsal funiculus (Bar: 100 μm). (*C*) E8 embryo. BEN protein expression is now down-regulated and progressively disappears in both motoneurons and DRG neurons (Bar: 100 μm). (*D*) Schematic representation of BEN expression during spinal nerve formation. The drawings show transverse sections of the spinal cord and the DRG at increasing stages of development expressed according to the developmental table of Hamburger and Hamilton (1951).

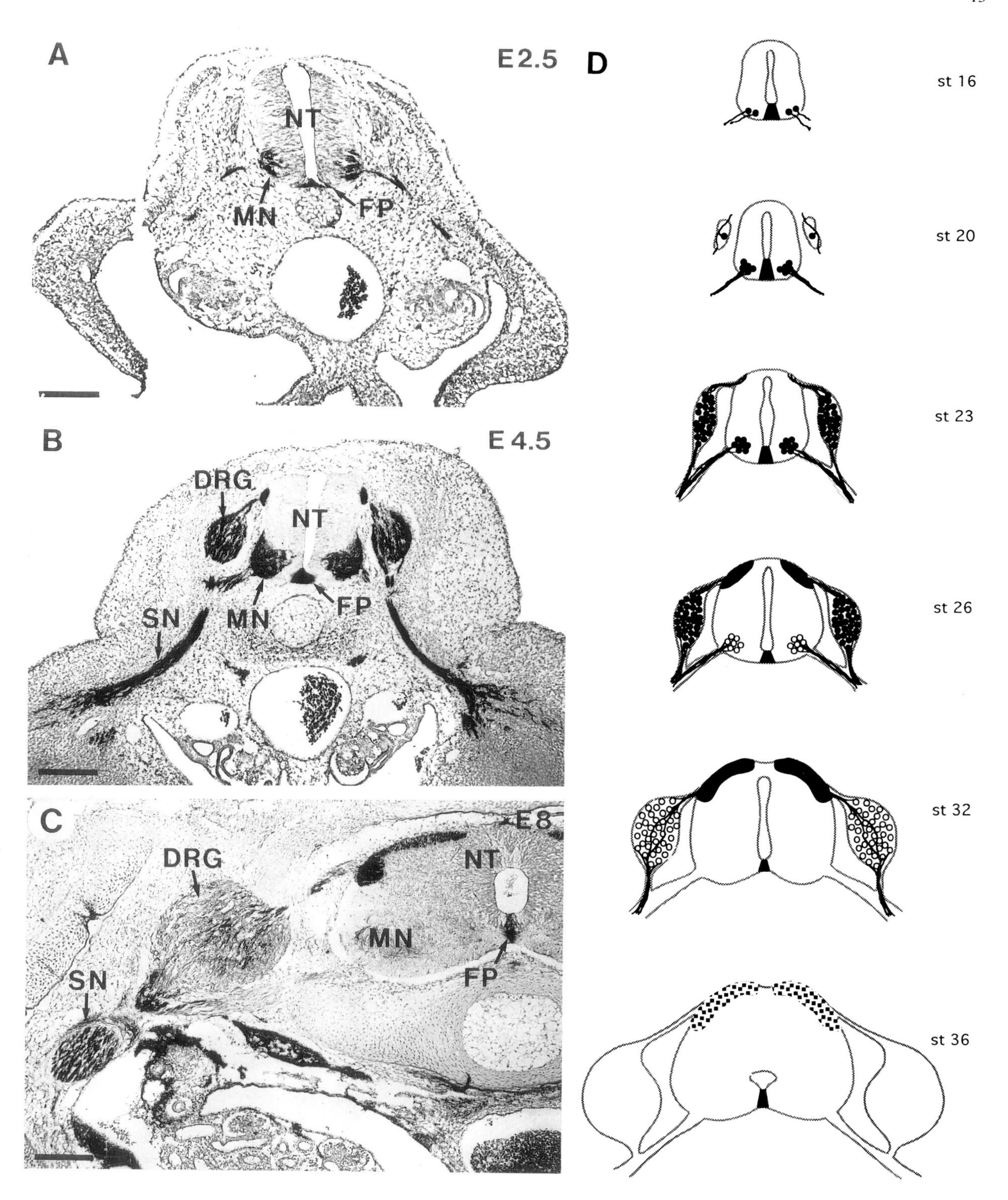

A
E2.5
NT
MN
FP
B
E 4.5
DRG
NT
SN
MN
FP
C
E8
DRG
NT
MN
SN
FP
D
st 16
st 20
st 23
st 26
st 32
st 36

chick embryo. BEN/DM-GRASP/SC1 has the typical structure of an integral membrane protein with a large external domain associated with a hydrophobic transmembrane portion and a short intracytoplasmic tail. The extracellular region of the BEN molecule is composed of five Ig-like domains, as in the human melanoma marker MUC18 (Lehmann et al., 1989) and in IrreC, a recently identified *Drosophila* molecule involved in retinal axon guidance (Ramos et al., 1993).

Although the overall homology at the amino acid level between BEN and MUC18 or IrreC is only 25%, the structure of these three molecules is basically similar. They may therefore constitute a new subgroup in the large Ig-superfamily (Fig. 8). Species homologs of BEN protein have now been discovered in other vertebrates such as goldfish, zebrafish and rat (Laessing et al., 1993; Kanki and Kuwada, 1993; Prince et al., 1992). As far as its putative function is concerned, this protein was shown by in vitro assays to mediate homophilic adhesive interactions (Burns et al., 1991; Tanaka et al., 1991; El Deeb et al., 1992).

We have first studied BEN protein expression in the peripheral nervous system. It is present during development on all neural crest derived neurons (Pourquié et al., 1990) soon after they start to extend their processes. In the sympathetic, parasympathetic and enteric nervous systems, the BEN antigen is detectable soon after gangliogenesis and remains present in PNS ganglia until at least hatching.

In the spinal cord, motoneurons start expressing the BEN gene when extending neurites. The floor plate, a specialized population of epithelial cells lying in the midline of the neural tube, is BEN-positive from E2 to E10. In the spinal ganglia, the expression of this protein starts in neurons soon after gangliogenesis and closely parallels its expression in motoneurons (Fig. 9). The immunoreactivity starts in both motor and sensory neurons cell bodies and spreads in the fibers fasciculating to form the spinal nerves. It finally disappears in a centrifugal manner at the time of synaptogenesis. The protein remains however present on the dorsal funiculi formed by the central afferent fibers of the DRG (up to E10) (Fig. 9). At E8, immunoreactivity is no longer detected in cell bodies of either sensory or motor neurons, but remains on their fibers. Later in development, the protein disappears from the spinal nerve roots. Interestingly, the BEN protein was shown to be expressed at the level of the functional neuromuscular synapse where it co-localizes with the acetycholine receptor (Fournier-Le Ray and Fontaine-Pérus, 1991). Therefore, at the level of the motor and sensory neurons, BEN gene expression is associated with axonal outgrowth and migration and is downregulated from cell body to synapse. These results, together with in vitro data suggesting a role for this molecule in homophilic adhesion, support the contention that it could play a role in nerve fiber fasciculation and synapse formation in the PNS.

Expression of BEN in the rhombencephalon

The earliest expression of the BEN gene was detected on basal plate derivatives of the rhombencephalon, namely the floor plate which becomes immunoreactive first and later in the nuclei formed by cranial nerves motoneurons. The onset of expression of the protein does not strictly follow a craniocaudal gradient in the cephalic region, as it does in the spinal cord. The protein first appears in somas of the facial nerve neurons, and slightly later, in the other motor nuclei (Chang et al., 1992; Guthrie and Lumsden, 1992). Expression was also first detectable on cell bodies and early pioneering axons soon after neurons become postmitotic. A few hours later, the BEN gene product was also detected in the cranial sensory ganglia. As observed for the spinal motoneurons and the DRG neurons, down regulation occurs from the cell body to the synapse around the time of synaptogenesis. Sensory neurons derived from the alar plate such as the *nucleus laminaris* start to express the BEN gene from about E6 (our unpublished observations). From E8 onwards, BEN expression is also detected on other nuclei, and particularly in the inferior olivary nucleus which has been the subject of further investigations.

Association of BEN with climbing fibers axonogenesis and synaptogenesis

The cerebellar cortex is connected to the rest of the brain by two main afferent systems. The mossy fibers

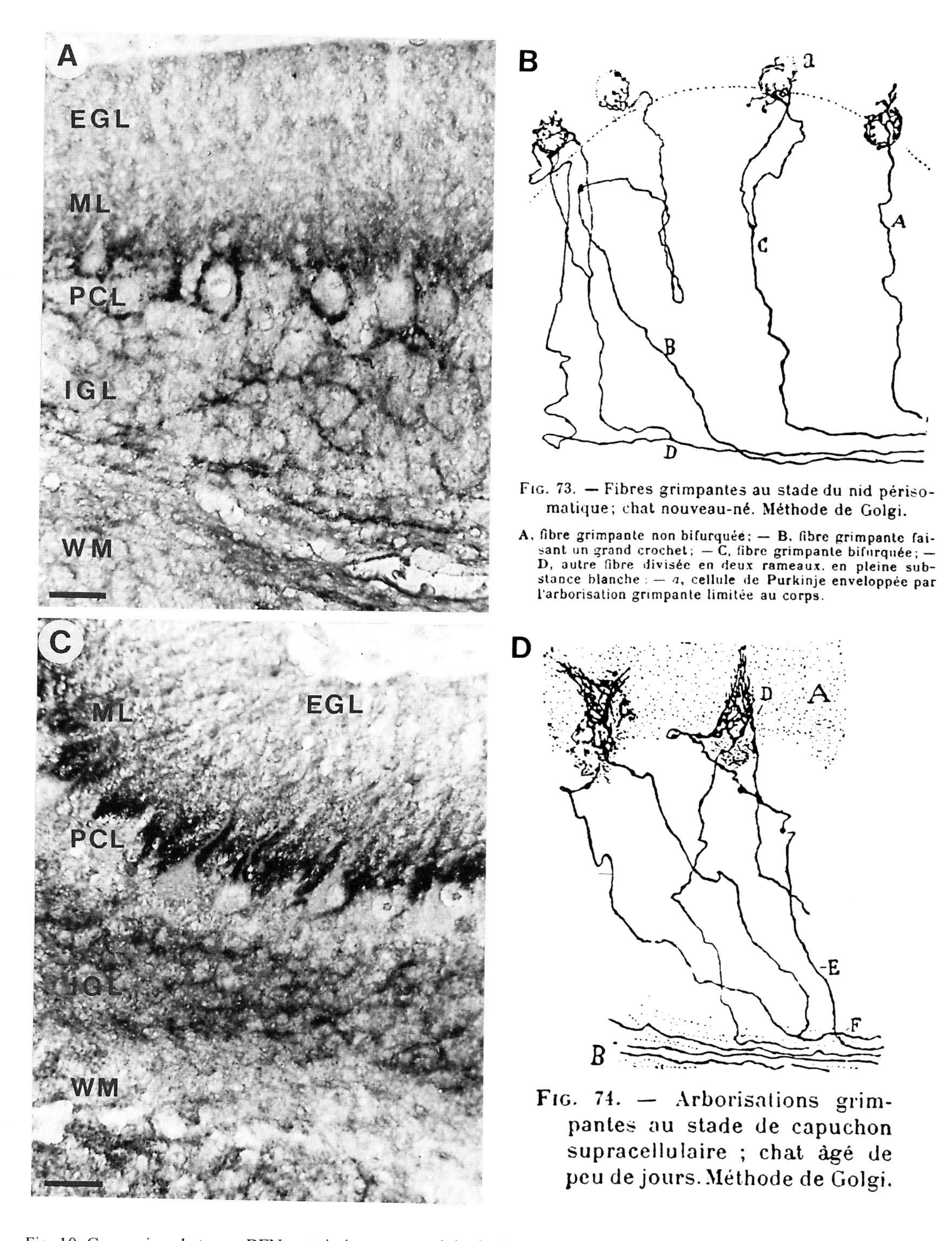
A
EGL
ML
PCL
IGL
WM
B
a
A
C
B
D
FIG. 73. — Fibres grimpantes au stade du nid périsomatique; chat nouveau-né. Méthode de Golgi.
A, fibre grimpante non bifurquée; — B, fibre grimpante faisant un grand crochet; — C, fibre grimpante bifurquée; — D, autre fibre divisée en deux rameaux, en pleine substance blanche ; — a, cellule de Purkinje enveloppée par l'arborisation grimpante limitée au corps.
C
ML
EGL
PCL
IGL
WM
D
D
A
E
F
B
FIG. 74. — Arborisations grimpantes au stade de capuchon supracellulaire ; chat âgé de peu de jours. Méthode de Golgi.

Fig. 10. Comparison between BEN protein immunoreactivity in the cerebellar cortex, and the classical descriptions of Ramon y Cajal (1911) of climbing fibers maturation. (*A*) Sagittal section of the cerebellar cortex at E15 showing the reactivity localized in the white matter (WM), the internal granular layer (IGL) and the Purkinje cell layer (PCL) evocating the nest stage described by Cajal illustrated in (*B*) (Bar: 20 μm). (*C*) Sagittal section of the cerebellar cortex a E16 (Bar: 20 μm). The maximal reactivity is now translocated at the bottom of the apical dendrite of Purkinje cells resembling the supracellular stage described by Cajal as shown in (*D*). EGL: external granular layer.

originate from various brain locations and establish synaptic connections with the granular and Golgi neurons known as glomeruli. The second type of afferents corresponds to the climbing fibers. They arise exclusively from the inferior olivary nucleus, which is derived from the myelencephalon as shown by the experiments reported above. Fibers from the inferior olivary nucleus establish connexions with the deep nuclei neurons and project to the cerebellar cortex to synapse on the Purkinje cell dendrites. As previously shown (Tan and Le Douarin, 1991) the inferior olivary nucleus neurons arise from the alar plate of the myelencephalon down to the level of somite 6. The climbing fibers contact the Purkinje cells according to a sequence of maturation stages of which Ramon y Cajal (1911) has provided a classical description (Fig. 10).

BEN expression in the developing cerebellar system turned out to be particularly interesting (Pourquié et al., 1992b). From E8–E9, BEN is strongly expressed by inferior olivary nucleus neurons. From E10 onward, the emergence and growth of climbing fibers arising from the inferior olivary nucleus can be followed. BEN immunoreactivity is successively detected on fibers in the cerebellar peduncles, the white matter and the Purkinje cell layer. The contacts established between the growing climbing fibers and the Purkinje cells in E14 and E18 embryos as revealed by BEN immunocytochemistry are strikingly similar to those seen by Cajal after Golgi staining (Fig. 10). The origin of the BEN positive fibers from the inferior olivary nucleus and thus their identity with the climbing fibers was further assessed by using the quail-chick chimera system.

From the fate mapping experiments of the rhombencephalon described previously, we knew that the presumptive territory of the inferior olivary nucleus is located in the myelencephalon and the rostral part of the spinal cord whereas the cerebellar cortex arises from the mesencephalon and the anterior metencephalon. Therefore chimeras were constructed by substituting the presumptive cerebellar territory (including its mes- and metencephalic components) of a chick by its quail counterpart. In such birds, the inferior olivary nucleus is composed of host cells, whereas the cerebellar cortex is derived from the graft.

We have taken advantage of this system and used species specific antibodies which recognize chick but not quail neuronal fibers. In these cerebellar chimeras, only the fibers afferent to the cerebellar cortex are of chick origin. Double staining experiments using these species specific antibodies together with anti-BEN antibodies showed that the immunoreactive chick fibers were also BEN positive. Together with the histological observation of BEN immunoreactivity during ontogenesis of the cerebellum, we conclude that BEN is expressed on climbing fibers during their axonogenesis and synaptogenesis.

Conclusions and discussion

This series of investigations demonstrates the use that can be made of the combination of two techniques: the quail-chick chimera system to trace morphogenetic movements of epithelia and migrations of embryonic cells and molecular methods to characterize the emergence of various cell types in the nervous system. The particular study carried out on the climbing fibers illustrates that such an approach can also be useful to illustrate and follow the onset of connexions between distant structures within the developing brain.

The investigations performed on the BEN/DM-GRASP/SC1 molecule suggest that its transient expression by definite subsets of neurons during development might be related to its role in axonal guidance and/or fasciculation and likely also to growth cone and target cell recognition during the establishment of certain types of synapses.

One striking fact in the biology of this molecule is that it is not specific for a particular type of neurons but rather of neurons emerging as homogenous groups. Although expression of BEN during development seems a general feature of motoneurons over the whole nervous system, other types of neurons also present this characteristic. Such is the case for virtually all the neurons arising from the neural crest and certain interneurons in the brain. Its distribution suggests that it could play a role in recognition and perhaps adhesion of homologous neurons together with a similar role in the projection of these neurons to restricted sets of targets.

Acknowledgements

We are indebted to Dr Françoise Dieterlen for critical reading of the manuscript. We thank Evelyne Bourson, Michelle Scaglia and Michelle Muesser for secretarial assistance. Our thanks go to Yann Rantier, Thierry Guérot and Sophie Gournet for photography and artwork. Financial support was provided by the Centre National de la Recherche Scientifique (CNRS), the Fondation pour la Recherche Médicale Française, and the Ligue Française pour la Recherche contre le Cancer and the Association Française contre les Myopathies (AFM).

References

Balaban, E., Teillet, M.A. and Le Douarin, N.M. (1988) Application of the quail-chick chimera system to the study of brain development and behavior. *Science*, 241: 1339–1342.

Burns, F.R., Von Kannen, S., Guy, L., Raper, J.A., Kamholz, J. and Chang, S. (1991) DM-GRASP, a novel immunoglobulin superfamily axonal surface protein that supports neurite extension. *Neuron*, 7: 209–220.

Chang, S., Fan, J. and Nayak, J (1992) Pathfinding by cranial nerve VII (facial) motoneurons in the chick hindbrain. *Development*, 114: 815–823.

El-Deeb, S., Thompson, S.C. and Covault, J. (1992) Characterization of a cell surface adhesion molecule expressed by a subset of developing chick neurons. *Dev. Biol.*, 149: 213–227.

Fournier Le Ray, C. and Fontaine-Pérus, J. (1991) Influence of spinal cord stimulation on the innervation pattern of muscle fibers in vivo. *J. Neurosci.*, 11: 3840–3850.

Fraser, S., Keynes, R. and Lumsden, A. (1990) Segmentation in the chick embryo hindbrain is defined by cell lineage restrictions. *Nature*, 344: 431–435.

Gans, C. and Northcutt, R.G. (1983) Neural crest and the origin of vertebrates: a new head. *Science*, 220: 268–274.

Guthrie, S. and Lumsden, A. (1992) Motor neuron pathfinding following rhombomere reversals in the chick embryo hindbrain. *Development*, 114: 663–673.

Guy, N., Teillet, M.A., Schuler, B., Lasalle, G.L., Le Douarin, N.M., Naquet, R. and Batini, C. (1992) Pattern of electroencephalographic activity during light induced seizures in genetic epileptic chicken and brain chimeras. *Neurosci. Lett.*, 145: 55–58.

Guy, N.T.M., Batini, C., Naquet, R. and Teillet, M.A. (1993) Avian photogenic epilepsy and embryonic brain chimeras – neuronal activity of the adult prosencephalon and mesencephalon. *Exp. Brain Res.*, 93: 196–204.

Hallonet, M.E.R. and Le Douarin, N.M. (1993) Tracing neuroepithelial cells of the mesencephalic and metencephalic alar plates during cerebellar ontogeny in quail-chick chimeras. *Eur. J. Neurosci.*, 5: 1145–1155.

Hallonet, M.E.R., Teillet, M.A. and Le Douarin, N.M. (1990) A new approach to the development of the cerebellum provided by the quail-chick marker system. *Development*, 108: 19–31.

Hamburger, V. (1948) The mitotic patterns in the spinal cord of the chick embryo and their relation to histogenetic processes. *J. Comp. Neurol.*, 88: 221–284.

Hamburger, V. and Hamilton, H.L. (1951) A series of normal stages in the development of the chick embryo. *J. Morphol.*, 88: 49–92.

His, W. (1891) Die Entwicklung des menschlichen Rautenhirns vom Ende des ersten bis zum Beginn des dritten Monats. I. Verlängertes Mark. *Abh. Sächs Ges. Wiss., Mat. Phys. K. I.* , 29: 1–74.

Kanki, J.P. and Kuwada, J.Y. (1993) Cloning and characterization of a Zebrafish cDNA similar to chick DM-GRASP, a neural cell surface protein in the immunoglobulin superfamily. *Soc. Neurosci. Abstr.*, 19: 1090.

Kinutani, M., Coltey, M. and Le Douarin, N.M. (1986) Postnatal development of a demyelinating disease in avian spinal cord chimeras. *Cell*, 45: 307–314.

Laessing, U., Giordano, S., Lottspeich, F. and Stuermer, C.A.O. (1993) Molecular cloning of neurolin and its expression in Goldfish and embryonic Zebrafish CNS. *Soc. Neurosci. Abstr.*, 19: 1090.

Lance-Jones, C.C. and Lagenaur, C.F. (1987) A new marker for identifying quail cells in embryonic avian chimeras: a quail-specific antiserum. *J. Histochem. Cytochem.*, 35: 771–780.

Le Douarin, N.M. (1969) Particularités du noyau interphasique chez la Caille japonaise (*Coturnix coturnix japonica*). Utilisation de ces particularités comme 'marquage biologique' dans les recherches sur les interactions tissulaires et les migrations cellulaires au cours de l'ontogenèse. *Bull. Biol. Fr. Belg.*, 103: 435–452.

Le Douarin, N.M. (1973) A Feulgen-positive nucleolus. *Exp. Cell. Res.*, 77: 459–468.

Le Douarin, N.M. (1982) *The Neural Crest*, Cambridge University Press, Cambridge, 259 pp.

Le Douarin, N.M. (1993) Embryonic neural chimaeras in the study of brain development. *Trends Neurosci.*, 16(2): 64–72.

Leber, S.M., Breedlove, S.M. and Sanes, J.R. (1990) Lineage, arrangements and death of clonally related motoneurons in chick spinal cord. *J. Neurosci.*, 10: 2451–2462.

Lehmann, J.M., Riethmuller, G. and Johnson, J.P. (1989) MUC18, a marker of tumor progression in human melanoma, shows sequence similarity to the neural cell adhesion molecules of the immunoglobulin superfamily. *Proc. Natl. Acad. Sci. USA*, 86: 9891–9895.

Martinez, S. and Alvarado-Mallart, R.M. (1989) Rostral cerebellum originates from the caudal portion of the so-called "mesencephalic" vesicle: a study using chick/quail chimeras. *Eur. J. Neurosci.*, 1: 549–560.

McMahon, A.P. and Bradley, A. (1990) The Wnt-1 (int-1) proto-oncogene is required for development of a large region of the mouse brain. *Cell*, 62: 1073–1085.

Pourquié, O., Coltey, M., Thomas, J.L. and Le Douarin, N.M. (1990) A widely distributed antigen developmentally regulated in the nervous system. *Development*, 109: 743–752.

Pourquié, O., Corbel, C., Le Caer, J.P., Rossier, J. and Le Douarin, N.M. (1992a) BEN, a surface glycoprotein of the immunoglobulin superfamily, is expressed in a variety of developing systems. *Proc. Natl. Acad. Sci. USA*, 89: 5261–5265.

Pourquié, O., Hallonet, M.E.R. and Le Douarin, N.M. (1992b) BEN glycoprotein expression is associated with climbing fibers axonogenesis in the avian cerebellum. *J. Neurosci.*, 12: 1548–1557.

Prince, J.T., Nishiyama, A., Healy, P.A., Beasley, L. and Stallcup, W.B. (1992). Expression of the F84-1 glycoprotein in the spinal cord and the cranial nerves of the developing rat. *Dev. Brain Res.*, 68: 193–202.

Ramon Y Cajal, S. (1911) *Histologie du Système Nerveux de l'Homme et des Vertébrés* (reprinted by Consejo Superior de Investigaciones Cientificas, Madrid, 1955), Maloine, Paris.

Ramos, R.G.P., Igloi, G.L., Lichte, B., Baumann, U., Maier, D., Schneider, T., Brandstätter, H.J., Fröhlich, A. and Fischbach, K.-F. (1993) The irregular chiasm C-roughest locus of *Drosophila*, which affects axonal projections and programmed cell death, encodes a novel immunoglobulin-like protein. *Genes Dev.*, 7: 2533–2547.

Simeone, A., Acampora, D., Gulisano, M., Stornaiuolo, A. and Boncinelli, E. (1992). Nested expression of four homeobox genes in developing rostral brain. *Nature*, 358: 687–690.

Takagi, S., Tsuji, T., Kinutani, M. and Fujisawa, H. (1989) Monoclonal antibodies against species-specific antigens in the chick central nervous system: putative application as transplantation markers in the chick-quail chimeras. *J. Histochem. Cytochem.*, 37: 177–184.

Tan, K. and Le Douarin, N.M. (1991) Development of the nuclei and cell migration in the medulla oblongata – application of the quail-chick chimera system. *Anat. Embryol.*, 183: 321–343.

Tanaka, H., Kinutani, M., Agata, A., Takashima, Y. and Obata, K. (1990) Pathfinding during spinal tract formation in Quail-Chick chimera analysed by species specific monoclonal antibodies. *Development*, 110: 565–571.

Tanaka, H., Matsui, T., Agata, A., Tomura, M., Kubota, I., Mcfarland, K.C., Kohr, B., Lee, A., Phillips, H.S. and Shelton, D.L. (1991) Molecular cloning and expression of a novel adhesion molecule, SC1. *Neuron.*, 7: 535–545.

Teillet, M.A., Naquet, R., Lasalle, G.L., Merat, P., Schuler, B. and Le Douarin, N.M. (1991) Transfer of genetic epilepsy by embryonic brain grafts in the chicken. *Proc. Natl. Acad. Sci. USA*, 88: 6966–6970.

Thomas, K.R. and Capecchi, M.R. (1990) Targeted disruption of the murine int-1 proto-oncogene resulting in severe abnormalities in midbrain and cerebellar development. *Nature*, 346: 847–850.

Wenger, C. (1950) An experimental analysis of relations between parts of the brachial spinal cord of the embryonic chick. *J. Exp. Zool.*, 114: 51–91.

Yamada, T., Placzek, M., Tanaka, H., Dodd, J. and Jessell, T.M. (1991) Control of cell pattern in the developing nervous system: polarizing activity of the floor plate and notochord. *Cell*, 64: 635–647.

F. Bloom (Editor)
Progress in Brain Research, Vol. 100

CHAPTER 2

Development of sympathetic neurons: neurotransmitter plasticity and differentiation factors

Story C. Landis

Department of Neurosciences, Case Western Reserve University, School of Medicine, Cleveland, OH 44106, USA

Introduction

Almost two decades ago, studies of sympathetic neurons developing in cell culture revealed that postmitotic neurons could alter their neurotransmitter phenotype (Landis and Patterson, 1981). By growing neurons dissociated from the superior cervical ganglia of newborn rats in the presence of non-neuronal cells, such as heart or skeletal muscle, noradrenergic neurons were induced to become functionally cholinergic. The induction of cholinergic properties by co-culture with non-neuronal cells does not require cell contact but occurs through the release of a soluble cholinergic inducing activity into the medium. That the cholinergic factor produced by non-neuronal cells causes individual neurons that express noradrenergic properties to become cholinergic was formally demonstrated in studies of neuron/heart microcultures in which the changing transmitter properties of single neurons were followed over time (Potter et al., 1986). While initial analyses in this system focussed on the classical transmitters, norepinephrine and acetylcholine, subsequent studies indicated neuropeptide expression by sympathetic neurons is also influenced by the same classes of environmental factors (Kessler, 1985; Nawa and Sah, 1990).

It is now clear that neurotransmitter plasticity analogous to that initially uncovered in vitro occurs in vivo, both during development and following injury. Further, significant progress has been made in identifying the differentiation factors that affect sympathetic neurotransmitter phenotype. Two of these, leukemia inhibitory factor (LIF) and ciliary neurotrophic factor (CNTF), are members of a recently described family of cytokines, and have actions on hematopoietic cells and neurons, including sympathetic, sensory and motor neurons (Patterson, 1992). The present challenge is to define where and how these cytokines and other differentiation factors normally act, not only to influence the properties of sympathetic neurons but also those of the other classes of responsive neurons. The recent generation of transgenic mice deficient in LIF and CNTF will aid in this endeavor (Rao et al., 1993a; Masu et al., 1993).

Transmitter plasticity occurs during development and after injury in vivo

To determine whether the neurotransmitter phenotype of sympathetic neurons is plastic during development in vivo as in vitro, we examined how sympathetic neurons acquire their adult complement of transmitters (Landis, 1990). Mature sympathetic neurons possess an extensive repertoire of neurotransmitter phenotypes: while many are noradrenergic, a minority are cholinergic and most contain one or more neuropeptides selected from the following: neuropeptide Y, enkephalin, vasoactive intestinal peptide (VIP), somatostatin or calcitonin gene-related peptide (CGRP). Initially most, if not all, sympathetic neurons express noradrenergic properties, including those that in the adult will be cholinergic. The best characterized cholinergic sympathetic neurons innervate sweat glands, concentrated in footpads. The properties expressed by the developing sweat gland innervation are strikingly different from those expressed by the mature innervation. When axons contact the developing glands, they exhibit noradrenergic properties and the cholinergic and peptidergic markers that characterize the mature innervation are undetectable. Choline acetyltransferase activity and VIP and CGRP immunoreactivities appear during the second and third postnatal weeks as the expression of noradrenergic properties decreases.

The changes in transmitter properties of sweat gland neurons are retrogradely specified by interactions with the target tissue (Landis, 1990). When sweat gland primordia in early postnatal rats are replaced with parotid gland, a target receiving noradrenergic sympathetic innervation, the sweat gland neurons that innervate the transplanted parotid gland retain catecholamines and fail to develop choline acetyltransferase or VIP. In addition, the terminals of presumptive sweat gland neurons that innervate the footpads of *Tabby* mutant mice that lack sweat glands fail to acquire the normal cholinergic and peptidergic phenotype. Conversely, when sweat gland-containing skin is transplanted to the lateral thorax so that sympathetic neurons that would normally innervate noradrenergic targets in hairy skin innervate sweat glands instead, the innervation of the transplanted glands exhibits properties appropriate for the novel target rather than hairy skin. Although catecholamine containing fibers initially form a plexus in the transplanted glands, by 6 weeks catecholamines have largely disappeared from the sympathetic axons and choline acetyltransferase activity and VIP have been induced.

Striking changes occur in neuropeptide expression in adult sympathetic neurons in vivo after injury. For example, the superior cervical ganglion normally contains little VIP and only an occasional neuron is immunoreactive. If the carotid nerves carrying postganglionic axons of neurons in the superior cervical ganglion are cut, however, there is a dramatic induction of VIP protein and mRNA which is accompanied by a significant increase in the number of intensely VIP immunoreactive neurons (Hyatt-Sachs et al., 1993). Similarly, little substance P (SP) is present, no immunoreactive cell bodies are detectable and only occasional fibers contain SP. After axotomy, SP content and tachykinin mRNA increase and numerous SP-immunoreactive cell bodies appear (Rao et al., 1993b). The very large increases observed after cutting postganglionic axons contrast with the modest increases seen after section of the superior cervical trunk carrying preganglionic input to the ganglion. Since some neurons in the superior cervical ganglion send axons through the cervical sympathetic trunk, the small increases observed after cutting the cervical sympathetic trunk may arise from axotomy rather than denervation.

Cell culture studies provide evidence that ganglionic non-neuronal cells play a role in mediating neuropeptide induction that occurs in response to axotomy. For example, co-culture with ganglionic non-neuronal cells, fibroblasts, satellite and Schwann cells, increases VIP content in dissociated cell cultures and this effect is mimicked by treatment of neuron-enriched cultures by medium conditioned by ganglionic non-neuronal cells (Sun et al., 1994). VIP is also induced when adult ganglia are placed in explant culture and medium conditioned by explanted ganglia induces VIP in neuron-enriched cultures (Sun et al., 1994). How axotomy causes ganglionic non-neuronal cells to produce peptide-inducing factors is unknown but it seems likely that both a retrograde signal initiated by axon transection and intercellular signalling between axotomized neuron cell bodies and ganglionic non-neuronal cells are involved.

The neurotransmitter plasticity displayed by sympathetic neurons is not unique. Examination of the development of transmitter properties has disclosed a number of examples of altered expression of transmitter synthetic enzymes and neuropeptides and suggests that not only quantitative, but also qualitative, changes in transmitter expression are common (Patterson and Nawa, 1993). The most thoroughly studied example is the transient catecholaminergic cells of the gut which give rise to many if not all the neurons, cholinergic, serotonergic and peptidergic, of the enteric nervous system (Baetge et al., 1990). Gershon and colleagues have put forward the hypothesis that the local environment, rather than the target, provides the instructive cues that result in the acquisition of the adult transmitter phenotype. Increases in neuropeptide expression, both peptide content and its mRNA, are also a common consequence of axotomy in peripheral neurons. As in sympathetic neurons, following nerve section or crush, VIP and galanin are induced in dorsal root ganglion neurons and CGRP and galanin increase in motor neurons.

Cholinergic differentiation factors are members of the neuropoietic cytokine family

Two proteins have been identified that induce cholinergic function in cultured sympathetic neurons. Cholinergic differentiation factor (CDF), an approximately 45 kDa glycoprotein, was purified from heart cell conditioned medium. It has subsequently been found to be identical to leukemia inhibitory factor, a multifunctional cytokine with effects of a variety of non-neuronal cells (Yamamori et al., 1989). The second cholinergic factor, ciliary neurotrophic factor, was purified from sciatic nerve on the basis of its ability to support the survival of chick ciliary neurons and was subsequently shown to induce cholinergic function in cultured sympathetic neurons. Pattern-based sequence comparison and homology modeling of protein fold suggest that CDF/LIF and CNTF are members of a family of hemopoietic cytokines that also includes interleukin-6, granulocyte colony-stimulating factor and oncostatin M (Patterson, 1992). Further evidence that the grouping of at least CDF/LIF, CNTF, interleukin-6 and oncostatin M together based on their predicted structure is correct comes from examination of their candidate receptor molecules (Ip et al., 1992). The evidence at present suggests that the CDF/LIF receptor is comprised of two subunits, gp 130 and LIFRβ, and that this receptor can be converted into a CNTF receptor by the addition of a CNTFRα subunit. At least one of these subunits, gp130, also contributes to the IL-6 and oncostatin M receptors. The development of a rapid reverse transcriptase polymerase chain reaction screen for the effects of candidate molecules on transmitter expression by sympathetic neurons will certainly add new differentiation factors to the present list (Fann and Patterson, 1993).

While the first neuronal action identified for LIF was the induction of cholinergic function in cultured sympathetic neurons, an ever increasing list of actions on neurons in cell cultures is being compiled. Further, to the extent that comparisons have been made, many of these activities are shared with CNTF. LIF and CNTF both alter the expression of neuropeptides in cultured sympathetic neurons, increasing VIP, SP and somatostatin and decreasing NPY. They increase the survival of motor neurons and choline acetyltransferase activity in cultured motor neurons (Arakawa et al., 1990; Martinou et al., 1992) while LIF decreases the expression of tyrosine hydroxylase in cranial sensory neurons (Fan and Katz, 1993). It also increases the production of neurons from spinal cord precursors and sensory neurons from neural crest (Murphy et al., 1991). Both LIF and CNTF support the survival of oligodendrocytes in cell culture (Barres et al., 1993). One interesting exception to the list of common activities is the ability of CNTF but not LIF to support the survival of chick ciliary neurons. In a small number of cases, these analyses of the actions of exogenous CNTF and LIF have been extended to in vivo systems. For example, CNTF has been shown to support the survival of axotomized neonatal motor neurons (Sendtner et al., 1990).

The identification of the cholinergic differentiation factor present in heart cell conditioned medium as LIF as well as other reports on the effects of interleukins on neuron survival and differentiation has resulted in the recognition that cytokines acting on hematopoietic cells may also act in the developing nervous system. Developing a catalogue of candidate effects using reduced culture systems should expedite elucidation of the roles of these factors in vivo.

Do the cholinergic differentiation factors influence sympathetic neurotransmitter phenotype in vivo?

Several lines of evidence suggest that the noradrenergic/cholinergic transmitter switch induced in the sweat gland innervation is not mediated by either LIF or CNTF, but by a novel member of the cytokine family. Extracts of developing and adult rat footpads contain an activity that induces choline acetyltransferase activity and VIP and reduces catecholamines and tyrosine hydroxylase in cultured sympathetic neurons (Rao et al., 1992; Rohrer, 1992). Sweat glands appear to be the source of this activity since significantly less cholinergic inducing activity is extracted from footpads of Tabby mutant mice that lack sweat glands than from footpads of normal mice. The cholinergic inducing activity in the extract is not immunoprecipitated or blocked by LIF antisera (Rao et al., 1992). Consistent with this finding, the neurotransmitter properties of the sweat gland innervation in transgenic mice which are deficient in LIF is indistinguishable from normal (Rao et al., 1993a). While a majority of the cholinergic inducing activity can be immunoprecipitated from footpad extracts with CNTF antisera (Rao et al., 1992; Rohrer, 1992), immunoblot and northern blot analysis did not reveal the presence of authentic CNTF protein or mRNA even though both were detectable in sciatic nerve which approximately equivalent amounts of cholinergic inducing activity as footpads (Rao et al., 1992). Recent data from transgenic mice that lack CNTF indicate, as in the case of the LIF deficient mice, that the neurotransmitter properties of the sweat gland innervation are normal (Masu et al., 1993). Taken together, these observations suggest that the sweat gland differentiation factor is related to, but distinct from, CNTF.

While LIF does not appear to be responsible for the adrenergic/cholinergic switch that takes place in the sweat gland innervation, it does play an important role in the induction of neuropeptides in adult sympathetic neurons after axotomy. LIF induces VIP and SP, peptides induced by axotomy in vivo, in sympathetic neurons in dissociated cell culture (Nawa and Patterson, 1990). LIF mRNA rises within sympathetic ganglia within several hours after axotomy. VIP and SP are induced when sympathetic ganglia are placed in explant culture, presumably because this entails axotomy. This induction is significantly suppressed when function-blocking LIF antibodies are included in the medium (Sun et al., 1994). Finally, the induction of VIP and neurokinin A, which like SP is derived from the β-preprotachykinin mRNA, is largely absent when superior cervical ganglia of LIF deficient mice are explanted into culture or axotomized in situ (Rao et al., 1993a). It is of interest that the role of LIF revealed in these studies while not that of directing the noradrenergic/cholinergic switch initially predicted by the cell culture studies is nonetheless consistent with them. LIF does mediates a change in neurotransmitter properties, peptide expression, in sympathetic neurons but does so in a different context than anticipated, in response to injury and not during development.

Small molecule transmitters, like the cytokines, function as differentiation factors

Analysis of the developing sweat gland innervation has revealed that establishing a functional synapse requires not only retrograde signalling but also anterograde. First, production of the cholinergic differentiation factor by sweat glands requires noradrenergic innervation (Habecker and Landis, 1993). In culture, sweat gland cells produce cholinergic inducing activity only when they are co-cultured with sympathetic neurons. This effect is blocked by adrenergic antagonists, suggesting that it is mediated by catecholamines. Similarly, the cholinergic inducing activity present in extracts of footpads of rat pups was reduced when the glands were surgically denervated (Rohrer, 1992) and eliminated when the sympathetic innervation was specifically lesioned at birth by treatment with 6-hydroxydopamine (Habecker and Landis, 1993). Second, cholinergic innervation is required for the development of secretory responsiveness (Grant and Landis, 1994). During development, the onset of sweat secretion in response to nerve stimulation or cholinergic agonists occurs after the appearance of cholinergic properties in the innervation and glands respond to agonists only if activated by nerve stimulation. When innervation and the transmitter switch is delayed, the development of physiological responsiveness is also delayed. Finally, adult rats, sympathectomized at birth, do not sweat. These data suggest that a factor(s) associated with the cholinergic innervation is required for gland function and acetylcholine is an excellent candidate. When we disrupted transmission in developing rats with the muscarinic antagonist, atropine, the acquisition of secretory function was prevented. After atropine was withdrawn, responsiveness developed. Thus, activation of muscarinic receptors is responsible for the induction and also maintenance of secretory responsiveness.

Our observations on the development of sympathetic neurons provide evidence that complex reciprocal interactions may be required to establish functional synapses between these neurons and their target tissues. Further, in addition to the instructive role(s) that member of the neuropoietic cytokine family play, we have found that the transmitters, norepinephrine and acetylcholine, act as differentiation signals.

References

Arakawa, Y., Sendtner, M. and Thoenen, H. (1990) Survival effects of ciliary neurotrophic factor (CNTF) on chick embryonic motorneurons in culture: comparison with other neurotrophic factors and cytokines. *J. Neurosci.*, 10: 3507–3515.

Baetge, G., Pintar, J.E. and Gershon, M.D. (1990) Transiently catecholaminergic (TC) cells in the bowel of the fetal rat: precursors of noncatecholaminergic enteric neurons. *Dev. Biol.*, 141: 353–380.

Barres, B., Schmid, R., Sendtner, M. and Raff, M. (1993) Multiple extracellular signals are required for long-term oligodendrocyte survival. *Development*, 118: 283–295.

Fan, G. Katz, D. (1993) Non-neuronal cells inhibit catecholaminergic differentiation of primary sensory neurons: role of leukemia inhibitory factor. *Development*, 118: 83–93.

Fann, M. and Patterson, P.H. (1993) A novel approach to screen for cytokine effects on neuronal gene expression. *J. Neurochem.*, 61: 1359–1355.

Grant, M. and Landis, S. (1994) Induction and maintenance of secretory responsiveness in sweat glands by acetylcholine. *J. Neurosci.*, submitted.

Habecker, B. and Landis, S. (1993) Noradrenergic transmission influences sweat gland cholinergic differentiation factor production. *Soc. Neurosci. Abstr.*, 19: 710.11.

Hyatt-Sachs, H., Schreiber, R., Bennett, T. and Zigmond, R. (1993) Phenotypic plasticity in adult sympathetic ganglia in vivo: effects of deafferentation and axotomy on the expression of vasoactive intestinal peptide. *J. Neurosci.*, 13: 1642–1653.

Ip, Y., Nye, S., Boulton, T., Davis, S., Taga, T., Li, Y., Birren, S., Yasukawa, K., Kishimoto, T., Anderson, D. and Yancopoulos, G. (1992) CNTF and LIF act on neuronal cells via shared signalling pathways that involve the IL-6signal transducing receptor component gp130. *Cell*, 69: 1121–1132.

Kessler, J.A. (1985) Differential regulation of peptide and catecholamine characters in cultured sympathetic neurons. *Neuroscience*, 15: 827–839.

Landis, S.C. (1990) Target regulation of neurotransmitter phenotype. *Trends Neurosci.*, 13: 344–350.

Landis, S.C. and Patterson, P.H. (1981) Neural crest cell lineages. *Trends Neurosci.*, 4: 1172–175.

Martinou, J., Martinou, I. and Kato, A. (1992) Cholinergic differentiation factor (CDF/LIF) promotes survival of isolated rat embryonic motoneurons in vitro. *Neuron*, 8: 737–744.

Masu, Y., Wolf, E., Holtmann, B., Sendtner, M., Brem, G. and Thoenen, H. (1993) Disruption of the CNTF gene results in motor neuron degeneration. *Nature*, 365: 27–32.

Murphy, M., Reid, K., Hilton, D. and Bartlett, P. (1991) Generation of sensory neurons is stimulated by leukemia inhibitory factor. *Proc. Natl. Acad. Sci. USA*, 88: 3498–3501.

Nawa, H. and Patterson , P. (1990) Separation and partial characterization of neuropeptide-inducing factors in heart cell conditioned medium. *Neuron*, 4: 269–277.

Nawa, H. and Sah, D.W. (1990) Different biological activities in conditioned media control the expression of a variety of neuropeptides in cultured sympathetic neurons. *Neuron*, 4: 279–287.

Patterson, P. (1992) The emerging neuropoietic cytokine family: first CDF/LIF, CNTF and IL-6; next ONC, MGF GCSF? *Curr. Opinions Neurobiol.*, 2: 94–97.

Patterson, P.H. and Nawa, H. (1993) Neuronal differentiation factors/cytokines and synaptic plasticity. *Cell*, 72: 123–137.

Potter, D.D., Landis, S.C., Matsumoto, S.G. and Furshpan, E.J. (1986) Synaptic functions in rat sympathetic neurons in microcultures. II. Adrenergic/cholinergic dual status and plasticity. *J. Neurosci.*, 6: 1080–1096.

Rao, M.S., Patterson, P.H. and Landis, S.C. (1992) Multiple cholinergic differentiation factors are present in footpad extracts: comparison with known cholinergic factors. *Development*, 116: 731–744.

Rao, M.S., Escary, J., Sun, Y., Perreau, J., Patterson, P.H., Zigmond, R.E., Brulet, P. and Landis, S.C. (1993a) Leukemia inhibitory factor mediates an injury response but not a target-mediated developmental transmitter switch in sympathetic neurons. *Neuron*, 11: 1–12.

Rao, M., Sun, Y., Vaidyanathan, U., Landis, S. and Zigmond, R. (1993b) Regulation of substance P is similar to that of vasoactive intestinal peptide after axotomy or explantation of the rat superior cervical ganglion. *J. Neurobiol.*, 24: 571–580.

Rohrer, H. (1992) Cholinergic neuronal differentiation factors: evidence for the presence of both CNTF-like and non-CNTF-like factors in developing footpad. *Development*, 114: 689–698.

Sendtner, M., Kreutzberg, G.W. and Thoenen, H. (1990) Ciliary neurotrophic factor prevents the degeneration of motor neurons after axotomy. *Nature*, *345*: 440–441.

Sun, Y., Rao, M.S., Zigmond, R.E. and Landis, S.C. (1994) Regulation of vasoactive intestinal peptide expression in sympathetic neurons in culture and after axotomy: the role of cholinergic differentiation factor/leukemia inhibitory factor. *J. Neurobiol.*, 25: in press.

Yamamori, T., Fukada, K., Aebersold, R., Korsching, S., Fann, M.J. and Patterson, P.H. (1989) The cholinergic neuronal differentiation factor from heart cells is identical to leukemia inhibitory factor. *Science*, 246: 1412–1416.

F. Bloom (Editor)
Progress in Brain Research, Vol. 100

CHAPTER 3

Lessons from genetic knockout mice deficient in neural recognition molecules

Melitta Schachner

Department of Neurobiology, Swiss Federal Institute of Technology, 8093 Zurich, Switzerland

Introduction

A variety of cell surface glycoproteins expressed by neurons and glia have been recognized as important mediators of recognition among neural cells, determining the specificity of cell interactions during development and during functional maintenance, regeneration and modification of synaptic activity in the adult. Most of the recognition molecules' functions have been derived from perturbation experiments in vitro using antibodies and the isolated recognition molecules themselves as functional blockers and ligands or competitors, respectively. As the ultimate test for a particular recognition molecule's function, however, its action in the intact organism should be observed. With the advent of recombinant DNA technology and availability of embryonic stem cells, it has been possible to ablate the genes encoding neural recognition molecules in the mouse and to study the nervous system in the molecule's absence, with the hope of eventually identifying its role in a complex cellular environment.

The present review summarizes our experience with genetic knockout mutants of the mouse deficient in three neural recognition molecules: (1) the major peripheral myelin glycoprotein of mammals, the immunoglobulin superfamily derived recognition molecule PO; (2) the adhesion molecule on glia AMOG, a recognition molecule and integral component of the Na,K-ATPase; and (3) the minor glycoprotein of central and peripheral nervous system myelin forming cells, the myelin-associated glycoprotein MAG. The three molecules are all mostly expressed in the nervous system and therein predominantly produced by glial cells at later stages of their development. It was hoped that this spatial and temporal restriction in expression would be beneficial for the production of a viable mutant animal. Analysis of the three mutants shows the range of phenotypes that can be expected from general considerations on knockout strategies. Our findings carry implications for future strategies intended to elucidate the functional role of a neural recognition molecule by genetic ablation.

All three mutants were generated by using homologous recombination in embryonic stem cells to replace the endogenous genes on the mouse chromosome by an insertionally inactivated gene (see references cited below).

The PO knockout mouse or the expected abnormal phenotype (Giese et al., 1992)

PO is the major protein of peripheral myelin of mammals. It is uniquely expressed by myelinating Schwann cells of the mammalian peripheral nervous system and accounts for 60% of the protein in the myelin sheath. It is also expressed in the compacted myelin throughout adulthood. A member of the immunoglobulin superfamily containing only one immunoglobulin-like domain, PO may engage in homophilic binding within the surface membrane of the same cell (cis-interaction) or between apposing surface membranes (trans-interaction) of myelinating Schwann cells (see Giese et al. (1992) for references).

The behaviour of the PO knockout mice was apparently normal until 2 weeks postnatally when the mice showed weak vibrations when lifted by the tail. Four-week-old mutants showed clasping of hindlimbs when lifted by the tail, uncoordinated swimming performance, slight tremors, and dragging or jerking movements of the hindlimbs. With increasing age, these behavioural traits became more pronounced, with some showing convulsions, and self-mutilation and consistently weak fore- and hindlimbs, but nevertheless without paralysis. Mice survived, with the oldest mouse maintained being now 14 months old. This abnormal mutant phenotype is reminiscent of some genetically transmitted peripheral neuropathies in humans, such as the polyneuropathy of the Charcot-Marie-Tooth or Dejerine-Sottas types. Indeed, mutations in PO have been found in these diseases (Kulkens et al., 1993; Hayasaka et al., 1993a,b; Su et al., 1993).

When peripheral nerves from 9 to l0-week-old mice were inspected for morphological abnormalities, a high degree of hypomyelination of larger calibre axons was conspicuous. Axon-Schwann cell units achieved a normal one-to-one ratio of association, but myelin-like sheaths formed fewer turns around axons with much less membrane compaction than in wild-type mice. Non-myelinating Schwann cells appeared normal. Some axons were only covered by a basal lamina, indicating the earlier presence of a Schwann cell, while still others contained myelin-like figures typically seen in Schwann cells undergoing Wallerian degeneration.

As an important prerequisite to understand the mutant phenotype, we investigated whether other neural recognition molecules known to be expressed during myelination in peripheral nerves are present in the mutant. The neural recognition molecule Ll is normally expressed by all premyelinating Schwann cells and downregulated when myelination starts. This downregulation was also observed in the mutant, in that myelin-like figures were Ll-negative. In contrast to Ll, NCAM, which is normally downregulated at the onset of myelination, remains highly expressed in the mutant in what would normally be myelinating Schwann cells. Other molecules showing a characteristic developmental regulation of expression in normal myelinating Schwann cells were also abnormally expressed; the low affinity nerve growth factor receptor, MAG and the extracellular matrix glycoprotein tenascin, all of which are downregulated with the onset of normal myelination, were instead highly expressed in the mutant. The proteolipid protein which is normally hardly detectable in the cytoplasm of myelinating Schwann cells is present in the myelin-like figures of the mutant. The myelin basic protein which is highly expressed in the compact myelin of normal mice is downregulated in the mutant. Two carbohydrate structures which are expressed by overlapping sets of neural recognition molecules (see Horstkorte et al. (1993) and Hall et al. (1993) for references) are abnormally absent, in the case of the L2/HNK-l carbohydrate, or normally expressed, in the case of the oligomannosidic L3 carbohydrate.

These results show that the majority of molecules that appear to be functionally involved in myelination and other molecules with as yet unknown functions in peripheral nervous system myelination are severely dysregulated in the mutant.

Insights into the function of PO derived from the PO knockout mutant

Although the histological defects observed in PO knockout mice could be called expected, since they are largely consistent with models of PO functions advanced previously on the basis of in vitro perturbation experiments, the interpretation of the mutant phenotype appears more complex. On the one hand, the erratic pattern of secondary effects on Schwann cell gene expression does not allow the conclusion that all abnormal features of the mutant are directly due to the absence of PO. Of the nine marker molecules investigated, only two, namely Ll and the L3 oligomannosidic carbohydrate, showed the expression expected from the wildtype. Given these multiple examples of abnormal gene expression in the mutant, it is plausible to assume that certain features of the mutant phenotype are not directly due to the absence of PO. For example, the continued expression of NCAM and MAG could be the cause rather than the consequence of one or all aspects of myelin abnormalities. On the other hand, it is also conceivable that the ability of some Schwann cell processes to engage in a limited amount of spiralling may be due to the continued expression of neural recognition molecules, such as NCAM and MAG. Possibly even other dysregulated molecules, such as proteolipid protein could at least partially compensate for the absence of PO. Our observations thus emphasize that it is not sufficient to analyze the behavioural and histological pheno-types of a mutant, but that a detailed analysis of secondary effects on gene expression may be necessary for a complete understanding of the mutant pheno-type.

The adhesion molecule on glia (AMOG) knockout mouse or the interpretable abnormal phenotype (Magyar et al., 1993)

AMOG was first described on the basis of in vitro experiments showing that antibodies against AMOG inhibited the migration of cerebellar granule cells along Bergmann glial processes (see Müller-Husmann et al. (1993) for references). AMOG is hardly detectable outside the central nervous system and is mainly expressed by glial cells, but also by certain types of neurons. Its expression is first detectable in the brain at late embryonic ages, increases during the first 2 weeks after birth along with the general maturation of the brain and reaches highest levels in the adult. Sequence analysis of AMOG revealed it to be a close homologue of the $\beta 1$ subunit of the Na,K-ATPase. The rat and human species homologues of AMOG were identified also as the $\beta 2$ subunit of the enzyme by low stringency hybridization using a $\beta 1$ subunit probe.

AMOG is a recognition molecule by several operational criteria. AMOG is tightly associated with the α subunits of the Na,K-ATPase, which co-purify with AMOG during stringent immunoaffinity chromatography purification procedures. The functional integration of AMOG into an ion pump introduces a new concept in the link between cell recognition and signal transduction: coupling of cell recognition with ion transport implicates cell interactions in the regulation of ionic homeostasis and the cellular parameters dependent on the ionic environment such as voltage-dependent ion channels, size of extracellular space volume and cell volume.

AMOG knockout mutants were behaviourally unremarkable during their first 2 weeks. At 14–15 days, mutant mice showed reduced righting behaviour and orientation when lifted by the tail. Within only 2–3 days, this motor incoordination rapidly worsened and AMOG mutant mice developed paralysis of their forelimbs and were unable to hold their heads upright. The hind limbs started to become tremorous and the animals were no longer able to stand. Shortly before their death at days 17 or 18, mutants were generally very weak and lay on their side but still showed normal grasping reflexes and responded to sound.

The abnormal phenotype was immediately obvious when inspecting the size of the lateral and third ventricles which were considerably enlarged in comparison to wild-type or heterozygous littermates. While the cortices of cerebellum and cerebrum, the hippocampus and optic nerve showed no histological abnormalities, the brain stem, thalamus, striatum and, to a lesser degree, spinal cord were all abnormal: Swollen cellular processes and vacuoles were detectable in frequent association with blood vessels. The most likely interpretation of these observations is that the vacuoles result from swelling and subsequent degeneration of astrocytic processes. In contrast to the PO mutant, other recognition molecules, such as NCAM, Ll or MAG were expressed at similar levels in AMOG mutants and wild-type littermates. Also, the $\beta 1$ subunit of Na,K-ATPase was normally expressed in the mutant. However, some dysregulation in the levels of the α subunits was seen, which were reduced in the mutant.

Insights into the function of AMOG derived from the mutant

The abnormal phenotype of the mutant begs the question as to the relation between cause and consequence of the lack of AMOG expression. Also, the question as to which of the two functional roles indicated for AMOG, namely cell-cell recognition on the one hand and pump activity on the other, needs consideration.

The mutant shows no evidence that AMOG plays a profound morphogenetic role during formation of the hippocampus, optic nerve, and cerebellar and cerebral cortices. Although antibodies against AMOG were initially found to interfere with granule neuron migration along Bergmann glial cells in vitro, this disturbance may be rather the consequence of an abnormal pump activity that can be elicited by the antibodies (Gloor et al., 1990) than caused by a defect in cell recognition between the two interacting partner cell types. It is also conceivable that other recognition molecules known to be involved in the migration process, such as tenascin, thrombospondin or Ll are able to compensate for the defect in AMOG (see Magyar et al. (1993) for references). From these observations it would appear necessary to analyze more systematically at different developmental stages whether AMOG plays a subtle, yet detectable role in morphogenetic cell interactions.

From the observations on the mutant phenotype it is evident that, without prior knowledge of the molecule's function, the mutant abnormalities would have been difficult to interpret on a molecular basis. Since spongiform encephalopathies resulting in neurodegeneration have been observed in many pathological situations, the non-uniform degeneration of certain cell types would have been impossible to rationalize. On the other hand, some of the functions deduced from in vitro perturbation experiments are less prominent in the animal than expected from these experiments, rendering a profound role of AMOG during morphogenesis more unlikely. Rather, it is conceivable that AMOG would be more instrumental in recognition-mediated triggering of pump activity in the adult, when expression of the molecule is highest. Despite these uncertainties in interpretation, the abnormalities of the mutant phenotype have allowed the design of further experiments to probe the complex functional roles of AMOG in the central nervous system.

The myelin-associated glycoprotein (MAG) knockout mouse or the enigmatic normal phenotype (unpublished observations)

MAG is a transmembrane glycoprotein of the immunoglobulin superfamily which is heavily glycosylated in its extracellular domain (see Schneider-Schaulies et al. (1991) for references). It occurs in developmentally regulated, alternatively spliced forms in the central and peripheral nervous system. MAG is also an adhesion molecule by several criteria. In addition to its adhesive role, MAG can also promote neurite outgrowth, again by a heterophilic mechanism. From these observations and the fact that MAG expresses the functionally important L2/HNK-l carbohydrate (Martini et al., 1992), a neurite outgrowth promoting role of MAG during regeneration in the peripheral nervous system could be envisaged.

No overtly abnormal phenotype of the mutant has been detected so far, through ages of several months. Behaviourally, the mutants show no gross defects. Histologically, myelin in the optic and sciatic nerves of 8-week-old mutants had an overall normal appearance at the light microscopic level.

Insights into the function of MAG derived from the mutant at the present stage of investigations

Although inferences about the function of MAG during the initiation of the myelination process were derived from the precise timing of its expression at the onset of myelination, the MAG mutant does not support the notion that MAG is an essential ingredient in myelin formation. It is at present difficult to reconcile the observations on the mutant with those resulting from in vitro experiments using anti-sense RNA approaches (Owens and Bunge, 1991). It is, however, conceivable that acute ablations in cell culture are less prone to evoke compensatory mechanisms than a chronic ablation as created by the knockout situation. What these compensatory mechanisms might be will need to be determined. Also, a more detailed analysis of myelin morphology in the adult and at early formative stages will need to be performed to be sure that subtle abnormalities may not have escaped detection. Again, knowledge of MAG's discrete temporal and spatial expression pattern and of some of its functions in vivo will make this search a focussed one. On the other hand, MAG may not have an essential function and may be an evolutionary vestige that has been retained during phylogeny for optimization of a function which could be carried out, in the absence of MAG, also by another or several other recognition molecules. However, in the absence of a presently detectable abnormal phenotype, all speculations about the possible causes of the apparent normality remain unfounded.

Conclusions

Ablating molecules that appear to be of significance in development and maintenance of nervous system functions appear to represent an intriguing, but also risky approach. In this short review, three mutants were characterized that cover a broad spectrum of expectations from the knowledge on the structure and in vitro functions of the molecules. On the one hand, the PO mutant presents itself indeed as expected, but with the caveat that other mechanisms related indirectly to the absence of PO may account for the observed abnormalities. Such indirect, either dysregulatory or compensatory effects will have to be taken more into consideration in the interpretation of any mutant phenotype.

The adhesion molecule on glia (AMOG) knockout mutant displays a very complex phenotype that is both expected and unexpected in the sense that the dual function of the molecule does not become apparent. An abnormality in pump activity was expected, but somewhat unexpected was that no morphogenetic aberrations have so far been observed. With the knowledge of the structure and function of the molecule and its spatial and temporal expression in the nervous system, the design of further experiments that could clarify the evolvement of the mutant's abnormal phenotype is now feasible.

The other extreme of the mutant's contribution to an understanding of a molecule's function is represented by the MAG mutant which so far has not yielded any abnormal phenotype. One could argue that, as an evolutionary vestige, MAG is not an essential ingredient in the formation and maintenance of myelin and that compensatory mechanisms could come into play that may attribute to MAG a superfluous or, at the most, an optimization role in the myelination process. On the other hand, it could very well be that the present level of investigations has not been detailed enough to allow subtle abnormalities to be recognized. Until such possibilities are exhausted, the gene's function in vivo will remain elusive.

Given these three examples of mutant phenotypes, the general usefulness of knockout strategies ablating an entire gene will need to be reconsidered. It seems that in future strategies, more subtle ablation methods should be introduced in order to minimize compensatory mechanisms. One will have to invest heavily into acute knockout possibilities as afforded by recombinant events that are inducible by external stimuli. As complementary experimental strategies, acute manipulations of the animal using blocking antibodies, antisense oligonucleotide and RNA approaches or competing soluble fragments of the recognition molecule under study should be undertaken. Although these methods also have their drawbacks, with these combined possibilities ahead, the analysis of the function of neural recognition molecules in the authentic environment in vivo will hopefully attain its full power.

References

Giese, P., Martini, R., Lemke, G., Soriano, P. and Schachner, M. (1992) Mouse PO gene disruption leads to hypomyelination, abnormal expression of recognition molecules and degeneration of myelin and axons. *Cell*, 71: 565–576.

Gloor, S., Antonicek, H., Sweadner, K.J., Pagliusi, S., Frank, R., Moos, M. and Schachner, M. (1990) The adhesion molecule on glia (AMOG) is a homologue of the β subunit of the Na,K-ATPase. *J. Cell Biol.*, 110: 165–174.

Hall, H., Liu, L., Schachner, M. and Schmitz, B. (1993) The L2/HNK-l carbohydrate mediates adhesion of neural cells to laminin. *Eur. J. Neurosci.*, 5: 34–42.

Hayasaka, K., Himoro, M., Sato, W., Takatta, G., Uyemura, K., Shimizu, M., Bird, T.D., Coneally, P.M. and Chance, P.F. (1993a) Charcot-Marie-Tooth neuropathy type IB is associated with mutations of the myelin P0 gene. *Nature Genet.*, 5: 31–34.

Hayasaka, K., Himoro, M., Swaishi, Y., Nanao, K., Takahashi, T., Takada, G., Nicholson, G.A., Ouvrier, R.A. and Tachi, N. (1993b) De novo mutation of the myelin PO gene in Dejerine-Sottas disease (hereditary motor and sensory neuropathy type III). *Nature Genet.*, 5: 266–268.

Horstkorte, R., Schachner, M., Magyar, J.P., Vorherr, T. and Schmitz, B. (1993) The fourth immunoglobulin-like domain of NCAM contains a carbohydrate recognition domain for oligomannosidic glycans implicated in association with Ll and neurite outgrowth. *J. Cell Biol.*, 121: 1409–1421.

Kulkens, T., Bolhuis, P.A., Wolterman, R.A., Kemp, S., te Nijenhuis, S., Valentijn, L.J., Hensels, G.W., Jennekens, F.G.I., de

Visser, M., Hoogendijk, J. and Baas, F. (1993) Deletion of the serine 34 codon from the major peripheral myelin protein P0 gene in Charcot-Marie-Tooth disease type IB. *Nature Genet.*, 5: 35–39.

Magyar, J.P., Bartsch, U., Wang, Z-Q., Howells, N., Aguzzi, A., Wagner, E. and Schachner, M. (1993) Degeneration of neural cells in the central nervous system of mice deficient in the gene for the adhesion molecule on glia (AMOG), the $\beta 2$ subunit of murine Na,K-ATPase. Submitted.

Martini, R., Xin, Y., Schmitz, B. and Schachner, M. (1992) The L2/HNK-1 carbohydrate epitope is involved in the preferential outgrowth of motor neurons on ventral roots and motor nerves. *Eur. J. Neurosci.*, 4: 628–639.

Müller-Husmann, G., Gloor, S. and Schachner, M. (1993) Functional characterization of β isoforms of murine Na,K-ATPase: The adhesion molecule on glia (AMOG/$\beta 2$), but not $\beta 1$, promotes neurite outgrowth. *J. Biol. Chem,*. 268: 26260–26267.

Owens, G.C. and Bunge, R. (1991) Schwann cells infected with a recombinant retrovirus expressing myelin-associated glycoprotein antisense RNA do not form myelin. *Neuron*, 7: 565–575.

Schneider-Schaulies, J., Kirchhoff, F., Archelos, J. and Schachner, M. (1991) Down-regulation of myelin-associated glycoprotein on Schwann cells by interferon-gamma and tumor necrosis factor-alpha affects neurite outgrowth. *Neuron*, 7: 995–1005.

Su, Y., Brooks, D.G., Li, L., Lepercq, J., Trofatter, J.A., Ravetch, J.V. and Lebo R.V. (1993) Myelin protein zero gene mutated in Charcot-Marie-Tooth type 1B patients. *Proc. Natl. Acad. Sci. USA*, 90: 10856–10860.

SECTION II

Molecular Brain Research

F. Bloom (Editor)
Progress in Brain Research, Vol. 100

CHAPTER 4

Quantitative analysis of neuronal gene expression

James L. Roberts

Dr. Arthur M. Fishberg Research Center for Neurobiology, Mount Sinai School of Medicine, One Gustave Levy Place, New York, NY 10029, USA

Overview

The vast diversity of neuronal and glial cell types derives primarily from the heterogeneity of gene expression in these two types of brain cells. However, a significant proportion of this diversity also resides in the quantitative aspect of the exact level of expression of a specific gene within neuronal or glial cell. For example, a given neuron may express the mRNAs encoding five different subunits for the $GABA_A$ receptor Cl^- channel, but the actual level of expression of each individual subunit mRNA will ultimately determine which types $GABA_A$ Cl^- channels will actually be expressed within that neuron. If four are expressed at 100 copies per cell and the fifth is present at 10 mRNA molecules per cell, then any special properties conferred by the latter subunit to the Cl^- channel function will be under-represented. Thus, the question arises, once a gene has been turned on within the neuron or glia, at what level will it be expressed as cytoplasmic mRNA.

The biosynthesis of a gene product in eukaryotes is a long and complicated process. It begins with the transcription of the gene by RNA polymerase II in the nucleus to form the primary transcript and the subsequent processing of introns from the primary transcript in order to produce a mature mRNA. The mRNA is then transported from the nucleus to the cytoplasm in an energy dependent process where that mRNA can then be translated into its encoded protein. Finally, the mRNA is degraded in the cytoplasm, a process which under certain circumstances can be regulated. Modulation of gene expression can occur at all of these levels. In general, transcriptional control is the major level at which gene expression is regulated, and through the action of various transcription factors upon cis-acting enhancer elements in the promoter region of the gene, the appropriate transcriptional rate is determined. Regulation can also occur at the nuclear RNA processing stage. Indeed, in many neural genes there is a modulation of RNA splicing or termination sites of transcription which can generate different mRNAs from a single gene transcript. This modulation, however, has more to do with the type of mRNA made, rather than the level. Finally, a mRNA is turned over in the cytoplasm at a specific rate, which in conjunction with its synthetic rate, will determine the overall level of that mRNA in the cytoplasm. The rate of cytoplasmic degradation can also be modulated by neuronal activity and other types of information input to the cell.

Not only will the level of expression of a particular gene help to dictate function within the cell, but the level of mRNA will also be able to define how rapidly changes in gene expression can occur. As a consequence of the different biochemical events taking place at different stages of the mRNA biosynthetic pathway there can be quite different response times to change. For example, levels of neuropeptide encoding mRNAs often change quite slowly, because those mRNAs are often present in tens of thousands of copies within individual neurons. On the other hand, one can rapidly observe changes in the mRNA for the transcription factor cFos after depolarization of a neuron with the induction of *cfos* gene transcription. In the basal state, there are only about 20–30 *cfos* mRNA molecules present in the cell, thus the synthesis of 200 new *cfos* mRNA molecules after 10 min of neuronal stimulation would have a major effect on the level of *cfos* gene expression. A similar change of 200 new mRNAs for the neuropeptide described above would not significantly alter the level of that mRNA. The change in new mRNA synthesis would have to persist for many hours to show changes in the cytoplasmic mRNA levels. On the other hand, however, one can rapidly affect the level of a neuropeptide mRNA by enhancing its degradation within the cytoplasm, accomplishing in less than 1 h what might take days to accomplish by lowering the level of transcription. Of course, this mechanism can only work in one direction. Thus, knowledge of the actual level of a specific mRNA provides information both as to the possible function of that gene in the cell as well as the mechanisms by which its change may be useful in modulating the function of that cell.

Methods of quantitative analysis of gene expression

While there are numerous methods of quantitating levels of a specific RNA transcript utilizing blot hybridization technologies, this chapter focuses on the solution hybridization/nuclease protection assay because of its unique properties. These properties give distinct advantages for analyzing gene expression in the brain, foremost being the extreme sensitivity of the assay. Using conventional autoradiography techniques, one can readily quantitate levels as small as 50 fg of a specific RNA transcript, and with the use of phosphoimagery techniques, sensitivity can be taken down at least another order of magnitude. Thus, this technique becomes quite useful for quantitating levels of transcripts which may be present at a very low copy number within the population of neurons under investigation, such as neurotransmitter receptors. This also makes it helpful in identifying the low abundance in nuclear transcripts involved in the biosynthesis of this specific mRNA. Secondly, because one can identify the size of the species being analyzed, numerous different RNAs of different size of protected fragments can be analyzed in a single sample. We have called this type of assay a "multiplex" solution hybridization assay and have recently reported the details and its application to quantitation of neuroendocrine gene expression (Jakubowski and Roberts, 1992). This also becomes an important issue in looking at multiple transcripts from a single gene. Ultimately, spliced RNAs can be identified based upon which portions of the probe get protected or on the other hand, multiple transcripts in the biosynthetic pathway can also be identified (Levin et al., 1989; Jakubowski and Roberts, 1994). This latter issue becomes quite important in analysis of gene expression in quantitating different nuclear transcripts along the mRNA biosynthetic pathway.

Another useful aspect quantitating levels of the primary transcript of a particular gene using this technique is the ability to determine mRNA turnover rates utilizing transcription inhibitors to block transcription acutely. In classic biochemical studies, the loss of mRNA from the cytoplasm is measured after blockade of new RNA synthesis. However, quite often artifacts are introduced since short-lived transcripts disappear rapidly and their products may affect the half-lives of longer lived transcripts. Utilizing the approach of "decay from steady state" at the primary transcript level, one need only analyze the first 20–30 min after blockade of transcription, avoiding this pitfall. Since the primary transcript turns over much more rapidly with a $t_{1/2}$ in the range of minutes, one can follow the loss of the primary transcript and determine an initial decay rate. Under steady state conditions, two principles apply; the synthetic rate of new primary transcript and the cytoplasmic mRNA turnover rate must be equal and the loss of primary transcript by processing must equal the rate of synthesis of the primary transcript. Hence, the rate of the loss of primary transcript must be equal to the rate of cytoplasmic mRNA turnover. We have verified this technique in a study of the GnRH gene (Yeo et al., 1994b). This becomes extremely useful in either in vivo situations or in in vitro cultures where sufficient amounts of labeled RNA cannot be incorporated into RNA to perform the more classical pulse chase labeling techniques for determining mRNA half-life.

Levels of neuroendocrine gene expression

Very crucial in all these arguments is the concept that, at least to a first approximation, the level of a specific mRNA is to some degree proportional to the level of protein translated from that mRNA. Thus, an mRNA which is present in thousands of copies in the cell will produce more protein within that cell than the mRNA that is present in only a few dozen copies. In general, this concept holds true. Neuropeptide encoding mRNAs, for example, POMC in the arcuate nucleus (16 000 copies) (Fremeau et al., 1989) or vasopressin in the paraventricular nucleus (30 000 copies) (Sherman et al., 1988; Sherman and Watson, 1988) are quite abundant. At the other end of the spectrum, the mRNAs which encode specific neurotransmitter receptors, such as dopamine D2 or $GABA_A$ receptor proteins are significantly less abundant in the range of tens to hundreds of copies of mRNA per cell (Autelitano et al., 1989; Berman et al., 1994). This difference in novel expression generally reflects a fact that you only have 10 000–50 000 molecules of receptor present on the surface of a cell, whereas there will be millions of molecules of a peptide in secretory granules. Thus, the cell needs to be able to synthesize larger amounts of neuropeptide, because it will be released and turned over extracellularly, whereas the receptor will be utilized many times before it is finally turned over.

Gonadotropin releasing hormone gene expression as an example

Utilizing the information obtained from performing quantitative analysis of gene expression, we were able to elucidate an interesting pathway for GnRH gene regulation in the rodent hypothalamus. As discussed above, the GnRH gene is expressed at approximately 12 000 copies of mRNA per GnRH neuron in the hypothalamus (Jakubowski and Roberts, 1994). As such, it falls into that class where one would presume it does not exhibit rapid regulation due to the large mass of the mRNA in the cytoplasm. Utilizing the quantitative assays described above, we had made the interesting observation that in the animal, the levels of nuclear RNA transcripts for GnRH are far higher than those seen for most genes, comprising approximately 20–40% of the total number of GnRH transcripts in the neuron (Jakubowski and Roberts, 1994). In particular, the primary transcript and its processing intermediates were present in very high levels. Initially we thought that this simply reflected inefficient or slow processing of the RNA, thus the precursor and its intermediates accumulated to a higher level. Another possible explanation was that transcription of the GnRH gene was occurring in an extremely high rate, and the steady state high level of primary transcript and processing intermediates reflected this very high level of synthesis. This would presume, however, that degradation was also at a very high rate.

In a separate set of studies, using an in situ hybridization approach, investigators found a very rapid, almost twofold rise in GnRH mRNA in the hypothalamus after only 60 min of NMDA treatment (Petersen et al., 1991). Because of the rapidity and magnitude of this effect, requiring the synthesis of about 10 000 new GnRH mRNA molecules in less than 1 h, our first reaction was that possibly this was a result of some type of in situ hybridization phenomenon where the NMDA treatment caused an increase in the accessibility of pre-existing GnRH mRNA to the probe, enhancing the observed signal. We subsequently investigated this phenomenon using the solution hybridization/nuclease protection assay and found that indeed there was an increase in the actual numbers of GnRH mRNA molecules present in the hypothalamus, exactly as reported by the in situ hybridization technique (Gore and Roberts, 1994). Interestingly, this dramatic rise in cytoplasmic mRNA was not accompanied by reciprocal changes in the levels of nuclear GnRH primary transcript, processing intermediates or mRNA, suggesting that the change was not due to a dramatic rise in the rate of the transcription of the GnRH gene or translocation of the nuclear transcripts to the cytoplasm. Only one explanation was left to account for the ability of the GnRH neuron to produce such rapid increase in the level of cytoplasmic GnRH mRNA. The GnRH gene would have to be transcribed at a very high rate and then turned over in the cytoplasm at an equivalently high rate, in essence yielding a system that has an extremely rapid "flow-through". By blocking the degradation of GnRH mRNA in the cytoplasm, the rapid flow-through rate then allows the cell to accumulate new GnRH mRNA quickly. This mechanism is also able to account for the high levels of precursor and processing intermediate in the nucleus; with an extremely high rate of synthesis in the nucleus and degradation in the cytoplasm, there are relatively more molecules reflecting the initial states of mRNA biosynthesis in the nucleus.

To test this hypothesis, we utilized an immortalized GnRH neuronal cell line, the GT1 cells, to determine the kinetic parameters of GnRH gene transcription, RNA processing and cytoplasmic GnRH mRNA turnover. Again, utilizing the solution hybridization/nuclease protection assay, we were able to exactly quantitate levels of primary transcript, processing intermediates as well as nuclear and cytoplasmic GnRH mRNA (Yeo et al., 1994a). Utilizing two different RNA synthesis inhibitors, actinomycin D which inhibits DA polymerase binding to DNA, and DRB, an ATP analog more specific to RNA Pol II activity, we addressed the issue of how rapidly the different GnRH RNA molecules were chased through their biosynthetic pathway when transcription was stopped. From these studies, we found that the primary transcript for GnRH was turned over rapidly, with a half-life of approximately 15 min, while the cytoplasmic mRNA has an extremely long half-life of 60–80 h. Thus, while our studies in the rat hypothalamus discussed above strongly argued that the GnRH mRNA was turning over very rapidly, the observations in the GT1 cultured cell line argued just the opposite, that GnRH mRNA had an extremely long half-life. Either our interpretation of the GnRH mRNA synthetic events in the animal were incorrect or the GT1 cells were not a good model for studying GnRH mRNA biosynthesis.

Shedding light on this quandary, recently Wray et al. (1993) reported experiments where they measured the half-life of GnRH mRNA in embryonic rat hypothalamic explant cultures, utilizing an in situ hybridization technique for quantitating the mRNA. In these cultures, which maintain many of the types of neuronal contacts on the GnRH neuron seen in the intact animal, the GnRH mRNA had a half-life of only 2–4 h after actinomycin D treatment, in stark contrast to the observations we made in the GT1 cells. However, these observations are in complete agreement with our interpretations based on the NMDA regulation studies discussed above. Thus, it appears that in the in vivo GnRH neuron, the turnover of GnRH mRNA is much more rapid than that seen in the cultured cell line, cells which are essentially devoid of all the normal inputs to the GnRH neuron. It appears that through contacts maintained in vivo, the turnover rate of GnRH mRNA in the cytoplasm is dramatically increased creating a biosynthetic pathway which is capable of sustaining rapid changes in GnRH mRNA to meet the demands of the GnRH neuron for production of its primary neuropeptide. Supporting this conclusion, we have recently discovered that in the GT1 cells, GnRH mRNA degradation is induced by protein kinase C activation, suggesting that in the basal state of GT1 cells, the GnRH mRNA turnover pathway is essentially off, giving the long half-life we observed.

Interestingly, this mechanism of gene regulation can also explain another phenomenon observed in several neuroendocrine systems, including GnRH (reviewed in King and Rubin, 1992); "new" neurons expressing the neuropeptide can be recruited at times of high demand. In the model where GnRH mRNA is rapidly turned over (see Fig. 1), some neurons may degrade the mRNA so quickly that they are in essence "negative" for GnRH. Upon proper stimulation, the GnRH mRNA degradation is suppressed, the neuron accumulates the mRNA and begins expressing GnRH

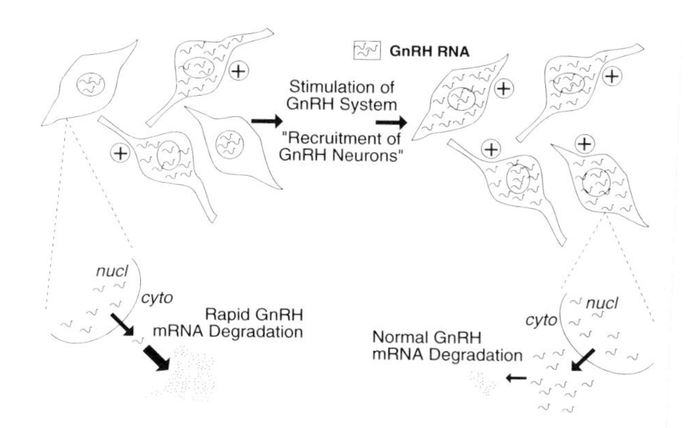

Fig. 1. Model of GnRH gene expression in GnRH neurons.

peptide. Possibly similar mechanisms are involved in other neuroendocrine systems.

Conclusions

Hopefully this chapter has highlighted the value of performing quantitative analysis of gene expression for the insights in elucidating the mechanisms by which neural cells can change the levels of gene expression. The absolute level of expression of a given gene can often yield useful information in determining the function that that gene will have within a neuron. In other cases, the relative levels of different RNA transcripts shed light on the mechanisms responsible for changing the levels and give the investigators clues as to which stages of the regulatory pathway are most likely to be involved in modulating the production of a gene transcript.

References

Autelitano, D.J., Snyder, L., Sealfon, S.C. and Roberts, J.L. (1989) Dopamine D2-receptor mRNA is differentially regulated by dopaminergic agents in rat anterior and neurointermediate pituitary. *Mol. Cell. Endocrinol.*, 67: 101–105.

Berman, J.A., Roberts, J.L. and Pritchett, D.B. (1994) Molecular and pharmacological characterization of $GABA_A$ receptors in the rat pituitary. *J. Neurochem.*, in press.

Fremeau, R.T. Jr., Autelitano, D.J., Blum, M., Wilcox, J. and Roberts, J.L. (1989) Intervening sequence-specific in situ hybridization: detection of the pro-opiomelanocortin gene primary transcript in individual neurons. *Mol. Brain Res.*, 6: 197–202.

Gore, A.C. and Roberts, J.L. (1994) Regulation of gonadotropin-releasing hormone gene expression by the excitatory amino acids kainic acid and *N*-methyl-D,L-aspartate in the male rat. *Endocrinology*, in press.

Jakubowski, M. and Roberts, J.L. (1992) Multiplex solution hybridization-RNase protection assay for quantitation of different RNA transcripts from snap-frozen neuroendocrine tissues of individual animals. *J. Neuroendocrinol.*, 4: 79–89.

Jakubowski, M. and Roberts, J.L. (1994) Processing of gonadotropin-releasing hormone gene transcripts in the rat brain. *J. Biol. Chem.*, 269: 4078–4083.

King, J.C. and Rubin, B.S. (1992) GnRH subgroups: a microarchitecture. In: W.F. Crowley and P.M. Conn (Eds.), *Modes of Action of GnRH and GnRH Analogs*, Springer-Verlag, Berlin, pp. 161–178.

Levin, N.J., Blum, M. and Roberts, J.L. (1989) Modulation of basal and corticotropin-releasing factor stimulated proopiomelanocortin gene expression by vasopressin in rat anterior pituitary. *Endocrinology*, 125: 2957–2966.

Petersen, S.L., McCrone, S., Keller, M. and Gardner, E. (1991) Rapid increases in LHRH mRNA levels following NMDA. *Endocrinology*, 129: 1679–1681.

Sherman, T.G. and Watson, S.J. (1988) Differential expression of vasopressin alleles in brattleboro heterozygote. *J. Neurosci.*, 8: 3797–3811.

Sherman, T.G., Day, R., Civelle, O., Douglass, J., Herbert, E., Akil, H. and Watson, S.L. (1988) Regulation of hypothalamic magnocellular neuropeptides and their mRNAs in the brattleboro rat: coordinate responses to further osmotic challenge. *J. Neurosci.*, 8: 3797–3811.

Wray, S., Key, S., Bachus, S. and Gainer H. (1993) Regulation of LHRH and oxytocin gene expression in CNS slice-explant cultures: effects of second messengers. Abstract 571.4, *23rd Annual Society for Neuroscience Meeting*, Washington, DC.

Yeo, T.S., Dong, K.-W., Zeng, Z., Blum, M. and Roberts, J.L. (1994a) Transcriptional and post-transcriptional regulation of gonadotropin releasing hormone gene expression by protein kinase C pathway in mouse hypothalamic GT1 cells. *Mol. Endocrinol.*, in press.

Yeo, T.S., Jakubowski, M., Dong, K.-W., Blum, M. and Roberts, J.L. (1994b) Characterization of gonadotropin releasing hormone gene transcripts in a mouse hypothalamic neuronal GT1 cell line. *J. Biol. Chem.*, in press.

F. Bloom (Editor)
Progress in Brain Research, Vol. 100

CHAPTER 5

Molecular pharmacology of NMDA receptors: modulatory role of NR2 subunits

Perry B. Molinoff, Keith Williams, Dolan B. Pritchett and Jie Zhong

Department of Pharmacology, University of Pennsylvania School of Medicine, Philadelphia, PA 19104-6084, USA

Introduction

Glutamate is the major fast excitatory neurotransmitter in the vertebrate central nervous system (CNS). The last decade has seen an explosion of interest in the pharmacology, physiology and pathophysiology of glutamatergic systems and of the cell surface receptors that mediate the effects of glutamate on CNS neurons. Ligand-gated ion channels that are sensitive to glutamate are classified on the basis of their sensitivity to the selective agonists *N*-methyl-D-aspartate (NMDA), AMPA and kainate (Fig. 1). Glutamate also activates receptors that are coupled to G proteins, the so-called "metabotropic" receptors (Fig. 1). Over the last several years, genes coding for subunits of AMPA/kainate, kainate and NMDA receptors and for a family of metabotropic receptors have been cloned (Fig. 1) (Nakanishi, 1992; Seeburg, 1993). The ion channel receptors are thought to be oligomeric complexes composed of combinations of two or more types of subunits similar to pentameric nicotinic acetylcholine receptors. Metabotropic receptors are composed of a single polypeptide with limited structural similarity to other G protein-linked receptors (Nakanishi, 1992).

NMDA receptors play a pivotal role in the generation of various forms of synaptic plasticity, including some types of associative long-term potentiation and long-term depression, and in defining neuronal architecture and synaptic connectivity including experience-dependent synaptic modifications in the developing nervous system (Collingridge and Lester, 1989). Excessive or abnormally prolonged activation of NMDA receptors has been implicated in a number of pathological states including ischemic neuronal cell death, epilepsy and chronic neurodegenerative diseases (Choi, 1988). Dysfunction or abnormal regulation of NMDA receptors may also be involved in the etiology of schizophrenia. Activation of NMDA receptors is antagonized by ethanol, and these receptors may mediate some of the acute and/or chronic effects of ethanol in the CNS.

The NMDA receptor contains an integral ion channel that gates Na^+, K^+ and Ca^{2+} and is blocked at resting membrane potentials by physiological concentrations of Mg^{2+}. The Mg^{2+} block of the ion channel is voltage-dependent and is relieved during membrane depolarization, allowing activation of the receptor by NMDA or glutamate. This conditional activation of the receptor, requiring both membrane depolarization and glutamate binding, may underlie the associative nature of induction of long-term potentiation. The NMDA receptor/channel complex contains a number of distinct recognition sites for endogenous and exogenous ligands (Fig. 2). These include binding sites for glutamate (or NMDA), glycine, Mg^{2+}, Zn^{2+}, polyamines and open-channel blockers such as phencyclidine (PCP) and MK-801. The receptor is also modulated by histamine, arachidonic acid, pH and redox reagents. There is an absolute requirement for glycine for the channel to be opened by NMDA or glutamate. Thus, glycine can be con-

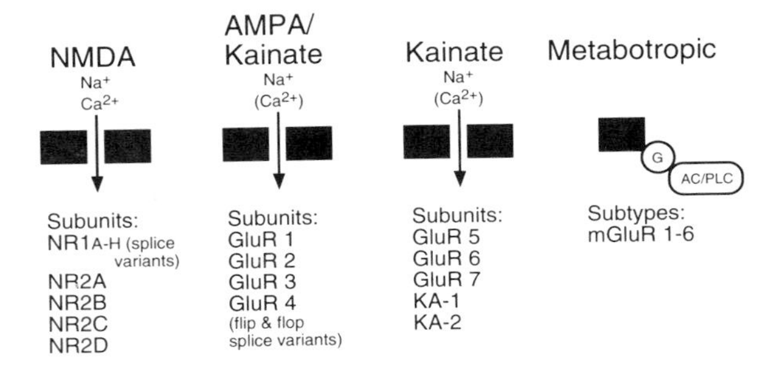

Fig. 1. Classification of mammalian glutamate receptors and corresponding cloned subunits.

sidered a "co-agonist" at the NMDA receptor complex.

Cloned subunits of the NMDA receptor: structure and function

A major advance in our understanding of the structural and functional properties of NMDA receptors has come with the cloning of cDNAs encoding subunits of the receptor. The first clone to be isolated, NMDAR1 (NR1), encodes a polypeptide with a protein molecular weight (105 kDa) and structural topography similar to that of the GluR subunits (Moriyoshi et al., 1991). The proposed topography includes four (or five) transmembrane regions and a large extracellular amino terminal domain (Fig. 3). Many of the properties of native NMDA receptors are seen with homomeric NR1 receptors expressed in *Xenopus* oocytes (Moriyoshi et al., 1991). For example, homomeric NR1 receptors have been shown to require glycine as a co-agonist to gate Ca^{2+}, and they are blocked in a voltage-dependent manner by Mg^{2+}. The receptors are also sensitive to Zn^{2+} and open-channel blockers such as MK-801.

Four related rat brain cDNA clones, designated NR2A, NR2B, NR2C and NR2D, were isolated by homology to NR1 and GluR1-4 (Fig. 3) (Monyer et al., 1992; Ishii et al., 1993). Equivalent cDNAs, termed $\zeta 1$ (NR1) and $\varepsilon 1$–4 (NR2A–D) were cloned from mouse brain (Kutsuwada et al., 1992; Meguro et al., 1992). The NR2 subunits are large polypeptides (≈160 kDa) and are 50–70% homologous to each other but only 15–20% identical to NR1. NR2 subunits do not form functional homomeric receptors. However, co-expression of NR1 and NR2 subunits generates channels that produce much larger whole-

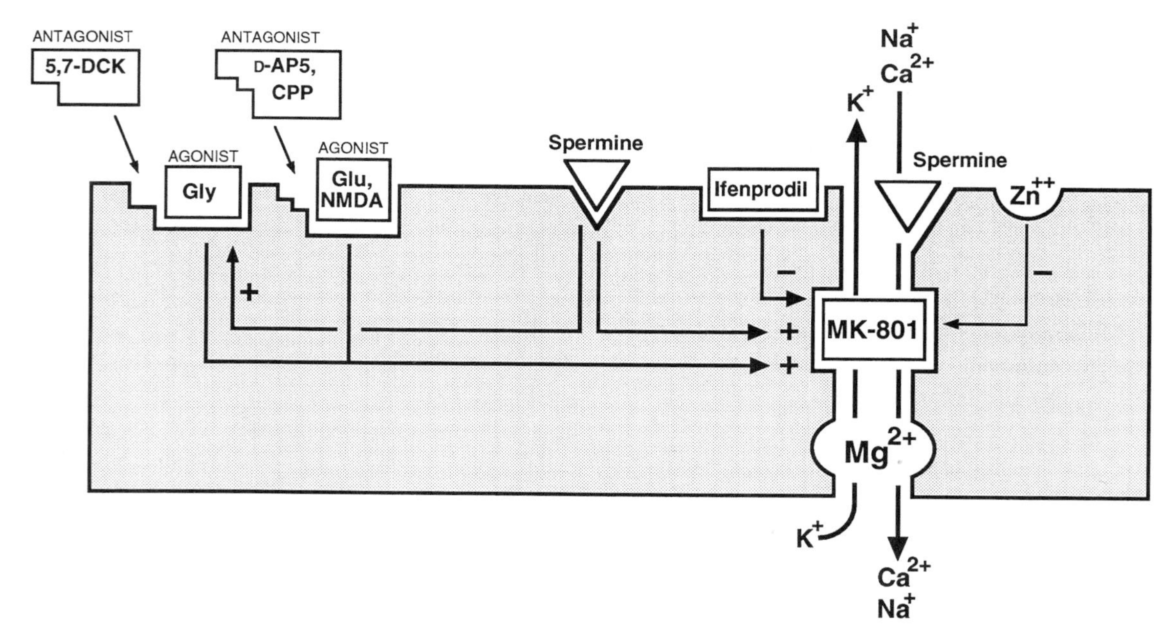

Fig. 2. Schematic model of the NMDA receptor.

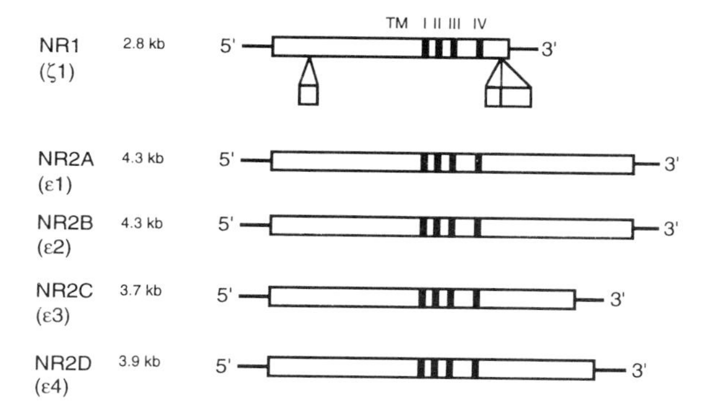

Fig. 3. Schematic of the cDNAs encoding subunits of the NMDA receptor. All subunits contain four putative transmembrane domains (TMI-IV). The NR1 gene is transcribed as eight alternatively spliced mRNAs by the inclusion or deletion of one 5′ and/or two 3′ exons.

cell currents than are seen with homomeric NR1 receptors. This suggests that the two kinds of subunit form heteromeric multisubunit complexes. Results of in situ hybridization histochemistry have shown that mRNAs encoding each of the NR2 subunits are selectively expressed in particular brain regions and their expression changes during development, whereas NR1 mRNA is expressed throughout the brain (Monyer et al., 1992; Nakanishi, 1992; Watanabe et al., 1992). The subunit composition of native NMDA receptors in identified brain regions has not yet been defined, but the receptors are likely to consist of combinations of NR1 and one or more NR2 subunits.

The TMII regions of NR1 and NR2A-D contain a conserved asparagine residue in a position analogous to the glutamine or arginine (Q/R) residue that controls the permeability of GluR1–4 channels to divalent cations. Mutation of this asparagine in NR1 or in NR2A or NR2C to glutamine or arginine reduces or abolishes Ca^{2+} permeability and voltage-dependent blockade by Mg^{2+} and decreases the affinity for MK-801 (Burnashev et al., 1992; Sakurada et al., 1993). These results suggest that the TMII region of NR1 and NR2 subunits may be involved in forming the ion-channel pore of NMDA receptors and that the conserved asparagine residues are critical for the control of permeability to divalent cations. Mutation of the asparagine in NR1 has somewhat different effects on Mg^{2+} block and Ca^{2+} permeability than does mutation of the equivalent residue in NR2 subunits (Burnashev et al., 1992). This suggests a non-equivalent or non-symmetrical contribution of TMII regions in NR1 and NR2 subunits to formation of the ion-channel pore.

The diversity of NMDA receptor subunits and potential subunit combinations has been increased by the discovery of splice variants of the original NR1 subunit. Eight variants have been described based on the alternative splicing of a 5′ exon and one or two adjacent 3′ exons (Fig. 3) (Sugihara et al., 1992; Hollmann et al., 1993). The inclusion of these exons changes the amino acid sequence in the presumed extracellular amino- and carboxy-terminal portions of the protein. The 5′ insert contains multiple positively charged residues while the inserts in the 3′ end of the molecule contain consensus sequences for phosphorylation catalyzed by protein kinase C (Tingley et al., 1993). The carboxy terminus may thus be located intracellularly and be preceded by five rather than four transmembrane regions (Seeburg, 1993).

Properties of recombinant heteromeric NMDA receptors

Although the NR1 subunit can form functional NMDA receptors when expressed in oocytes, it is likely that native NMDA receptors are heterooligomers composed of combinations of NR1 and NR2 subunits. A number of recent studies have shown that the inclusion of different NR2 subunits in heteromeric NMDA receptors can markedly alter the functional and pharmacological properties of the receptors. Some of these differences are strikingly similar to those seen in studies of native NMDA receptors in different brain regions or during various stages of development.

Receptors containing the NR2C subunit are less sensitive to blockade by Mg^{2+} and MK-801 than are receptors containing NR2A or NR2B subunits (Kutsuwada et al., 1992; Monyer et al., 1992; Ishii et al., 1993). Since the NR2C subunit is expressed predominantly in the cerebellum, this is consistent with the hypothesis that NR2C may be a major determinant of the properties of cerebellar NMDA receptors,

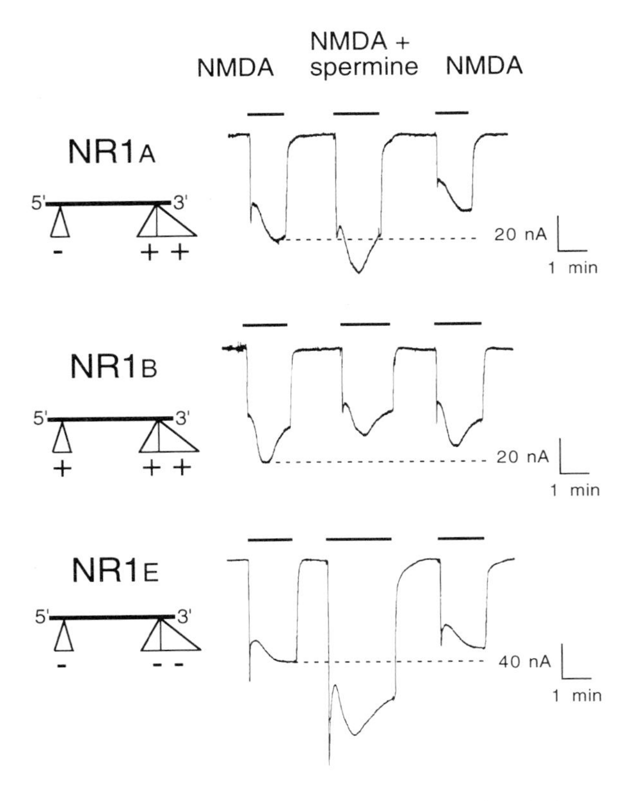

Fig. 4. Spermine stimulation is seen at NR1A and NR1E but not NR1B receptors. Effects of 100 μM spermine on inward currents induced by NMDA were measured in oocytes expressing homomeric NR1A, NR1B, and NR1E receptors and voltage-clamped at –70 mV. The splicing patterns of the subunit cDNAs are shown.

which have a lower affinity for open-channel blockers, such as MK-801, than do receptors in the cerebral cortex and hippocampus.

The single channel properties of NR1/NR2A and NR1/NR2B receptors are different from those of NR1/NR2C receptors (Stern et al., 1992). Both NR1/NR2A and NR1/NR2B receptors exhibit conductance levels and opening patterns that are very similar to those of native NMDA receptors on hippocampal neurons. In contrast, NR1/NR2C receptors have a lower unitary conductance than NR1/NR2A and NR1/NR2B. The properties of NR1/NR2C receptors resemble those of native NMDA receptors on large cerebellar neurons (Stern et al., 1992).

Differences in sensitivity to glycine and to glutamate-site antagonists have also been reported for receptors containing different NR2 subunits. For example, NR1/NR2B receptors have a 10-fold higher affinity for glycine than do NR1/NR2A receptors (Kutsuwada et al., 1992). The consequences of a difference in sensitivity to glycine of native receptors are not known, but if the concentration of glycine in the synapse is not saturating, then changes in the concentration of glycine could selectively alter the activity or the activation threshold of some subtypes of NMDA receptor.

Polyamines such as spermine have a variety of effects on native NMDA receptors. These include “glycine-dependent” stimulation, which involves an increase in the affinity of the receptor for glycine, “glycine-independent” stimulation, which is seen in the presence of saturating concentrations of glycine, voltage-dependent inhibition, and a decrease in the affinity of the receptor for NMDA and glutamate. Variability in the effects of spermine on native NMDA receptors studied electrophysiologically has been observed. Results of studies using recombinant NMDA receptors are beginning to explain the inconsistent results observed in studies of native receptors on cultural neurons. Homomeric NR1 receptors expressed from splice variants such as NR1A or NR1E that do not contain a 5′ insert exhibit glycine-independent stimu-

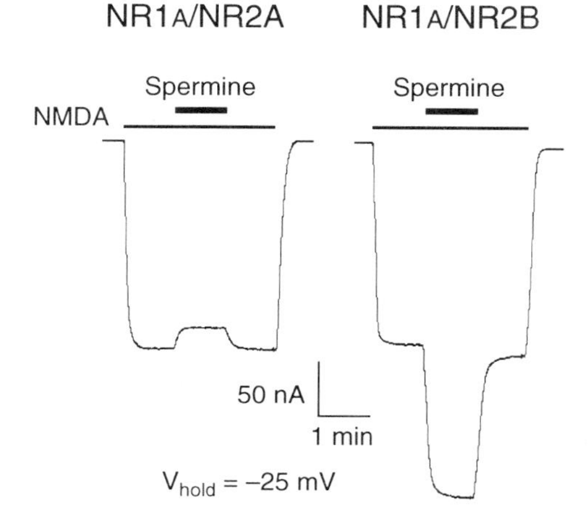

Fig. 5. Spermine stimulation occurs at NR1A/NR2B but not NR1A/NR2A receptors. The effects of 100 μM spermine on responses to 100 μM NMDA (with 10 μM glycine) were measured in oocytes expressing NR1A/NR2A and NR1A/NR2B receptors and voltage-clamped at –25 mV.

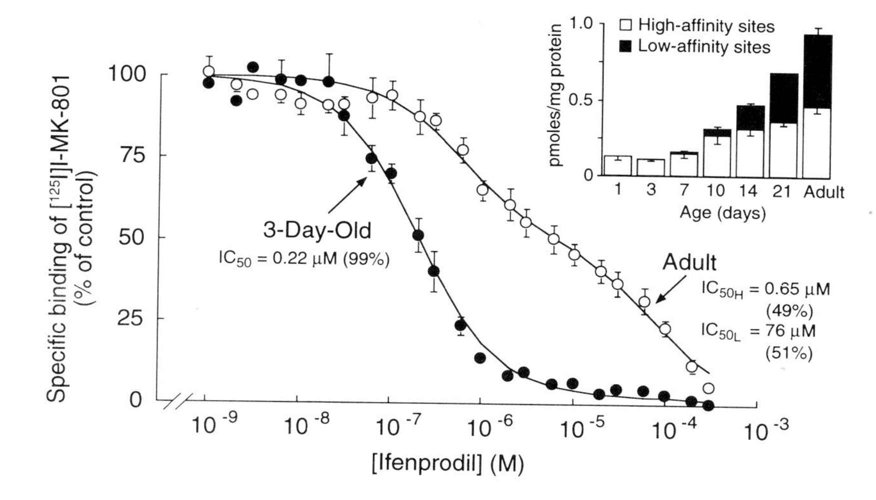

Fig. 6. Inhibitory effects of ifenprodil on NMDA receptors in developing rat brain. The effects of ifenprodil on the binding of [^{125}I]I-MK-801 were determined using membranes prepared from 3-day-old and adult rat forebrain. Inset: The number of binding sites having a high or a low affinity for ifenprodil was determined using membranes prepared from rats of different ages. The density of receptors increased by approximately 8-fold between postnatal day 1 and adult. Data are from Williams et al., 1993.

lation by spermine. In contrast, homomeric receptors expressed from variants such as NR1B, containing a 5′ insert, do not show spermine stimulation (Fig. 4) (Durand et al., 1993). Moreover, in studies of heteromeric receptors, it was found that NR1A/NR2B but not NR1A/NR2A receptors show glycine-independent stimulation by spermine (Fig. 5). Voltage-dependent inhibition is seen at both types of receptor (data not shown). Thus, inclusion in a heteromeric receptor of an NR1 variant such as NR1A is necessary for stimulation by polyamines, but the manifestation of this stimulatory effect is controlled by the type of NR2 subunit present in the receptor complex (Williams et al., 1994).

The atypical antagonist ifenprodil discriminates two subtypes of native NMDA receptors present in equal proportions in adult rat forebrain, one having a high affinity and the other a low affinity for ifenprodil.

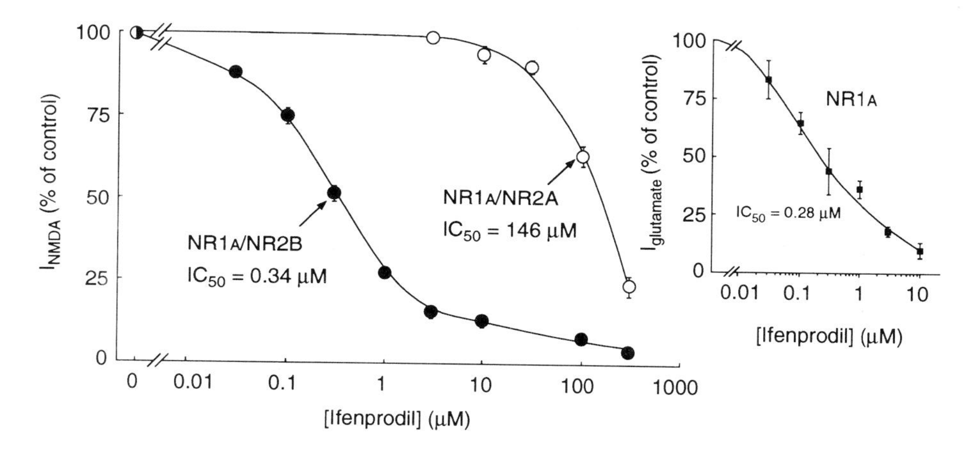

Fig. 7. Effects of ifenprodil on recombinant NMDA receptors. The effects of ifenprodil on responses to NMDA or glutamate were studied in oocytes expressing heteromeric NR1A/NR2B and NR1A/NR2A receptors and (inset) homomeric NR1A receptors. Data are from Williams, (1993) and Williams et al. (1993).

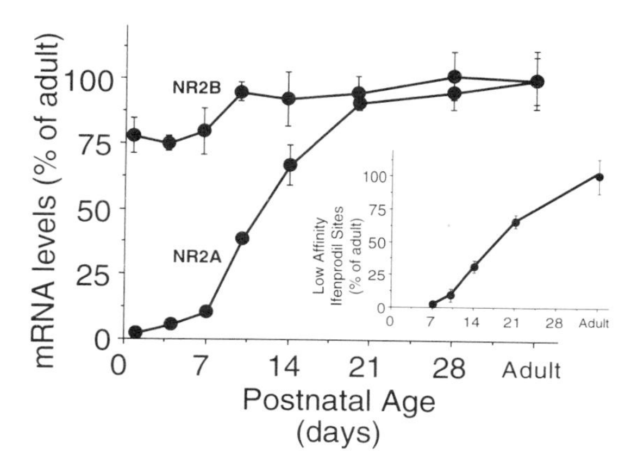

Fig. 8. Expression of receptor subtypes and subunit mRNAs in developing rat brain. Levels of mRNAs encoding NR2A and NR2B were measured by solution hybridization/RNase protection assays in rat cerebral cortex. Inset: The number of receptors having a low affinity for ifenprodil (data from Williams et al., 1993) is expressed as a percentage of the density in adult rat forebrain.

The two subtypes are differentially expressed during postnatal development. In neonatal rats only the form of the receptor with a high affinity for ifenprodil is expressed (Fig. 6) (Williams et al., 1993). Receptors having a low affinity for ifenprodil are expressed after postnatal day 7 (Fig. 6, inset). In studies of recombinant NMDA receptors, ifenprodil has a high affinity at homomeric NR1A receptors (Fig. 7, inset) and at heteromeric NR1A/NR2B receptors but a low affinity at NR1A/NR2A receptors (Fig. 7) (Williams, 1993). The time course of expression of NR2 mRNAs has been defined in the developing brain. NR2B mRNA is expressed at high levels in neonatal and adult rat forebrain. In contrast, NR2A mRNA is found at very low levels in neonates, and expression increases markedly between postnatal days 7 and 21 with a time course that is similar to the expression of receptors having a low affinity for ifenprodil (Fig. 8) (see also Watanabe et al., 1992). Thus, inclusion of NR2A in native NMDA receptors is probably responsible for the delayed development of receptors with a low affinity for ifenprodil.

Future directions

An appreciation of the functional and pharmacological diversity of NMDA receptors is leading to an understanding of the structural features of these receptors and the way in which particular features relate to function. There will undoubtedly be rapid progress in the identification of the subunit composition of at least some native NMDA receptors. Furthermore, site-directed mutagenesis and the use of chimeric subunits will allow tentative identification of regions of NR1 and NR2 subunits that control sensitivity to agonists, antagonists, and modulators of these receptors and will reveal more about the gating and permeability properties of the corresponding channels. The regulation of receptor subunit expression in both the developing and mature nervous systems and the effects of post-translational modifications are also areas of considerable interest and potential importance. The expression of different forms of neurotransmitter receptors, regulated ultimately by the expression of the genes coding for the receptors, may control much of the functional diversity and specialization of neurons in the CNS. The potential array of NMDA receptor subtypes, like that of other ligand-gated channels, is experimentally daunting. However, the diversity of genes and splice variants makes it possible to anticipate the development of therapeutic agents directed at particular subtypes of NMDA receptor. This may lead to the development of improved therapeutic approaches for the treatment of epilepsy and/or reduc-tion of excitotoxic cell death in CNS ischemia. Understanding the repertoire and molecular properties of glutamate receptors will also contribute to a better understanding of phenomena such as long-term poten-tiation and long-term depression that may underlie higher brain functions including learning and memory and may lead to the development of agents that facilitate the development and/or retention of higher brain functions.

Acknowledgment

Supported by USPHS grants GM 34781 and NS 30000

References

Burnashev, N., Schoepfer, R., Monyer, H., Ruppersberg, J.P., Günther, W., Seeburg, P.H. and Sakmann, B. (1992) Control by asparagine residues of calcium permeability and magnesium blockade in the NMDA receptor. *Science,* 257: 1415–1419.

Choi, D.W. (1988) Glutamate neurotoxicity and diseases of the nervous system. *Neuron,* 1: 623–634.

Collingridge, G.L. and Lester, R.A.J. (1989) Excitatory amino acid receptors in the vertebrate central nervous system. *Pharmacol. Rev.,* 41: 143–210 .

Durand, G.M., Bennett, M.V.L. and Zukin, R.S. (1993) Splice variants of the *N*-methyl-D-aspartate receptor NR1 identify domains involved in regulation by polyamines and protein kinase C. *Proc. Natl. Acad. Sci. USA,* 90: 6731–6735.

Hollmann, M., Boulter, J., Maron, C., Beasley, L., Sullivan, J., Pecht, G. and Heinemann, S. (1993) Zinc potentiates agonist-induced currents at certain splice variants of the NMDA receptor. *Neuron,* 10: 943–954.

Ishii, T, Moriyoshi, K., Sugihara, H., Sakurada, K., Kadotani, H., Yokoi, M., Akazawa, C., Shigemoto, R., Mizuno, N., Masu, M. and Nakanishi, S. (1993) Molecular characterization of the family of the *N*-methyl-D-aspartate receptor subunits. *J. Biol. Chem.,* 268: 2836–2843.

Kutsuwada, T., Kashiwabuchi, N., Mori, H., Sakimura, K., Kushiya, E., Araki, K., Meguro, H., Masaki, H., Kumanishi, T., Arakawa, M. and Mishina, M. (1992) Molecular diversity of the NMDA receptor channel. *Nature,* 358: 36–41.

Meguro, H., Mori, H., Araki, K., Kushiya, E., Kutsuwada, T., Yamazaki, M., Kumanishi, T., Arakawa, M., Sakimura, K. and Mishina M. (1992) Functional characterization of a heteromeric NMDA receptor channel expressed from cloned cDNAs. *Nature,* 357: 70–74.

Monyer, H., Sprengel, R., Schoepfer, R., Herb, A., Higuchi, M., Lomeli, H., Burnashev, N., Sakmann, B. and Seeburg, P.H. (1992) Heteromeric NMDA receptors: molecular and functional distinction of subtypes. *Science,* 256: 1217–1221.

Moriyoshi, K., Masu, M., Ishii, T., Shigemoto, R., Mizuno, N. and Nakanishi, S. (1991) Molecular cloning and characterization of the rat NMDA receptor. *Nature,* 354: 31–37.

Nakanishi, S. (1992) Molecular diversity of glutamate receptors and implications for brain function. *Science,* 258: 597–603.

Sakurada, K., Masu, M. and Nakanishi, S. (1993) Alteration of Ca^{2+} permeability and sensitivity to Mg^{2+} and channel blockers by a single amino acid substitution in the *N*-methyl-D-aspartate receptor. *J. Biol. Chem.,* 268: 410–415.

Seeburg, P.H. (1993) The molecular biology of mammalian glutamate receptor channels. *Trends Neurosci.,* 16: 359–365.

Stern, P., Béhé, P., Schoepfer, R. and Colquhoun, D. (1992) Single-channel conductances of NMDA receptors expressed from cloned cDNAs: comparison with native receptors. *Proc. R. Soc. London Ser. B,* 250: 271–277.

Sugihara, H., Moriyoshi, K., Ishii, T., Masu, M. and Nakanishi, S. (1992) Structures and properties of seven isoforms of the NMDA receptor generated by alternative splicing. *Biochem. Biophys. Res. Commun.,* 185: 826–832.

Tingley, W.G., Roche, K.W., Thompson, A.K. and Huganir, R.L. (1993) Regulation of NMDA receptor phosphorylation by alternative splicing of the C-terminal domain. *Nature,* 364: 70–73.

Watanabe, M., Inoue, Y., Sakimura, K. and Mishina, M. (1992) Developmental changes in distribution of NMDA receptor channel subunit mRNAs. *NeuroReport,* 3: 1138–1140.

Williams, K. (1993) Ifenprodil discriminates subtypes of the *N*-methyl-D-aspartate receptor: selectivity and mechanisms at recombinant heteromeric receptors. *Mol. Pharmacol.,* 44: 851–859.

Williams, K., Zappia, A.M., Pritchett, D.B., Shen, Y.M. and Molinoff, P.B. (1994) Sensitivity of the *N*-methyl-D-aspartate receptor to polyamines is controlled by NR2 subunits. *Mol. Pharmacol.,* in press.

Williams, K., Russell, S.L., Shen, Y.M. and Molinoff, P.B. (1993) Developmental switch in the expression of NMDA receptors occurs in vivo and in vitro. *Neuron,* 10: 267–278.

F. Bloom (Editor)
Progress in Brain Research, Vol. 100

CHAPTER 6

Glutamate receptors and the induction of excitotoxic neuronal death

Dennis W. Choi

Department of Neurology and Center for the Study of Nervous System Injury, Box 8111, Washington University School of Medicine, 660 S. Euclid Avenue, St. Louis, MO 63110, USA

Introduction

Glutamate or related excitatory amino acids probably mediate the death of central neurons in several human pathological conditions, for example after toxic food ingestion, or after acute insults such as hypoxia-ischemia, trauma, or prolonged seizures (Olney, 1986; Choi, 1988). In addition, intriguing clues have emerged suggesting that this glutamate-mediated neuronal death, "excitotoxicity", may contribute to the pathogenesis of certain neurodegenerative disorders, such as Huntington's disease, Alzheimer's disease, or motor neuron disease.

To facilitate consideration of underlying mechanisms, we have proposed that excitotoxic neuronal death might be considered in three stages analogous to the stages of long-term potentiation: induction, amplification and expression (Choi, 1992). In this scheme, induction consists of the initial cellular changes immediately attributable to glutamate exposure. Amplification consists of subsequent modulatory events that amplify these initial derangements, increasing their intensity and promoting the injury of additional neurons. Expression consists of the cytotoxic cascades directly responsible for neuronal disintegration.

The most active area of investigation in the excitatory amino acid field at present is probably the study of glutamate receptors. Glutamate activates three major families of ionophore-linked receptors classified by their preferred agonists: *N*-methyl-D-aspartate (NMDA), kainate and α-amino-3-hydroxy-5-methyl-4-isoxazolepropionic acid (AMPA) (Watkins et al., 1990). Multiple functional receptor subunits from each family have been cloned (Hollmann et al., 1989; Nakanishi, 1992; Sommer and Seeburg, 1992). Glutamate also activates a family of metabotropic receptors that activate second messenger systems rather than directly gating ion channels (see below). Historically, NMDA receptors have received the greatest attention with regard to excitotoxicity, but more recently it has become apparent that non-NMDA receptors may play an important role. This brief review presents an overview of the participation of glutamate receptors in excitotoxic induction, with mention of some recent developments with regard to AMPA and metabotropic receptors.

Induction of excitotoxicity: why are NMDA receptors often prominently involved?

The induction of glutamate neurotoxicity consists of the development of an initial set of intracellular derangements, to a great extent resulting directly from glutamate receptor activation. These initial derangements serve as triggers for subsequent amplification and expression events. Although potentially lethal, induction events precede irreversible injury. Neurons can be rescued following full induction by removing extracellular Na^+ and Ca^{2+} for 30 min. Induction is most simply accomplished by receptor overstimulation, but normal physiological levels of receptor activation may become neurotoxic if neuronal energy levels are compromised (Beal et al., 1993).

The channels gated by NMDA, or AMPA/kainate receptors (AMPA and kainate receptors are overlapping populations, and difficult to distinguish pharmacologically) are permeable to both Na^+ and K^+. Channels gated by NMDA receptors, but only a minority subset of channels gated by AMPA/kainate receptors (see below), additionally possess high permeability to Ca^{2+}.

If glutamate exposure is intense, widespread cortical neuronal death can be induced by exposure times as short as 2–3 min, a phenomenon we have termed "rapidly-triggered excitotoxicity". Two components of injury are distinguishable: (1) an acute component, marked by immediate neuronal swelling and dependent on the presence of extracellular Na^+ and Cl^-; and (2) a delayed component marked by neuronal disintegration occurring over a period of hours after exposure, dependent on the presence of extracellular Ca^{2+}. The first component probably reflects the influx of extracellular Na^+, accompanied passively by the influx of Cl^- and water, resulting in cell volume expansion. The second component is likely triggered by excessive Ca^{2+} influx. Although either the acute Na^+-dependent component or the delayed Ca^{2+}-dependent component of glutamate neurotoxicity can alone produce irreversible neuronal injury, the latter component normally predominates.

Experiments with glutamate antagonists suggest that both AMPA/kainate and NMDA type glutamate receptors contribute to acute neuronal swelling, but that most delayed disintegration requires NMDA receptor activation. Death following brief intense glutamate exposure can be almost completely blocked by selective blockade of NMDA receptors, but selective blockade of AMPA/kainate receptors has only a small effect on late neuronal death. Only when both NMDA and AMPA/kainate receptors are blocked is acute glutamate-induced neuronal swelling eliminated. However, selective AMPA/kainate receptor activation can cause the widespread death of cortical neurons if exposure time is extended for several hours. With 24-h exposure, 10 μM of either kainate or AMPA are highly lethal, a phenomenon we have termed "slowly-triggered excitotoxicity" to emphasize the requirement for prolonged receptor activation.

Abnormal entry of extracellular Ca^{2+} may be the primary factor responsible for the induction of both rapidly triggered and slowly triggered excitotoxicity. The dependence of rapidly triggered toxicity upon extracellular Ca^{2+} and NMDA receptor activation is consistent with the idea that it is initiated by excessive Ca^{2+} influx through the Ca^{2+}-permeable NMDA receptor- gated channel. Slowly triggered, AMPA/kainate receptor-mediated excitotoxicity may also be initiated by excessive Ca^{2+} influx. Most channels gated by AMPA or kainate receptors have limited Ca^{2+} permeability (but see below), so Ca^{2+} influx induced by these receptors may occur mainly via indirect routes, such as voltage-gated Ca^{2+} channels, reverse operation of the Na^+-Ca^{2+} exchanger, or membrane stretch-activated conductances.

A key role of Ca^{2+} entry in rapidly triggered excitotoxicity is supported by a quantitative correlation between the extent of cortical neuronal death induced by exposure to glutamate receptor agonists and the amount of extracellular $^{45}Ca^{2+}$ that accumulates in neurons during the exposure period (Hartley et al., 1993). During brief intense glutamate exposure, NMDA receptors mediate neuronal $^{45}Ca^{2+}$ accumulation which is several-fold greater than that induced by comparable exposure to high concentrations of K^+, kainate, or AMPA. The critical reason why NMDA receptor-mediated excitotoxicity is rapidly triggered may be this high rate of Ca^{2+} influx.

Variations in AMPA/kainate receptor behavior may have important implications for excitotoxic death

While as noted above, most AMPA/kainate receptors gate channels permeable only to monovalent cations, it has become recently recognized that a minority subset of these receptors do gate channels permeable to several divalent cations, including Ca^{2+} (Iino et al., 1990). Expression studies with cloned AMPA receptor subunits suggests that expression of an edited form of the GluR2 (alternatively termed GluR-B) subunit dominantly confers Ca^{2+}-impermeability to channels formed by combinations of GluR1/GluR-A, GluR3/ GluR-C, or GluR4/GluR-D subunits (Sommer and Seeburg, 1992). Early studies with high affinity kainate receptors formed from GluR6 subunits suggest that RNA editing may also determine the Ca^{2+} permeability of these receptors (Köhler et al., 1993).

If indeed Ca^{2+} influx triggered by glutamate receptor activation is a critical mediator of excitotoxic injury, the prediction arises that the presence of AMPA/kainate receptors gating Ca^{2+}-permeable channels should confer enhanced vulnerability to death induced by AMPA or kainate. Specifically, neurons bearing substantial numbers of such atypical AMPA/ kainate receptors should be destroyed by exposures to AMPA or kainate too brief to destroy neurons lacking these receptors.

To test this idea, we have utilized kainate-activated Co^{2+} uptake as a histochemical marker for cells bearing Ca^{2+}-permeable AMPA receptors (Pruss et al., 1991). While most cultured cortical neurons exhibit little kainate-activated Co^{2+} uptake, about 15% of the neuronal population showed high levels of uptake. This minority subpopulation was selectively destroyed after AMPA or kainate exposures of only 10–60 min, exposure times too brief to cause much death in the general neuronal population (Turetsky et al., 1992). Studies by Reid et al. (1993) showed that this subpopulation expresses a distinctive profile of AMPA receptor subunits as determined by immunostaining. Neurons exhibiting kainate-activated Co^{2+} uptake were much less likely to express GluR2/GluR3, and much more likely to express GluR1 or GluR4, than the general cortical neuronal population. Thus, expression of AMPA receptors lacking the GluR2 subunit may account for the divalent cation permeability properties of Co^{2+} uptake-positive cells. Of note, most of these cells also stain for glutamic acid decarboxylase, indicating that they are GABAergic (H. Yin, D. Turetsky, J. Weiss and D. Choi, unpublished observations). Our findings fit with another recent study showing that single inhibitory neurons in layer 4 of visual cortex have reduced levels of GluR2/ GluR-B mRNA compared to layer 5 pyramidal neurons (H. Monyer and P. Seeburg, personal communication). It is intriguing to speculate that the preferential loss of cortical GABAergic neurons may occur in disease states associated with the excitotoxic overstimulation of AMPA receptors, an occurrence that could have important functional implications, for example the development of a seizure focus.

Another feature of AMPA/kainate receptor behavior that may influence participation in excitotoxicity is desensitization. If AMPA receptor desensitization is blocked with cyclothiazides (Yamada and Rothman, 1992), an enhanced contribution to glutamate neurotoxicity results (Bateman et al., 1993). Presumably, desensitization is critical in limiting net influx of Na^{+} through AMPA receptor-gated channels; reduction of desensitization presumably permits this influx (and thus resultant secondary Ca^{2+} influx) to reach lethal proportions. While kainate has been considered in the past to be a non-desensitizing agonist on AMPA receptors, more recent work with rapid perfusion techniques suggests that some desensitization still occurs with kainate stimulation (Patneau et al., 1993), and cyclothiazide does potentiate kainate-induced membrane current (Patneau et al., 1993) and toxicity (M. Goldberg and K. Yamada, unpublished observations). The identification of drugs capable of increasing AMPA receptor desensitization may be a useful pathway in the future for the development of new neuroprotective agents based on non-competitive AMPA receptor inhibition.

How does metabotropic receptors activation influence excitotoxicity

The role of metabotropic receptor in excitotoxic injury has been difficult to define because of limitations in available pharmacology. Recently, the cloning of several metabotropic receptor subtypes (and splice variants), and the emergence of partially selective agonists and antagonists has permitted some progress.

Original descriptions of metabotropic receptor focused on the activation of inositol phosphate metabolism, leading to the formation of inositol-1,4,5-trisphosphate and release of Ca^{2+} from intracellular stores (Schoepp and Conn, 1993). Extrapolating from the idea that intracellular Ca^{2+} overload was a key early step in excitotoxic cell death, an injury promoting role of metabotropic receptor activation seemed most likely.

However, studies by Koh et al. (1991) revealed that the broad spectrum selective metabotropic receptor agonist, *trans*-1-aminocyclopentane-1,3-dicarboxylic acid (tACPD), was not intrinsically excitotoxic; rather, it reduced rapidly triggered excitotoxicity in cortical cultures. Subsequent in vivo studies have indicated that the situation is complex. Injection of tACPD into rat hippocampus produced seizures and neuronal loss (Schoepp and Conn, 1993), and injection into striatum potentiated NMDA-induced toxicity (McDonald and Schoepp, 1992).

How can these disparate results be reconciled? The answer likely lies in the multiplicity of cellular actions mediated by various metabotropic receptor subtypes. While some subtypes increase phosphoinositide hydrolysis, others increase or decrease cAMP levels. In addition to releasing intracellular Ca^{2+} stores, metabotropic receptor activation has several actions likely to promote excitotoxic injury, including: (1) increasing NMDA and AMPA/kainate receptor-mediated membrane current; (2) promoting slow onset potentiation or long term potentiation at excitatory synapses; (3) reducing GABAergic inhibition; and (4) reducing inhibitory K^+ currents, I_M and I_{AHP}. On the other hand, metabotropic receptor activation also has several actions that may protect neurons from excitotoxic injury, including (1) reducing synaptic glutamate release; (2) promoting long term depression; and (3) reducing Ca^{2+} influx through voltage-gated Ca^{2+} channels. Thus, the net effect of metabotropic receptor activation on excitotoxic injury may depend critically on the balance between injury promoting, and injury attenuating effects.

One key variable may be the presence of organized excitatory circuits. In intact systems, where these circuits are preserved, pro-excitant effects of metabotropic receptor activation may predominate, leading to increased glutamate release and injury enhancement. In simplified systems such as cell culture, attenuation of circuit inhibition may be less important than attenuation of Ca^{2+} currents entering directly through voltage-gated channels, resulting in a net reduction of excitotoxic injury.

A challenge for the future will be the definition of specific metabotropic receptor subtypes responsible for various actions. It may turn out that certain subtypes predominantly mediated pro-excitant effects (for example, mGluR1 or mGluR5), whereas other subtypes (for example mGluR2, mGluR3, or mGluR4) predominantly mediate pro-inhibitory effects (Nakanishi, 1992). If so, this would represent auspicious circumstances for therapeutic manipulation of the receptor system.

Conclusions

Activation of either NMDA or AMPA/kainate receptors can induce excitotoxic neuronal death. NMDA receptors in particular play a key role in the induction of excitotoxic neuronal death by high concentrations of glutamate, probably reflecting their ability to mediate high levels of Ca^{2+} influx. Recent advances in the pharmacology and molecular biology of AMPA/kainate and glutamate metabotropic receptors have strengthened arguments that these receptors may also contribute importantly to, or influence, the induction of excitotoxicity. Specific manipulation of glutamate receptor subtypes may constitute a useful clinical therapeutic strategy aimed at reducing certain types of pathological neuronal death.

Acknowledgments

Supported in part by NIH grant NS 30337.

References

Bateman, M.C., Bagwe, M.R., Yamada, K.A. and Goldberg, M.P. (1993) Cyclothiazide potentiates AMPA neurotoxicity and oxygen-glucose deprivation injury in cortical culture. *Soc. Neurosci. Abstr.*, 19: 1643.

Beal, M.F., Hyman, B. and Koroshetz, W. (1993) Do defects in mitochondrial energy metabolism underlie the pathology of neurodegenerative diseases? *Trends Neurosci.*, 16: 125–131.

Choi, D.W. (1988) Glutamate neurotoxicity and diseases of the nervous system. *Neuron*, 1: 623–634

Choi, D.W. (1992) Excitotoxic cell death. *J. Neurobiol.*, 23: 1261–1276.

Hartley, D.M., Kurth, M., Bjerkness, L., Weiss, J.H. and Choi, D.W. (1993) Glutamate receptor-induced $^{45}Ca^{2+}$ accumulation in cortical cell culture correlates with subsequent neuronal degeneration. *J. Neurosci.*, 13: 1993–2000

Hollmann, M., O'Shea-Greenfield, A., Rogers, S.W. and Heinemann, S. (1989) Cloning by functional expression of a member of the glutamate receptor family. *Nature*. 342: 643–648.

Iino, M., Ozawa, S. and Tsuzuki, K. (1990) Permeation of calcium through excitatory amino acid receptor channels in cultured rat hippocampal neurones. *J. Physiol.*, 424: 151–165.

Koh, J.Y., Palmer, E. and Cotman, C.W. (1991) Activation of the metabotropic glutamate receptor attenuates *N*-methyl-D-aspartate neurotoxicity in cortical cultures. *Proc. Natl. Acad. Sci. USA*, 88: 9431–9435.

Köhler, M., Burnashev, N., Sakmann, B. and Seeburg, P. (1993) Determinants of Ca^{++} permeability in both TM1 and TM2 of high affinity kainate receptor channels: diversity by RNA editing. *Neuron*, 10: 491–500.

McDonald, J.W. and Schoepp, D.D. (1992) The metabotropic excitatory amino acid receptor agonist 1S,3R-ACPD selectively potentiates *N*-methyl-D-aspartate-induced brain injury. *Eur. J. Pharmacol.*, 215: 353–354.

Nakanishi, S. (1992) Molecular diversity of glutamate receptors and implications for brain function. *Science*, 258: 597–603.

Olney, J.W. (1986) Inciting excitotoxic cytocide among central neurons. *Adv. Exp. Med. Biol.*, 203: 631–645.

Patneau, D.K., Vyklicky Jr., L. and Mayer, M.L. (1993) Hippocampal neurons exhibit cyclothiazide-sensitive rapidly desensitizing responses to kainate. *J. Neurosci.*, 13: 3496–3509

Pruss, R.M., Akeson, R.L., Racke, M.M. and Wilburn, J.L. (1991) Agonist-activated cobalt uptake identifies divalent cation-permeable kainate receptors on neurons and glial cells. *Neuron*, 7: 509–18

Reid, S., Yin, H. and Weiss, J.H. (1993) Cortical neurons subject to kainate triggered Co^{2+} accumulation display differential AMPA/kainate receptor immunoreactivity. *Soc. Neurosci. Abstr.*, 19: 474

Schoepp, D. and Conn, P.J. (1993) Metabotropic glutamate receptors in brain function and pathology. *Trends Pharmacol. Sci.*, 14: 13–20.

Sommer, B. and Seeburg, P.H. (1992) Glutamate receptor channels: novel properties and new clones. *Trends Pharmacol. Sci.*, 13: 291–296.

Turetsky, D.M., Goldberg, M.P. and Choi, D.W. (1992) Kainate-activated cobalt uptake identifies a subpopulation of cultured cortical cells that are preferentially vulnerable to kainate-induced damage. *Soc. Neurosci. Abstr.*, 18: 81.

Watkins, J.C., Krogsgaard-Larsen, P. and Honore, T. (1990) Structure-activity relationships in the development of excitatory amino acid receptor agonists and competitive antagonists. *Trends Pharmacol. Sci.*, 11: 25–33.

Yamada, K.A. and Rothman, S.M. (1992) Diazoxide blocks glutamate desensitization and prolongs excitatory postsynaptic currents in rat hippocampal neurons. *J. Physiol. (London)*, 458: 409–423

F. Bloom (Editor)
Progress in Brain Research, Vol. 100

CHAPTER 7

Sodium/potassium-coupled glutamate transporters, a "new" family of eukaryotic proteins: do they have "new" physiological roles and could they be new targets for pharmacological intervention?

Niels C. Danbolt, Jon Storm-Mathisen, Ole P. Ottersen

Anatomical Institute, University of Oslo, P.O.B. 1105 Blindern, N-0317 Oslo, Norway

Introduction

Chemical signalling is critically dependent on effective mechanisms for terminating transmitter action and for maintaining an extracellular concentration of transmitter that is low enough to avoid undue activation of the appropriate receptors. In the central nervous system, the major transmitters are removed from the extracellular space by uptake into the presynaptic element or glial cells. This uptake is effected by plasma membrane transporters that are driven by transmembrane ion gradients and that are selective for a single transmitter or a few closely related transmitter species (Amara and Kuhar, 1993). These transporters have attracted much interest since they are in the position to regulate the efficacy of synaptic transmission and thus provide potential targets of pharmacological intervention. Several important drugs in current use are thought to exert their effects through action on transmitter transporters. The tricyclic antidepressants, acting on monoamine transporters, are notable examples (Amara and Kuhar, 1993).

The first neurotransmitter transporter to be cloned was a GABA transporter (Guastella et al., 1990). Glycine transporters and several monoamine transporters were cloned subsequently and it became clear that they all belong to a family of molecules with similar ion dependence (sodium and chloride) and molecular structure (12 putative transmembrane domains) (Amara and Kuhar, 1993). Only recently did attempts to clone a glutamate transporter meet with success. The breakthrough came with three independent reports, each using a different approach and identifying a separate transporter, that were published almost simultaneously in late autumn 1992 (Kanai and Hediger, 1992; Pines et al., 1992; Storck et al., 1992). The three transporters displayed about 50% sequence identity, but had no significant primary structural homology to the superfamily of Na^+ and Cl^- coupled transporters or to any other known eukaryotic protein. Thus, glutamate transporters represent a "new" family of molecules. The number of transmembrane domains is not known as the hydropathy plots give room for several interpretations (see below).

A perturbed function at glutamate synapses has been implicated in many disease states, including ischemia and epilepsy, and several neurodegenerative disorders such as amyotrophic lateral sclerosis and Huntington's chorea (Whetsell and Shapira, 1993). Given the recent advances in the understanding of the molecular biology and function of the glutamate transporters, it is time to ask whether these have potentials as clinically beneficial targets of pharmacological intervention. For rational drug design it is essential to know: (1) the physiological roles of the glutamate transporters; (2) whether the transporters are subject to regulation; and (3) whether the transporters are heterogeneous in terms of functional properties and cellular and regional distribution. These issues are addressed in the present overview.

Glutamate transporters: physiological roles

It has long been known, primarily through the work of Kanner and collaborators, that glutamate uptake is electrogenic and dependent on external Na^+ and internal K^+ (Kanner and Schuldiner, 1987; Danbolt et al., 1990). Detailed studies on the stoichiometry of glutamate uptake have been carried out on Müller cells, a specialized form of glial cells, isolated from the salamander retina (Attwell et al., 1991; Bouvier et al., 1992). These cells are not equipped with glutamate receptors and the current elicited by glutamate therefore reflects electrogenic uptake. Analyses based on whole cell patching indicate that glutamate (carrying one net negative charge) is transported into the cell together with two sodium ions and that the return of the carrier to the outside of the cell is coupled to an outward transport of one potassium and one hydroxyl ion. This stoichiometry provides glutamate uptake with a substantial driving force which on theoretical grounds would be sufficient to maintain an external glutamate concentration as low as 0.6 μM (i.e. less than 1:10 000 of the assumed intracellular concentration). This value is in good agreement with microdialysis data from brain.

The maintenance of a low level of external glutamate is one obvious function of the glutamate transporter. This function is crucial as glutamate becomes neurotoxic when its extracellular concentration exceeds a certain level. The magnitude of this level is hard to determine, due to the efficiency of the uptake and the lack of good uptake blockers. It may amount to a few hundred μM (Nicholls and Attwell, 1990) or be as low as 1 μM (Frandsen and Schousboe, 1990). The importance of glial glutamate transporters for reducing glutamate neurotoxicity is illustrated by experiments showing that neurons are much more sensitive when grown alone than in co-culture with glia (Rosenberg et al., 1992) and that cells on the surface of tissue slices are much more sensitive than cells situated deeper into the slices (Garthwaite et al., 1992). The toxic effects are due to excessive depolarization associated with swelling, and an accumulation of calcium recruited from the extracellular space or from intracellular stores. In conditions such as anoxia and ischemia, the driving force of glutamate uptake collapses due to the breakdown of the electrochemical sodium and potassium gradients, and glutamate transport will be compromised or reversed. In fact, there is increasing evidence to suggest that the extracellular overflow of glutamate in ischemic and anoxic conditions primarily reflects inadequate or reversed glutamate transport and that exocytotic release plays a minor role (Nicholls and Attwell, 1990).

As pointed out in the Introduction, neurotransmitter transporters are generally regarded as instrumental for terminating the synaptic action of the respective transmitters. This view is supported by numerous studies demonstrating a prolonged and enhanced transmitter action after inhibition of uptake (Amara and Kuhar, 1993). Surprisingly, it is still not clear to what extent this holds true in the case of the glutamate transporters. The issue is whether the diffusion of glutamate away from the synaptic cleft is so fast that the decay of the synaptic current is determined simply by the kinetics of the receptor channels. This appears to be the case for the NMDA receptor channels, which have a long activated lifetime compared to the estimated time course of glutamate in the synaptic cleft, and which may be saturated following glutamate release from a single vesicle (Clements et al., 1992). Compared to the NMDA receptors, the AMPA receptor channels have a lower affinity to glutamate and a shorter duration, leading Clements et al. (1992) to predict that the decay of the current through these could in part depend on the rate of transmitter clearance. If so, changes in uptake activity would be expected to alter the synaptic efficacy.

Due to the lack of adequate uptake blockers, this prediction is difficult to verify experimentally. All the known compounds that inhibit glutamate uptake with "high" affinity, are competitive inhibitors which have merely about the same affinity as glutamate, are themselves transported and can exchange with intracellular excitatory amino acids. Because the extracellular space is narrow and the glutamate concentration in the synaptic cleft after a single event may amount to mM levels (Clements et al., 1992), it is not clear whether the inhibitor gets into the synaptic cleft at concentrations sufficient to compete successfully with synaptically released glutamate. Isaacson and Nicoll (1993) recently studied the effects of a new glutamate uptake blocker, L-*trans*-pyrrolidine-2,4-dicarboxylate (L-*trans*-PDC), on the time course and kinetics on NMDA and non-NMDA receptor mediated synaptic currents in the hippocampal slice preparation. They concluded that the decay time course of both types of current could be explained by the kinetics of the receptors and that clearance of glutamate from the synaptic cleft by lateral diffusion was faster. Similar results were obtained by using L-*trans*-PDC at the cerebellar mossy fibre to granular cell synapses (Sarantis et al., 1993) and previously at various sites by use of other less potent glutamate uptake blockers. The implication would be that the role of the glutamate transporters is restricted to that of maintaining a favorable gradient for glutamate diffusion from the synaptic cleft.

However, this conclusion may not be valid for all systems or in all experimental situations. Recently, Kovalchuk and Attwell (1994) and Barbour et al. (1994) have found that L-*trans*-PDC, D-aspartate and other glutamate uptake inhibitors do prolong the decay time of the synaptic currents at the parallel and climbing fibre synapses onto Purkinje cells in the cerebellum. The latter authors provide calculations indicating that the restricted diffusion dictated by the geometry of the cerebellar axo-spinous synapses explains the long normal timecourse of the synaptic current and its prolongation by inhibition of glutamate uptake. The "outlet" of the space lateral of the synaptic cleft in this type of synapse is through the narrow surround of the spine necks and the entering and leaving parallel fibre axons. This space is confined by glial lamellae imbued with glutamate transporters (Fig. 1). Previously, Eliasof and Werblin (1993) have reported that inhibition of glutamate uptake abolished the light response in horizontal cells of the salamander retina, indicating that deficient uptake compromised the rapid decrease in synaptic glutamate concentration that signals light onset through the cone-horizontal

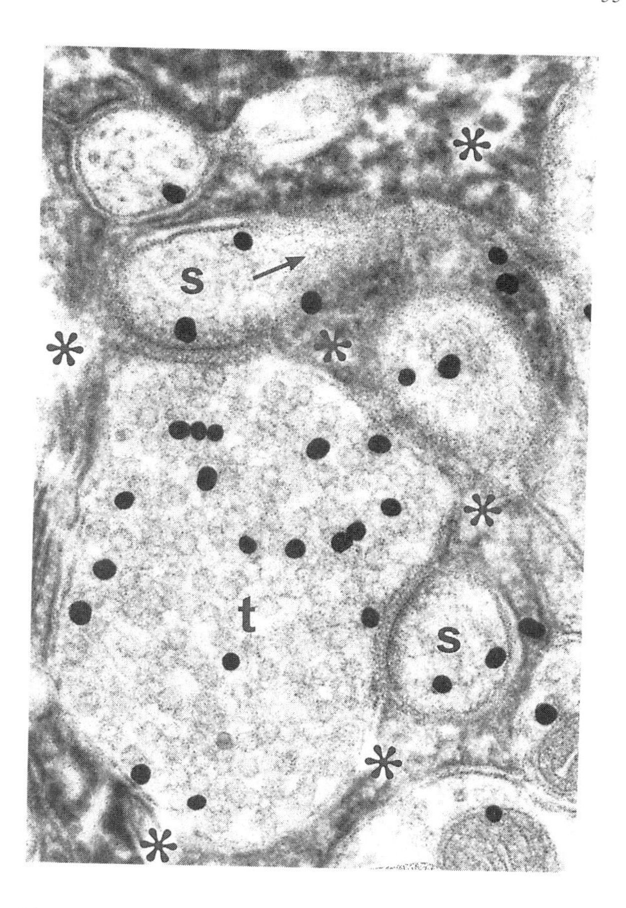

Fig. 1. A parallel fibre bouton (t) in the cerebellar molecular layer forming synapses with asymmetric membrane specializations onto two dendritic spines (s). The arrow points in the direction of the narrow spine neck (tangentially cut). The space surrounding the synapse is completely ensheathed with processes (*) of Bergmann fibres, a type of astrocyte, immunoperoxidase labelled for an intracellular epitope of the glutamate transporter GLAST. Similar localization is seen for GLT1. The bouton is enriched with glutamate (30 nm particles) as shown by postembedding immunogold labelling with antibody to the glutaraldehyde fixed amino acid (see Zhang et al., 1993). Other neuronal structures contain some "metabolic" glutamate. Quantitation in similar material has shown the particle density in parallel fibre boutons to be about 3.5 times higher than in Purkinje cell dendrites and 4 times higher than in astrocytic processes (Ottersen et al., 1992). Modified from Lehre et al. (1994).

cell synapse. Uptake blockers that bind irreversibly with high affinity will be required to settle the question of whether glutamate uptake affects the timecourse of synaptic events at glutamatergic synapses in general.

Another question that pertains to the physiological roles of glutamate transporters is whether the basal level of glutamate is sufficiently high to exert a background activation of glutamate receptors. This is probably not the case for the AMPA receptors, which have low affinity for glutamate. The NMDA, receptors, on the other hand, exhibit a K_d in the low micromolar range and a proportion of these receptors may well be activated by the basal concentration of glutamate. The same may be true of presynaptic metabotropic glutamate receptors, which also show a high affinity for glutamate. Activation of the latter receptors facilitates glutamate release (Herrero et al., 1992), providing a possible feedback link between glutamate uptake activity and synaptic efficacy.

Regulation of glutamate transport

All glutamate transporters that have been cloned so far are equipped with potential protein kinase C dependent phosphorylation sites. Direct evidence that PKC phosphorylation is involved in the regulation of glutamate transport was recently provided by Casado et al. (1993). Phorbol esters were found to produce an increase in glutamate transport in C6 cells (a cell line of glial origin) and a concomitant increase in the phosphorylation level of the glutamate transporter (isolated by antibodies to GLT1). By site directed mutagenesis of GLT1 and transfection to HeLa cells, it could be shown that the effect of the phorbol ester was dependent on serine 113, which thus appears to be the biologically relevant phosphorylation site. The physiological stimulus for this regulatory mechanism remains to be identified, but $alpha_1$ and beta-adrenergic receptors may be involved, as activation of both types of receptor has been found to modulate glutamate uptake (for review, see: Amara and Kuhar, 1993). Phosphorylation may possibly underlie early observations of an increased glutamate uptake in the striatum after in vivo electrical stimulation of the frontal cortex (Nieoullon et al., 1983).

There is solid evidence that arachidonic acid has an inhibitory effect on glutamate uptake in glial cells and nerve terminals (Barbour et al., 1989) although it is still unclear whether this effect is caused by a direct action on the transporter molecules or, indirectly, through an interaction with their lipidic environment. This effect may be of physiological relevance during the induction of long term potentiation, which is associated with an increased production of arachidonic acid (Bliss and Collingridge, 1993). It has previously been reported (Herrero et al., 1992) that arachidonic acid may increase synaptic strength by stimulating glutamate release, acting presynaptically in concert with metabotropic glutamate receptors. A simultaneous inhibition of glutamate uptake would have a synergistic effect. Interestingly, glutamate uptake has been found to be inhibited also by NO (Pogun and Kuhar, 1993), which, like arachidonic acid, has been proposed to act as a "retrograde messenger" during the induction of LTP.

Another class of endogenous compounds that may influence glutamate uptake is the glucocorticoids. Glucocorticoids have been found to inhibit glutamate uptake in astrocytes (Virgin et al., 1991), apparently by reducing the affinity to glutamate. It is conceivable that this mechanism may contribute to the neurotoxic effects of high glucocorticoid levels.

It was recently shown, by quantitative immunoblotting, that the levels of GLT1 and GLAST in the striatum were decreased after decortication (Levy et al., 1993). This finding raises the possibility that the expression of the glial transporters is regulated by glutamate itself or by another factor that is released from corticostriatal terminals. The finding also casts doubt on the common assumption that the decrease in glutamate uptake after nerve transection is fully explained by a loss of presynaptic uptake sites.

The fact that glutamate transport thus appears to be subject to elaborate regulatory mechanisms strongly suggests an important role for the glutamate transporters in brain function, maybe through mechanisms that yet remain to be discovered. The notion of important regulatory roles is reinforced by the findings of multiple types of glutamate transporters, with different regional and cellular localizations.

Heterogeneity of glutamate transporters

The three first cloned glutamate transporters exhibit an

amino acid sequence identity of about 50% (Kanai et al., 1993b; Danbolt, 1994). The predicted number of membrane spanning domains was different for each transporter, but this may reflect differences in the interpretation of the hydropathy plots which differ only marginally between GLT1, GLAST, and EAAC1. Information is still scarce with regard to possible functional heterogeneities between the three transporters. Estimations of the affinities to glutamate have given lower K_m values for GLT1 than for GLAST and EAAC1 (Kanai et al., 1993b), but a comparison under identical experimental conditions by Fairman et al. (1993) on the human counterparts of GLAST, GLT and EAAC (termed EAAT-1, -2 and -3) expressed in COS-7 cells gave K_m values of 57, 101 and 70 μM, respectively.

Information is more plentiful with regard to the differential localization of the three transporters. The distributions of two of these (GLT1 and GLAST) have been compared in the same material by use of immunocytochemistry (Lehre et al., 1994) and in situ hybridization (Torp et al., 1994). The two techniques have produced consistent results and indicate that there are pronounced regional differences in the expression of the two transporters. The highest concentration of GLAST is found in the cerebellum while GLT1 predominates in telencephalic structures including the hippocampus, neocortex and striatum. However, the cellular distribution is very similar: both GLAST and GLT1 are expressed in glial cells (Fig. 1), but in different proportions depending on the region. Indeed, analyses of consecutive ultrathin sections labelled postembedding with antibodies to the respective transporters have revealed that GLAST and GLT1 are co-localized in the same glial cells in the hippocampus and cerebellum (Chaudhry et al., 1994). One possible interpretation of this finding is that the glial glutamate transporter is a heterooligomer and that the ratio between GLAST and GLT1 (and possibly other) subunits differs among different brain regions. This would further increase the possibilities for functional variation and would be in agreement with the observation that antibodies to GLT1 are capable of immunoprecipitating more than 90% of the glutamate uptake activity in rat brains (Danbolt et al., 1992). In view of the latter results and the apparent absence of GLT1 and GLAST from nerve terminals, it is interesting that Na^+-dependent high affinity excitatory amino acid transport has been demonstrated directly in hippocampal glutamatergic nerve terminals by ultrastructural immunogold localization of exogenous D-aspartate (Gundersen et al., 1993).

The distribution of EAAC1, which was cloned from another species (rabbit), has not been studied under conditions similar to those used for GLAST and GLT1. In their original report, Kanai and Hediger (1992) concluded on the basis of the distribution of hybridizing mRNA that this transporter was primarily neuronal. Recently, the same authors succeeded in cloning a rat homologue of EAAC1 (Kanai et al., 1993a), but as yet no information is available on its distribution. Most likely, several additional excitatory amino acid transporters remain to be identified.

Perturbations of the extracellular glutamate level: pathological consequences

An inappropriate action of glutamate is thought to be a pathogenetic factor in many neurological diseases (Whetsell and Shapira, 1993). Overt neurotoxicity, caused by an extracellular glutamate overflow in the range of 100 μM or more, is implicated in catastrophic events such as anoxia, ischemia and severe brain trauma. It should be emphasized, however, that neurotoxic effects could become apparent at much lower glutamate concentrations if the energy status of the tissue is suboptimal. In such conditions, the efficiency of glutamate transport may decrease, causing an elevation (maybe slight) of the basal glutamate level, which in turn will produce an increased metabolic demand on the neurons. Thus, the stage would be set for a vicious circle that could ultimately lead to cell death. This mechanism could be reinforced by a reduction of glutathione synthesis secondary to a glutamate induced inhibition of the cellular uptake of cysteine through the cysteine-glutamate antiporter (Murphy et al., 1989). A mismatch between the metabolic demand imposed by glutamate and the ability to withstand metabolic stress (Beal et al., 1993) may contribute to cell death, e.g. in Parkinson's disease and amyotrophic lateral sclerosis (ALS). The latter disease (Rothstein et al., 1992), and a mouse counterpart (Battaglioli et al., 1993), are associated with a decreased glutamate uptake, notably in the regions that are commonly affected by this disease. Interestingly, a decreased superoxide dismutase activity has been discovered in a familial form of ALS (Rosen et al., 1993). While these two defects have yet to be identified in the same patients, the findings highlight the possibility that a low level of glutamate excitotoxicity and reduced antioxidative capacity may act in concert to cause neurodegeneration (Coyle and Puttfarcken, 1993).

Although neurodegenerative diseases are bound to have a predominant role in any discussion of glutamate-related pathology, it should be emphasized that an inappropriate action of glutamate is likely to be relevant in other diseases as well. Epilepsy is one important example which is highlighted by the recent findings of During and Spencer (1993). These authors demonstrated by use of microdialysis probes that epileptic seizures in humans were associated with an increase in the extracellular level of glutamate. Interestingly, this increase appeared to precede seizure onset, suggesting that it could be responsible for triggering the attack.

On the other hand, reduced glutamatergic activity has been implicated in the pathogenetic mechanisms of schizophrenia and other psychoses (Riederer et al., 1992). Localized or generalized disturbances in the regulation of glutamate uptake could possibly be involved in such mechanisms.

Glutamate transporters: possible targets of pharmacological intervention?

Since perturbations of the concentration of extracellular glutamate may be involved in many diseases, there is a need for drugs that can counteract such perturbations. As outlined above, the glutamate transporters play a central role in the maintenance of the extracellular glutamate level and would be appropriate targets for such drugs. It is conceivable that a modulation of uptake activity could be effected without undue interference with synaptic transmission through AMPA receptors. The recognition that glutamate transporters are subject to various forms of regulation (see above) opens possible avenues for therapies aimed at increasing or decreasing the uptake activity. It is also of importance that neuronal and glial uptake are handled by different transporter molecules and that the two glial transporters differ substantially with regard to their regional distribution. These features provide the possibility of designing drugs that are selective for a given cell type or region. The availability of region-selective drugs would be of importance as most neurodegenerative diseases show a predilection for specific brain areas. It even implies the possibility of carefully targeted inhibition of uptake with a view to augment deficient glutamatergic activity. Clearly, the glutamate transporters hold promise as targets of pharmacological intervention.

Future developments

Where is the work on glutamate transporters heading? In the near future, we can foresee the identification of additional members of the excitatory amino acid transporter family, cloned on the basis of homology with GLAST, GLT1 and EAAC1, and new knowledge of the structure and localization of the genes encoding the transporters. Site directed mutagenesis and other methods will uncover the molecular anatomy of the transporting and regulatory sites, and insight will be gained in how the proteins are folded with respect to the phospholipid membrane. Eventually, the production of large amounts and the crystallization of the molecules will allow their three-dimensional structure to be described. The availability of the cloned transporters will aid studies of their function and the development of pharmacological tools (ligands of transport sites, modulators, oligonucleotides) for blockade and regulation of activity. Such tools may prove useful for probing the functional state of the transporters and of glutamatergic synapses, inter alia by imaging techniques. Transgenic animals with selective lack of transporter types or expressing mutated transporters may form models of human disease. The exact localizations of the transporter types and changes in these, depending on functional state and pathology, can now be studied by specific antibodies and nucleotide probes. Based on this work, it will be possible to uncover the functional role(s) of the transporters in the intact organism and their involvement in disease processes. It is likely that this will eventually lead to innovation of therapy and diagnostic procedures for several different neurological and psychiatric disorders, such as stroke, motor neuron disease, dementia and schizophrenia.

Acknowledgements

We are grateful to David Attwell, Boris Barbour and Roger Nicoll for making preprints of unpublished work available to us.

References

Amara, S.G. and Kuhar, M.J. (1993) Neurotransmitter transporters – recent progress. *Annu. Rev. Neurosci.*: 16: 73–93.

Attwell, D., Sarantis, M., Szatkowski, M., Barbour, B. and Brew, H. (1991) Patch-clamp studies of electrogenic glutamate uptake: ionic dependence, modulation and failure in anoxia. In: H. Wheal and A. Thomson (Ed.), *Excitatory Amino Acids and Synaptic Transmission,* Academic Press, London, pp 223–237.

Barbour, B., Szatkowski, M., Ingledew, N. and Attwell, D. (1989) Arachidonic acid induces a prolonged inhibition of glutamate uptake into glial cells. *Nature*, 342: 918–920.

Barbour, B., Keller, B.U., Llano, I. and Marty, A. (1994) Prolonged presence of glutamate during excitatory synaptic transmission to cerebellar Purkinje cells. Submitted.

Battaglioli, G., Martin, D.L., Plummer, J. and Messer, A. (1993) Synaptosomal glutamate uptake declines progressively in the spinal cord of a mutant mouse with motor neuron disease. *J. Neurochem.*, 60: 1567–1569.

Beal, M.F., Hyman, B.T. and Koroshetz, W. (1993) Do defects in mitochondrial energy metabolism underlie the pathology of neurodegenerative diseases? *Trends Neurosci.*, 16: 125–131.

Bliss, T.V. and Collingridge, G.L. (1993) A synaptic model of memory: long-term potentiation in the hippocampus. *Nature*, 361: 31–39.

Bouvier, M., Szatkowski, M., Amato, A. and Attwell, D. (1992) The glial cell glutamate uptake carrier countertransports pH-changing anions. *Nature,* 360: 471–474.

Casado, M., Bendahan, A., Zafra, F., Danbolt, N.C., Aragón, C., Giménez, C. and Kanner, B.I. (1993) Phosphorylation and modulation of brain glutamate transporters by protein kinase C. *J. Biol. Chem.*, 268: 27313–27317.

Chaudhry, F.A., Lehre, K.P., Danbolt, N.C., Ottersen, O.P. and Storm-Mathisen, J. (1994) Localization of glutamate transporters on the plasma membrane of astrocytes: electron microscopic post-embedding immunogold observations on freeze-substituted tissue. In preparation.

Clements, J.D., Lester, R.A.J., Tong, G., Jahr, C.E. and Westbrook, G.L. (1992) The time course of glutamate in the synaptic cleft. *Science,* 258: 1498–1501.

Coyle, J.T. and Puttfarcken, P. (1993) Oxidative stress, glutamate, and neurodegenerative disorders. *Science*, 262: 689–695.

Danbolt, N.C. (1994) The high affinity uptake system for excitatory amino acids in the brain. *Prog. Neurobiol.*, in press.

Danbolt, N.C., Pines, G. and Kanner, B.I. (1990): Purification and reconstitution of the sodium- and potassium-coupled glutamate transport glycoprotein from rat brain. *Biochemistry,* 29: 6734–6740.

Danbolt, N.C., Storm-Mathisen, J. and Kanner, B.I. (1992): A $[Na^{+}+K^{+}]$coupled L-glutamate transporter purified from rat brain is located in glial cell processes. *Neuroscience*, 51: 295–310.

During, M.J. and Spencer, D.D. (1993) Extracellular hippocampal glutamate and spontaneous seizure in the conscious human brain. *Lancet,* 341: 1607–1610.

Eliasof, S. and Werblin, F. (1993) Characterization of the glutamate transporter in retinal cones of the tiger salamander. *J. Neurosci.*, 13: 402–411.

Fairman, W.A., Arriza, J.L. and Amara, S.G. (1993) Pharmacological characterization of cloned human glutamate transporter subtypes. *Soc. Neurosci. Abstr.*, 19: 496.

Frandsen, A. and Schousboe, A. (1990) Development of excitatory amino acid induced cytotoxicity in cultured neurons. *Int. J. Dev. Neurosci.*, 8: 209–216.

Garthwaite, G., Williams, G.D. and Garthwaite, J. (1992) Glutamate toxicity – an experimental and theoretical analysis. *Eur. J. Neurosci.*, 4: 353–360.

Guastella, J., Nelson, N., Nelson, H., Czyzyk, L., Keynan, S., Miedel, M.C., Davidson, N., Lester, H.A. and Kanner, B.I. (1990) Cloning and expression of a rat brain GABA transporter. *Science*, 249: 1303–1306.

Gundersen, V., Danbolt, N.C., Ottersen, O.P. and Storm-Mathisen, J. (1993) Demonstration of glutamate/aspartate uptake activity in nerve endings by use of antibodies recognizing exogenous D-aspartate. *Neuroscience*, 57: 97–111.

Herrero, I., Miras-Portugal, M.T. and Sanchez-Prieto, J. (1992) Positive feedback of glutamate exocytosis by metabotropic presynaptic receptor stimulation. *Nature*, 360: 163–166.

Isaacson, J.S. and Nicoll, R.A. (1993) The uptake inhibitor L-trans-PDC enhances responses to glutamate but fails to alter the kinetics of excitatory synaptic currents in the hippocampus. *J. Neurophysiol.*, 70: 2187–2191.

Kanai, Y. and Hediger, M.A. (1992) Primary structure and functional characterization of a high-affinity glutamate transporter. *Nature,* 360: 467–471.

Kanai, Y., Lee, W.-S., Bhide, P.G. and Hediger, M.A. (1993a) Functional analysis and distribution of expression of the neuronal high affinity glutamate transporter. *Soc. Neurosci. Abstr.,* 19: 496.

Kanai, Y., Smith, C.P. and Hediger, M.A. (1993b) The elusive transporters with a high affinity for glutamate. *Trends Neurosci.*, 16: 365–370.

Kanner, B.I. and Schuldiner, S. (1987) Mechanism of transport and storage of neurotransmitters. *CRC Crit. Rev. Biochem.*, 22: 1–38.

Kovalchuk, Y. and Attwell D. (1994) Effects of adenosine and a glutamate uptake blocker on excitatory synaptic currents at two synapses in isolated rat cerebellar slices. *J. Physiol. (London) Proc.*, 475: 153P–154P.

Lehre, K.P., Levy, L.M., Ottersen, O.P., Storm-Mathisen, J. and Danbolt, N.C. (1994) Differential expression of two glial glutamate transporters in the rat brain: quantitative and immunocytochemical observations. Submitted.

Levy, L.M., Lehre, K.P., Walaas, I., Storm-Mathisen, J. and Danbolt, N.C. (1993) Down regulation of a glial glutamate transporter in striatum after destruction of the glutamatergic corticostriatal projection. *J. Neurochem.*, 61: S208.

Murphy, T.H., Miyamoto, M., Sastre, A., Schnaar, R.L. and Coyle, J.T. (1989) Glutamate toxicity in a neuronal cell line involves inhibition of cystine transport leading to oxidative stress. *Neuron*, 2: 1547–1558.

Nicholls, D. and Attwell, D. (1990) The release and uptake of excitatory amino acids. *Trends Pharmacol. Sci.*, 11: 462–468.

Nieoullon, A., Kerkerian, L. and Dusticier, N. (1983) Presynaptic dopaminergic control of high affinity glutamate uptake in the striatum. *Neurosci. Lett.*, 43: 191–196.

Ottersen, O.P., Zhang, N. and Walberg, F. (1992) Metabolic compartmentation of glutamate and glutamine: morphological evidence obtained by quantitative immunocytochemistry in rat cerebellum. *Neuroscience*, 46: 519–534.

Pines, G., Danbolt, N.C., Bjørås, M., Zhang, Y., Bendahan, A., Eide, L., Koepsell, H., Seeberg, E., Storm-Mathisen, J. and Kanner, B.I. (1992) Cloning and expression of a rat brain L-glutamate transporter. *Nature*, 360, 464–467.

Pogun, S. and Kuhar, M.J. (1993) Glutamic acid (Glu) uptake inhibition by nitric oxide (NO). *Soc. Neurosci. Abstr.*, 19: 1351.

Riederer P., Lange K. W., Kornhuber J. and Danielczyk W. (1992) Glutamatergic-dopaminergic balance in the brain. Its importance in motor disorders and schizophrenia. *Arzneimittelforschung*, 42: 265–268.

Rosen, D.R., Siddique, T., Patterson, D., Figlewicz, D.A., Sapp, P., Hentati, A., Donaldson, D., Goto, J., O'Regan, J.P., Deng, H.X., Rahmani, Z., Krizus, A., McKenna-Yasek, D., Cayabyab, A., Gaston, S.M., Berger, R., Tanzi, R.E., Halperin, J.J., Herzfeldt, B., Van den Bergh, R., Hung, W.Y., Bird, T., Deng, G., Mulder, D.W., Smyth, C., Laing, N.G., Soriano, E., Pericak-Vance, M.A., Haines, J., Rouleau, G.A., Gusella, J.S., Horvitz, H.R. and Brown, R.H. (1993) Mutations in Cu/Zn superoxide dismutase gene are associated with familial amyotrophic lateral sclerosis. *Nature*, 362: 59–62.

Rosenberg, P.A., Amin, S. and Leitner, M. (1992) Glutamate uptake disguises neurotoxic potency of glutamate agonists in cerebral cortex in dissociated cell culture. *J. Neurosci.*, 12: 56–61.

Rothstein, J.D., Martin, L J. and Kuncl, R.W. (1992) Decreased glutamate transport by the brain and spinal cord in amyotrophic lateral sclerosis. *N. Engl. J. Med.*, 326: 1464–1468.

Sarantis, M., Ballerini, L., Miller, B., Silver, R.A., Edwards, M. and Attwell, D. (1993) Glutamate uptake from the synaptic cleft does not shape the decay of the non-NMDA component of the synaptic current. *Neuron*, 11: 541–549.

Storck, T., Schulte, S., Hofmann, K. and Stoffel, W. (1992) Structure, expression, and functional analysis of a Na^+-dependent glutamate/aspartate transporter from rat brain. *Proc. Natl. Acad. Sci. USA*, 89: 10955–10959.

Torp, R., Danbolt, N.C., Babaie, E., Bjørås, M., Seeberg, E., Storm-Mathisen, J. and Ottersen, O.P. (1994) Differential expression of two glial glutamate transporters in the rat brain: an in situ hybridization study. *Eur. J. Neurosci.*, in press.

Virgin, C.E., Ha, T.P., Packan, D.R., Tombaugh, G.C., Yang, S.H., Horner, H.C. and Sapolsky, R.M. (1991) Glucocorticoids inhibit glucose transport and glutamate uptake in hippocampal astrocytes: implications for glucocorticoid neurotoxicity. *J. Neurochem.*, 57: 1422–1428.

Whetsell, W.O. and Shapira, N.A. (1993) Biology of disease – neuroexcitation, excitotoxicity and human neurological disease. *Lab. Invest.*, 68: 372–387.

Zhang, N., Storm-Mathisen, J. and Ottersen, O.P. (1993) A model system for specificity testing and antigen quantitation in single and double labelling postembedding electron microscopic immunocytochemistry. *Neurosci. Protocols*, 93-050-13: 1–20.

F. Bloom (Editor)
Progress in Brain Research, Vol. 100

CHAPTER 8

GABA$_C$ receptors

Graham A.R. Johnston

Department of Pharmacology, The University of Sydney, NSW, 2006, Australia

Introduction

Much progress has been made in recent years in our understanding of the rich variety of receptors for the inhibitory neurotransmitter GABA (γ-aminobutyric acid, 4-aminobutanoic acid). The current classification of GABA receptors was introduced by Hill and Bowery in 1981 and is based on pharmacological characteristics, GABA$_A$ receptors being antagonised by bicuculline, and GABA$_B$ receptors being insensitive to bicuculline antagonism and activated selectively by baclofen (2-*p*-chlorophenyl-4-aminobutanoic acid). It is becoming increasingly clear, however, that GABA can activate receptors that are insensitive to both bicuculline and baclofen and that the GABA$_A$/GABA$_B$ classification represents an oversimplification of the range of receptors available to this inhibitory neurotransmitter.

GABA receptors insensitive to both bicuculline and baclofen may represent a major class of GABA receptors in the animal kingdom. These receptors have been called GABA$_C$ or GABA$_{NANB}$ ("non-A, non-B") receptors. Such GABA receptors have been described in vertebrate retina, cerebellum, cerebral cortex, optic tectum, spinal cord, many insect species and bacteria. Molecular biological studies have shown that mammalian retinal and insect mRNA injected into *Xenopus* oocytes leads to expression of bicuculline-insensitive, baclofen-insensitive GABA receptors. Cloning of cDNAs indicates that a specific protein subunit, designated ρ_1, may be associated with the retinal receptors.

Early studies

The discovery, in 1970, of the GABA antagonist action of the convulsant alkaloid bicuculline provided vital pharmacological evidence for the role of GABA as an inhibitory neurotransmitter in the CNS. Bicuculline-sensitive synaptic inhibition was observed all over the brain and spinal cord providing a convenient pharmacological means to support other evidence for GABA being the major inhibitory neurotransmitter (Johnston, 1991). By 1981, it was clear that not all GABA receptors could be antagonised by bicuculline, and that a specific class of bicuculline-insensitive GABA receptors were activated by the GABA analogue, baclofen (Hill and Bowery, 1981) leading to the GABA$_A$/GABA$_B$ classification of GABA receptors. GABA$_A$ receptors were insensitive to baclofen, antagonised by bicuculline and gated chloride ion channels in neuronal membranes. GABA$_B$ receptors were insensitive to bicuculline, activated by baclofen and were linked to second messenger systems. By 1984, a range of GABA analogues had been described that could inhibit neuronal activity in a bicuculline-insensitive manner but did not interact with GABA$_B$ receptors as indicated by a lack of effect on [^{3}H]baclofen binding (Drew et al., 1984), giving rise to the possibility of a third class of GABA receptors, tentatively labelled GABA$_C$. The lead compound was *cis*-4-aminocrotonic acid (CACA), an unsaturated analogue of GABA in a partially folded conformation; CACA was approximately one-quarter as potent as GABA in inhibiting the firing of spinal neurones in

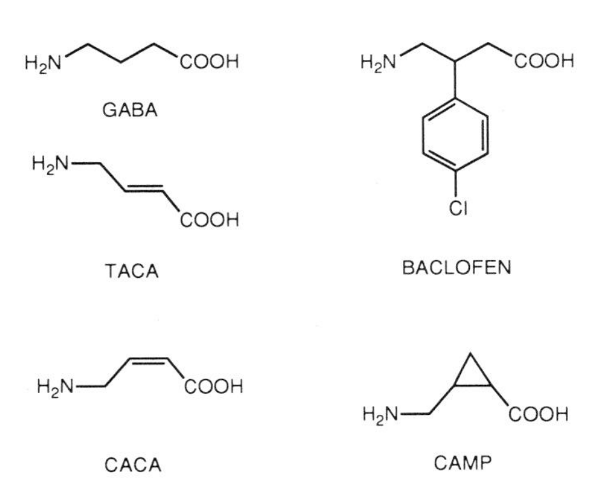

Fig. 1. Structures of GABA, baclofen, TACA (*trans*-4-aminocrotonic acid), CACA (*cis*-4-aminocrotonic acid) and CAMP (*cis*-2-(aminomethyl)-cyclopropane carboxylic acid. CACA and CAMP may be selective ligands for $GABA_C$ receptors.

cats under pentobarbitone anaesthesia, whereas the corresponding *trans*-isomer, TACA, was a bicuculline-sensitive inhibitor equal in potency to GABA (Johnston et al., 1975). The structures of these and some related compounds are shown in Fig. 1. TACA but not CACA inhibited the binding of [^{3}H]baclofen to rat cerebellar membranes (Drew et al., 1984). These studies showed that CACA was a neuronal inhibitor whose action did not appear to be linked to either bicuculline-sensitive ($GABA_A$) or baclofen-sensitive ($GABA_B$) receptors, but only the structural similarity between CACA and GABA linked the inhibitory action of CACA to a possible third class of GABA receptors.

Novel pharmacology of GABA receptors in the optic tectum and retina

Sivilotti and Nistri (1989) found GABA effects in frog optic tectum which exhibited a novel pharmacology. GABA and TACA were equipotent (ED_{50} 110 μM) in enhancing excitatory postsynaptic field potentials in a chloride dependent manner. CACA was some five times less potent (ED_{50} 500 μM). The effects of GABA, TACA and CACA were relatively insensitive to bicuculline (100 μM) but could be blocked by picrotoxin (IC_{50} 78 μM). The benzodiazepine midazolam did not influence this action of GABA. Pentobarbitone acted as a partial agonist, enhancing the field potentials in a picrotoxin-sensitive manner, and antagonising the GABA effects.

A number of studies on retina, published in 1991, led to further interest in a possible third major class of GABA receptors. Matthews et al. (1991) published an abstract entitled "Inhibition of presynaptic calcium current via $GABA_C$ receptors" describing their work on the large synaptic terminals of bipolar neurones in goldfish retina which receive feedback innervation from GABAergic amacrine cells. They reported that 5 μM CACA, like GABA, suppressed calcium currents in these terminals. Baclofen was ineffective. CACA did not activate $GABA_A$ chloride conductances in bipolar neurones. These results suggested "a physiological role for $GABA_C$ in modulation of presynaptic Ca current".

Polenzani et al. (1991) published an extensive paper suggesting that "mammalian retina contains RNAs encoding GABA receptors with distinct pharmacology". They injected poly(A)$^+$ RNA from bovine retina into *Xenopus* oocytes which resulted in the expression of GABA receptors which were insensitive to bicuculline and baclofen. These receptors could not be modulated by benzodiazepines or barbiturates that modulate classic $GABA_A$ receptors. Activation of the receptors produced a chloride current that could be blocked by picrotoxin.

Cutting et al. (1991) cloned a cDNA for a GABA receptor subunit, ρ_1, the mRNA for which is highly expressed in the retina. On injection into *Xenopus* oocytes, ρ_1 mRNA expressed GABA receptors insensitive to bicuculline. Activation of these receptors produced a picrotoxin-sensitive chloride conductance.

Further studies by this group (Shimada et al., 1992) on the pharmacology of these expressed ρ_1 subunit homooligomeric receptors was considered to provide "substantial evidence for a distinct and unique physiologic and pharmacologic role for this newest member of the GABA receptor gene family". The authors go on to state "if further evidence supports a self-associating role for this subunit, these genetic, pharmacologic and physiologic distinctions may be sufficiently unique to merit naming the receptors formed from these subunits $GABA_C$".

Patch clamp studies

Two papers published in the same issue of *Nature* in January 1993 described patch clamp studies on novel GABA responses in retina. Feigenspan et al. (1993) described GABA gated bicuculline-insensitive chloride channels in cultured rod bipolar cells of rat retina, which were not modulated by flunitrazepam, pentobarbitone and alphaxalone and were only marginally blocked by picrotoxin. CACA (100 μM) evoked small but consistent responses, comprising about 10% of the current induced by 20 μM GABA and TACA. Bicuculline (100 μM) reduced the currents induced by GABA and TACA by about 50% without influencing the currents induced by CACA. Feigenspan et al. (1993) concluded "the bicuculline- and baclofen-insensitive GABA receptors were activated selectively by the GABA analogue *cis*-4-aminocrotonic acid (CACA). Hence they may be similar to those receptors termed $GABA_C$ receptors." The accompanying paper of Qian and Dowling (1993) reported similar responses in dissociated rod-driven horizontal cells but not in bipolar cells from white perch retina. GABA responses in the retinal horizontal cells were insensitive to 500 μM bicuculline, 10 μM diazepam, 100 μM pentobarbitone, 500 μM phaclofen and 500 μM 2-hydroxysaclofen (the last two compounds being $GABA_B$ antagonists). The responses could be blocked by 500 μM picrotoxin. CACA produced similar bicuculline-insensitive, picrotoxin-sensitive responses to GABA but was considerably weaker, with EC_{50} values of 48.5 μM for CACA and 1.87 μM for GABA. CACA appeared to behave as a partial agonist, its maximal effect being about half that produced by GABA.

Novel pharmacology of receptors expressed in *Xenopus* oocytes

Extensive studies of the pharmacology of GABA ρ_1 receptors expressed in *Xenopus* oocytes have been reported. Woodward et al. (1993) found the following EC_{50} agonist values: TACA 0.6 μM, GABA 1.3 μM, and CACA 75 μM. ZAPA, THIP, 3-aminopropylphosphinic acid, 3-aminopropyl(methyl)-phosphinic acid, and δ-aminovaleric acid acted as antagonists. Bicuculline had some weak antagonist activity but was at least 5000 times less potent than at $GABA_A$ receptors. Kusama et al. (1993) found the following agonist K_d values: TACA 0.6 μM, GABA 1.7 μM and CACA 74 μM in excellent agreement with those of Woodward et al. (1993). Kusama et al. (1993) found, however, that ZAPA, THIP and 3-aminopropyl phosphinic acid were essentially inactive. TACA, CACA and CAMP (*cis*-2-(aminomethyl)-cyclopropanecarboxylic acid, another conformationally restricted GABA analogue) showed the most selectivity between ρ_1 and $GABA_A$ receptors resulting from the expression of $\alpha_5\beta_1$ subunits in *Xenopus* oocytes. CAMP was more than 150 times more potent against ρ_1 than $\alpha_5\beta_1$ receptors. CAMP has been shown to be a bicuculline-insensitive inhibitor of the firing of cat spinal neurones in vivo with a potency varying between one-twentieth and equipotent with GABA (Allan et al., 1980).

Binding studies

Some studies on the binding of [^{3}H]GABA to rat brain membranes reveal binding that is insensitive to bicuculline and to baclofen. Balcar et al. (1986) reported that the binding of [^{3}H]GABA to cerebral cortical membranes prepared from newborn and adult rats could be partially inhibited by 100 μM CACA and that this inhibition produced by CACA was additive to that produced by 100 μM bicuculline and 100 μM baclofen. Drew and Johnston (1992) reported studies on [^{3}H]GABA binding to rat cerebellar membranes and described a calcium-independent [^{3}H]GABA component that was insensitive to bicuculline and baclofen. They termed this binding $GABA_{NANB}$, to indicate that the binding sites involved were different from classically described $GABA_A$ and $GABA_B$ binding sites. Scatchard analysis indicated two components for the $GABA_{NANB}$ binding with K_d values of 42 nM and 9 μM, respectively. Up to 60% of the [^{3}H]GABA bound to rat cerebellar membranes appeared to bind to $GABA_{NANB}$ binding sites. This binding was inhibited by CACA (IC_{50} 2 μM), TACA (IC_{50} 22 μM) and CAMP (42% at 1 μM), and was insensitive to 2-hydroxysaclofen, securinine, gabapentin and 3-

aminopropylphosphonic acid (Drew and Johnston, unpublished). We have begun studying the binding of [^{3}H]CACA to rat cerebellar membranes (Drew, Duke and Johnston, unpublished). Both GABA and TACA are potent inhibitors of [^{3}H]CACA binding ($IC_{50} < 25$ nM) with CACA itself showing moderate potency (IC_{50} 0.5 μM).

Studies on insects and bacteria

The majority of GABA receptors described in insects appear to be insensitive to bicuculline and baclofen (Lummis, 1992). They gate chloride channels that can be weakly antagonised by picrotoxin. The function of these insect GABA receptors can be enhanced by benzodiazepines, although the pharmacology of the benzodiazepine enhancement seems to follow that of the vertebrate peripheral benzodiazepine sites that are not linked to $GABA_A$ receptors more closely than the vertebrate CNS benzodiazepines sites that are linked to such bicuculline-sensitive receptors. Many insecticides act on GABA receptors and part of their selective action between insects and vertebrates may be due to differences in GABA receptors. Recently, a series of insecticidal 1,2,3-triazoles have been described that block bicuculline-insensitive GABA responses in muscle cells of the nematode *Ascaris* which "may also prove to be useful antagonists of the $GABA_C$ receptor subtype" (Holden-Dye et al., 1994)

GABA binds to receptors in bacteria that appear to respond to mammalian $GABA_A$ agonists, such as muscimol, but not to mammalian $GABA_A$ antagonists, such as SR95531 (Balcar, 1990). The benzodiazepine, diazepam, binds to receptors in bacteria that can be modulated by GABA (Lummis et al., 1991). These sites are similar to those found in insects in that they resemble vertebrate peripheral binding sites but, as in insects, these sites can be modulated by GABA.

Conclusions

There is increasing evidence for the existence of classes of GABA receptors that are not covered by $GABA_A$ receptors, pharmacologically defined as bicuculline-sensitive, baclofen-insensitive receptors, and $GABA_B$ receptors, defined as baclofen-sensitive, bicuculline-insensitive receptors. $GABA_C$ receptors which are bicuculline-insensitive, baclofen-insensitive, linked to chloride channels and selectively activated by CACA and CAMP, may represent just one of such a class of receptors. $GABA_C$ receptors may be more widespread in insects and bacteria than in vertebrates where they may be localised in certain parts of the nervous system such as the retina.

$GABA_C$ receptors may represent receptors that subsequently evolved into $GABA_A$ receptors by gaining bicuculline-sensitivity, together with altered agonist and modulator specificity. The ρ_1 protein, which is highly expressed in retina and has considerable sequence homology with $GABA_A$ sub-unit proteins, yields homooligomeric receptors in *Xenopus* oocytes which have many but not all of the pharmacological properties of $GABA_C$ receptors as deduced from electrophysiological and neurochemical studies on neuronal membranes. A combination of molecular biology, electrophysiology, pharmacology and medicinal chemistry will be needed to provide more comprehensive information on $GABA_C$ and other receptors which do not fit the classic definitions of $GABA_A$ and $GABA_B$ receptors.

Acknowledgements

The author is grateful to the Australian NH&MRC for financial support and to Dr Robin Allan, Ms Muallâ Akinci, Dr Colleen Drew, Dr Rujee Duke, Dr Frances Edwards and Dr Ken Mewett for their collaboration on studies of $GABA_C$ receptors.

References

Allan, R.D., Curtis, D.R., Headley, P.M., Johnston, G.A.R., Lodge, D. and Twitchin, B. (1980) The synthesis and activity of *cis*- and *trans*-2-(aminomethyl)cyclopropanecarboxylic acid as conformationally restricted analogues of GABA. *J. Neurochem.*, 34: 652–654.

Balcar, V.J. (1990) Presence of a highly efficient "binding" to bacterial contamination can distort data from binding studies. *Neurochem Res.*, 15: 1239–1240.

Balcar, V.J., Joó, F., Kása, P., Dammasch, I.E. and Wolff, J.R. (1986) GABA receptor binding in rat cerebral cortex and superior cervical ganglion in the absence of GABAergic synapses. *Neurosci. Lett.*, 66: 269–274.

Cutting, G.R., Lu, L., O'Hara, B., Kasch, L.M., Donovan, D., Shimada, S., Antonarakis, S.E., Guggino, W.B., Uhl, G.R. and Kazazian, H.H. (1991) Cloning of the GABA ρ_1 cDNA: a novel GABA subunit highly expressed in retina. *Proc. Natl. Acad. Sci.*, 88: 2673–2677.

Drew, C.A. and Johnston, G.A.R. (1992) Bicuculline- and baclofen-insensitive γ-aminobutyric acid binding to rat cerebellar membranes. *J. Neurochem.*, 58: 1087–1092.

Drew, C.A., Johnston, G.A.R. and Weatherby, R.P. (1984) Bicuculline-insensitive GABA receptors: studies on the binding of (–)-baclofen to rat cerebellar membranes.

Feigenspan, A., Wössle, H. and Bormann, J. (1993) Pharmacology of GABA receptor Cl^- channels in rat retinal bipolar cells. *Nature*, 361: 159–162.

Hill, D.R. and Bowery N.G. (1981) ^{3}H-Baclofen and ^{3}H-GABA bind to bicuculline-insensitive $GABA_B$ sites in rat brain. *Nature*, 290: 149–152.

Holden-Dye, L., Willis, R.J. and Walker, R.J. (1994) Azole compounds antagonise the bicuculline insensitive GABA receptor on the cells of the parasitic nematode *Ascaris suum*. *Br. J. Pharmacol.*, in press.

Johnston, G.A.R. (1991) $GABA_A$ antagonists. *Semin. Neurosci.*, 3: 205–210.

Johnston, G.A.R., Curtis, D.R., Beart, P.M., Game, C.J.A., McCulloch, R.M. and Twitchin, B. (1975) *cis*- and *trans*-4-aminocrotonic acid as GABA analogues of restricted conformation. *J. Neurochem.*, 24: 157–160.

Kusama, T., Spivak, C.E., Whiting, P., Dawson, V.L., Schaeffer, J.C. and Uhl, G.R. (1993) Pharmacology of GABA ρ_1 and GABA α/β receptors expressed in *Xenopus* oocytes and COS cells. *Br. J. Pharmacol.*, 109: 200–206.

Lummis, S.C.R. (1992) Insect GABA receptors: characterization and expression in *Xenopus* oocytes following injection of cockroach CNS mRNA. *Mol. Neuropharmacol.*, 2: 167–172.

Lummis, S.C.R., Nicoletti, G., Johnston, G.A.R. and Holan G. (1991) Gamma-aminobutyric acid-modulated benzodiazepine binding sites in bacteria. *Life Sci.*, 49: 1079–1086.

Matthews, G., Ayoub, G. and Heidelberger, R. (1991) Inhibition of presynaptic calcium current via GABAC receptors. *Soc. Neurosci. Abstr.*, 17: 900.

Polenzani, L., Woodward, R.M. and Miledi R. (1991) Expression of mammalian g-aminobutyric acid receptors with distinct pharmacology in *Xenopus* oocytes. *Proc. Natl. Acad. Sci. USA*, 88: 4318–4322.

Qian, H. and Dowling, J.E. (1993) Novel GABA responses from rod-driven retinal horizontal cells. *Nature*, 361:162–164.

Shimada, S., Cutting, G. and Uhl, G.R. (1992) γ-Aminobutyric acid A or C receptor? γ-Aminobutyric acid ρ_1 receptor RNA induces bicuculline-, barbiturate-, and benzodiazepine-insensitive γ-aminobutyric acid responses in *Xenopus* oocytes. *J. Pharmacol. Exp. Ther.*, 41: 683–687.

Sivilotti, L. and Nistri, A. (1989) Pharmacology of a novel effect of γ-aminobutyric acid on the frog optic tectum in vitro. *Eur. J. Pharmacol.*, 164: 205–212.

Woodward, R.M., Polenzani, L. and Miledi, R. (1993) Characterization of bicuculline/baclofen-insensitive (ρ-like) γ-aminobutyric acid receptors expressed in *Xenopus* oocytes. 2. Pharmacology of γ-aminobutyric $acid_A$ and γ-aminobutyric $acid_B$ receptor agonists and antagonists. *Mol. Pharmacol.*, 43: 609–625.

F. Bloom (Editor)
Progress in Brain Research, Vol. 100

CHAPTER 9

The central cholinergic system during aging

Giancarlo Pepeu and Lisa Giovannelli

Department of Preclinical and Clinical Pharmacology, University of Florence, Viale Morgagni 65, 50134 Florence, Italy

Introduction

After the seminal papers of Drachman and Leavitt (1974), suggesting a relationship between impairment of the cholinergic system and memory deficits in aging, and of Davies and Maloney (1976) showing a decrease in cortical choline acetyltransferase (ChAT) activity in the cerebral cortex of patients affected by Alzheimer's disease, Bartus et al. (1982) marshalled the data available at that time and proposed the "cholinergic hypothesis of geriatric memory dysfunction". Discussing the aging of the cholinergic system today means, unavoidably, assessing the impact of this hypothesis on research and clinical practice, and its present importance. The recent admission by the FDA of tacrine, a potent reversible cholinesterase inhibitor (Freeman and Dawson, 1991), for the treatment of senile dementia of Alzheimer's and Alzheimer's type disease (AD) is a consequence of this hypothesis and further reason for its critical appraisal.

The hypothesis proposed by Bartus was mostly based on observation that in patients affected by AD, ChAT activity is strongly reduced in the cerebral cortex and hippocampus as a consequence of the degeneration of forebrain cholinergic nuclei (Bigl et al., 1990). Inconsistent data existed at that time on the extent of cholinergic dysfunction in non-pathological aging. Since much evidence indicates that forebrain cholinergic pathways play a role in cognitive processes (Collerton, 1986), the hypothesis offered a rationale for searching for drugs active on memory impairment, which is a predominant symptom of AD, as well as a frequent cause of complaint in normal aging. Undoubtedly, the hypothesis exerted an important heuristic effect by stimulating investigations aimed at defining the extent of cholinergic hypofunction, its relationship to cognitive impairment, and the development of therapeutic agents. However, its limits became rapidly evident since it was soon demonstrated that aging and particularly AD affect many neuronal systems besides the cholinergic (Hardy et al., 1985).

In this review, the question of whether the cholinergic hypothesis of memory impairment is still viable will be addressed by examining the recent evidence of cholinergic hypofunction associated with pathological and non-pathological aging, and its behavioral correlates. Furthermore, the possibility of correcting the hypofunction is discussed. For more information on aging and the cholinergic system, see Decker (1987), Sherman and Friedman (1990) and Pepeu et al. (1993a).

Loss and morphological changes of the cholinergic neurons

While the loss of neurons in the forebrain cholinergic nuclei in AD patients has been repeatedly confirmed (Bigl et al., 1990), investigations on their loss and morphological changes in normal aging in man, non-human primates, and rodents have generated controversial results. In aging men, a reduced density of the cortical cholinergic network has been demonstrated (Geula and Mesulam, 1989), and 50% of the total neuronal population of the nucleus basalis, including cholinergic and non-cholinergic neurons, has been found to be lost by 90 years of age, in comparison to the number found between 16 and 29 years (deLacalle et al., 1991). Investigations in old rats have demonstrated either a decrease or no change in the number of cholinergic neurons, a decrease and even an increase in size, a loss of dendritic spines or no morphological changes. Armstrong et al. (1993) point out that the lack of uniformity of the studies, with respect to sex, strain, age and histochemical methods, may explain the differences. Nevertheless, a decrease in the number of ChAT immunopositive cells in discrete forebrain nuclei is found in old behaviorally impaired rats as compared with non-behaviorally impaired animals of the same age. However, the finding that the loss of ChAT immunopositive neurons is not matched, in the same area, by that of neurons immunolabeled with antibodies against $p75^{NGF}$ or counterstained for Nissl substance suggests that the cholinergic neurons in aging rats do not actually die but only lose their ChAT immunoreactivity (Armstrong et al., 1993). Information on whether a decrease in ChAT immunopositive neurons also occurs in the striatum and the brainstem ascending cholinergic systems is still scarce.

Age-associated changes in ACh synthesis and release

Here there is a clear difference between the findings in AD and in non-pathological aging. In the first case, a decrease in cortical and hippocampal ChAT activity has been constantly reported (Bartus et al., 1982; Hardy et al., 1985). In the latter, it appears from the reviews of Decker (1987) and Sherman and Friedman (1990) that, in the cerebral cortex and hippocampus of aging humans and rodents, either no or small disparate changes in ChAT activity have generally been found. Moreover, it cannot be excluded that the decrease in ChAT activity, sometimes observed in normal aged humans, might be attributable to undetected AD cases. Since ChAT activity of the cholinergic neurons is normally very high, and ChAT does not catalyze a rate-limiting reaction, it is possible that determination under optimal substrate and co-factor conditions may mask small age-associated decreases in activity.

ACh synthesis also depends on high affinity choline uptake (HACU) whose rate is directly coupled to neuronal activity. Either no change or only a small decrease in HACU has been found in aging rats, with remarkable differences between rat strains and cerebral regions investigated (Decker, 1987; Sherman and Friedman, 1990). Nevertheless, these findings represent uptake under resting conditions. The possibility that a decrease in HACU activity might actually occur during aging is supported by the findings of a marked reduction in old rats in [^{3}H]hemicholinium binding , a marker of HACU sites (Forloni and Angeretti, 1992), and total tritium content, after incubation with [^{3}H]choline of electrically stimulated cortical slices (Vannucchi et al., 1990).

The marked reduction in cortical ChAT activity occurring in AD patients results in a decrease in acetylcholine (ACh) formation and release. This was shown by Sims et al. (1980) in their conclusive study on cortical biopsies. Also in non-pathological aging, most ex vivo and in vivo studies have demonstrated a decrease in ACh release from the brain. This finding is direct evidence of age-associated cholinergic hypofunction. In cortical slices, the basal efflux is not affected, but a marked reduction has been found in the evoked ACh release (Pedata et al., 1983) in rodents and in man (Feuerstein et al., 1992). The decrease usually begins between 11 and 14 months of age (Vannucchi and Pepeu, 1987), long before either morphological changes of the cholinergic neurons or ChAT reduction have ever been detected. In vivo, experiments with the microdialysis technique have confirmed the findings on brain slices. A decrease in ACh release, ranging from 35 to 60% was found in the cortex, hippocampus and striatum of 18–22-month old rats in comparison with that found in 2 to 3-month old rats (see references in Pepeu et al., 1993b). In the striatum, a 30% decrease occurred in 9-month-old rats. Interestingly, a decrease in choline efflux was also reported, suggesting that age-related changes in choline availability may underly the decrease in ACh release. The only discordant finding was obtained by Fischer et al. (1991b) who demonstrated no change in ACh release from the hippocampus of 24-month-old cognitively impaired female rats with a significant decrease in the number of ChAT-positive neurons in the septal-diagonal band.

This result and the finding that the decrease in ACh release may even occur at 9–14 months of age, long before the loss of ChAT immunopositive neurons, suggest that the decrease in ACh release and morphological changes are not directly correlated. The possibility should be considered that the decrease in ACh release may also depend on age related modifications of the choline pools used for ACh synthesis (Vannucchi et al., 1990), and on presynaptic modulation. Giovannelli et al. (1988) demonstrated a decrease of the inhibitory action of adenosine A_1 receptors on ACh release from old rats. Crawley and Wenk (1989) claim that the inhibitory effect of galanine on ACh release also is enhanced in AD.

Age associated changes in cholinergic receptors

Nordberg et al. (1992) demonstrated a significant decrease in the number of cortical M1 and M2 receptors and nicotinic receptors in the human brain during normal aging. In AD there is an increase in M1 and M2 receptors and a decrease in nicotinic receptors which was also observed by positron emission tomography (Nordberg et al., 1990). In aged rats, the reports of either a decrease or no change in the density of muscarinic binding sites were reviewed by Sherman and Friedman (1990). The reasons for the differences are not only the usual lack of uniformity of the studies, but also the presence of five different muscarinic receptor subtypes, and their plasticity. It may be assumed that a loss of presynaptic receptors, resulting from the disappearance of cholinergic nerve endings, induces a compensatory upregulation of postsynaptic receptors. Changes in density and affinity of the binding sites, and dysfunctions of their transducing mechanisms could be responsible for the limited therapeutic usefulness of cholinomimetics.

Relationship between the age-associated cholinergic hypofunction and cognitive impairment

Although much evidence demonstrates a cholinergic modulation of information processing (Warburton and Rusted, 1993), the relationship between cholinergic dysfunction and cognitive impairment in aging and AD is far from clear. An association between cognitive impairment and decrease in ACh release has been shown in aging rats (Vannucchi et al., 1990). However, association does not necessarily indicate causal relationship. Experimental lesions of the forebrain cholinergic nuclei are followed by deficits in the acquisition and performance of learned behaviors (Smith, 1988). However, caution should be exerted in attributing the deficits solely to cholinergic hypofunction since none of the lesion procedures so far used is strictly specific for cholinergic neurons (Fibiger, 1991). The same caution should hold for the aging process. Nevertheless, in animals, the recovery of the cholinergic hypofunction induced by drugs and transplants is always associated with improvement of the cognitive deficit (Fischer et al., 1987, 1991a; Gage et al., 1988; Vannucchi et al., 1990).

Can age-associated cholinergic hypofunction be improved?

Many drugs have been proposed for correcting the cholinergic hypofunction associated with normal aging and AD (Becker and Giacobini, 1991). The main approach has been to inhibit brain cholinesterase in order to increase ACh concentration in the synaptic cleft. This approach has led to the introduction of tacrine in the therapy of AD, after the unsuccessful trials with physostigmine. The true effectiveness and tolerability of tacrine in a long-term treatment, and whether its efficacy depends on cholinesterase inhibition or on other actions such as potassium channel inhibition, or muscarinic agonistic properties (Freeman and Dawson, 1991) will soon be demonstrated by its widespread use, and by comparison with new cholinesterase inhibitors. However, the most promising approach is presented by the trophic factors. It has been shown that nerve growth factor (NGF) promotes maintenance of function and survival of adult cholinergic neurons of the basal forebrain on which its specific receptors are located (for references see Hefti et al., 1991). In aging rats, a relationship exists between hippocampal NGF levels and spatial learning (Henriksson et al.,

1992), and intraventricular administration of NGF for 2 weeks increases the size of ChAT immunopositive forebrain neurons, stimulates ChAT activity in specific cholinergic nuclei, and improves spatial memory (Fischer et al., 1987, 1991a; Williams, 1991). These findings have led Olson et al. (1992) to test NGF on a patient affected by AD. Even if the observed improvement has been limited, the trial is important from a heuristic viewpoint, and the discovery of new trophic factors, along with means to manipulate their expression, may soon offer new possibilities to this therapeutic approach.

In aged rats, correction of age-associated cholinergic hypofunction has also been obtained with short-term intraperitoneal, or long-term oral administration of phosphatidylserine (PtdSer). In clinical trials, PtdSer treatments were beneficial in age-associated memory impairment and AD, even though its effects were not as rapid and clearcut as in rats. The mechanism of action, involving an effect on choline utilization, is still not understood (see references in Pepeu et al., 1993a,b)

Conclusions

The available experimental and clinical data make it possible to conclude that brain cholinergic dysfunction is actually associated with the aging process. This dysfunction may develop at different ages, involve few or several cholinergic nuclei, and vary in severity, being more severe in AD. However, to define the extent to which this cholinergic dysfunction is responsible for cognitive deficits is still a matter of research, and the hypothesis of Bartus et al. (1982) still needs the final demonstration that should come from the therapeutic results obtained with cholinergic agents. Unfortunately, we are all aware that so far the results have been somewhat disappointing (Kumar and Calache, 1991). It will be interesting to see whether the use of tacrine will change the picture. Finally, we would like to close on an optimistic note. As shown by the work with NGF and PtdSer, the aging of the cholinergic neurons appears to be reversible in the rat. Whether this is true in other animal species, including man, needs to be investigated. However, to understand the molecular basis for the recovery of cholinergic dysfunction in the rat may offer new leads for therapeutic intervention.

Acknowledgements

This work was supported by a grant from C.N.R., Target Project on Aging.

References

Armstrong, D.M., Sheffield, R., Buszaki, G., Chen, K.S., Hersh, L.B., Nearing, B. and Gage, F.H. (1993) Morphological alterations of choline acetyltransferase-positive neurons in the basal forebrain of aged behaviorally characterized Fisher 344 rats. *Neurobiol. Aging*, 14: 457–470.

Bartus, R.T., Dean, R.L., Beer, B. and Lippa, A.S. (1982) The cholinergic hypothesis of geriatric memory dysfunction. *Science*, 217: 408–410.

Becker, R.E. and Giacobini E., Eds. (1991) *Cholinergic Basis for Alzheimer Therapy*, Birkhauser, Boston.

Bigl, V., Arendt, T. and Biesold, D. (1990) The nucleus basalis of Meynert during ageing and in dementing neuropsychiatric disorders. In M. Steriade and D. Biesold (Eds.), *Brain Cholinergic System*, Oxford University Press, Oxford, pp. 364–386.

Collerton , D. (1986) Cholinergic function and intellectual decline in Alzheimer's disease. *Neuroscience*, 19: 1–28.

Crawley, J.N. and Wenk, G.L. (1989) Co-existence of galanin and acetylcholine: is galanin involved in memory processes and dementia? *Trends Neurosci.*, 12: 278–281.

Davies, P. and Maloney, A.J.R. (1976) Selective loss of cholinergic neurons in Alzheimer's disease. *Lancet*, 2: 1403.

Decker, M.W. (1987) The effects of aging on hippocampal and cortical projections of the forebrain cholinergic system. *Brain Res. Rev.*, 12: 423–438.

de Lacalle, S., Iraizos, I. and Gonzalo, L.M. (1991) Differential changes in cell size and number in topographic subdivisions of human basal nucleus in normal aging. *Neuroscience*, 43: 445–456.

Drachman, D.A. and Leavitt, J. (1974) Human memory and the cholinergic system: a relationship to ageing? *Arch. Neurol.*, 30: 113–121.

Feuerstein, T.J., Lehman, J., Sauermann, W., Van Velthoven, V. and Jackisch, R. (1992) The autoinhibitory feedback control of acetylcholine release in human neocortex tissue. *Brain Res.*, 572: 64–71.

Fibiger, H. (1991) Cholinergic mechanisms in learning, memory and dementia: a review of recent evidence. *Trends Neurosci.*, 14: 220–223.

Fischer, W., Wictorin, K., Bjorklund, A., Williams, L.R., Varon, S. and Gage, F.H. (1987) Amelioration of cholinergic neurons atrophy and spatial memory impairment in aged rats by nerve growth factor. *Nature*, 329: 65–68.

Fischer, W., Bjorklund, A., Chen, K. and Gage, F.H. (1991a) NGF improves spatial memory in aged rodents as a function of age. *J. Neurosci.*, 11: 1889–1906.

Fischer, W., Nilsson, O.G. and Bjorklund, A. (1991b) *In vivo* acetylcholine release as measured by microdialysis is unaltered in the hippocampus of cognitively impaired aged rats with degenerative changes in the basal forebrain. *Brain Res.*, 556: 44–52.

Forloni, G. and Angeretti , N. (1992) Decreased ^{3}H-hemicholinium binding to high-affinity choline uptake sites in aged rat brain. *Brain Res.*, 570: 354–357.

Freeman, S.E. and Dawson, R.M. (1991) Tacrine: a pharmacological review. *Prog. Neurobiol.*, 36: 257–277.

Gage, F.H., Armstrong, D.M., Williams, L.R. and Varon, S. (1988) Morphologic response of axotomized septal neurons to nerve growth factor. *J. Comp. Neurol.*, 269: 147–155.

Geula, C. and Mesulam, M.M. (1989) Cortical cholinergic fibers in ageing and Alzheimer's disease: a morphometric study. *Neuroscience*, 33: 469–476.

Giovannelli, L., Giovannini, M.G., Pedata, F. and Pepeu, G. (1988) Purinergic modulation of cortical acetylcholine release is decreased in aging rats. *Exp. Gerontol.*, 23: 175–181.

Hardy , J., Adolfsson, R., Alafuzoff, I., Bucht, G., Marcusson, J., Nyberg, P., Perdahl, E., Wester, P. and Winblad, B. (1985) Transmitter deficits in Alzheimer's disease. *Neurochem. Int.*, 7: 345–363.

Hefti, F., Brachet, P., Will, B. and Christen, Y. (1991) *Growth Factors and Alzheimer's Disease*. Springer, Berlin.

Henriksson, B.G., Soderstrom, S., Gower, A.J., Ebendal, T., Winblad, B. and Mohammed, A.H. (1992) Hippocampal nerve growth factor levels are related to spatial learning ability in aged rats. *Behav. Brain Res.*, 48: 15–20.

Kumar, V. and Calache N. (1991) Treatment of Alzheimer's disease with cholinergic drugs. *Int. J. Clin. Ther. Toxicol.*, 29: 23–37

Nordberg, A., Hartvig, P., Lilja A., Viitanen M., Amberla K., Lundqvist, H., Andersson, Y., Ulin, J., Winblad, B. and Langstrom, B. (1990) Decreased uptake and binding of ^{11}C-nicotine in brain of Alzheimer patients as visualized by positron emission tomography. *J. Neural Transmission*, 2: 215–224.

Nordberg, A., Alafuzoff I. and Winblad, B. (1992) Nicotinic and muscarinic subtypes in the human brain: changes with aging and dementia. *J. Neurosci. Res.*, 31: 103–111.

Olson, L., Nordberg, A., Von Holst, H., Backman, L., Ebendahl, T., Alafuzoff, I., Amberla, K., Hartvig, P., Herlitz, A., Lilja, A., Lundqvist, H., Langstrom, B., Meyersson, B., Persson, A., Viitanen, M. and Winblad, B. (1992) Nerve growth factor affects ^{11}C-nicotine binding, blood flow, EEG, and verbal episodic memory in an Alzheimer patient. *J. Neural Transmission*, 4: 79–95.

Pedata, F., Slavikova, J., Kotas, A. and Pepeu, G. (1983) Acetylcholine release from rat cortical slices during postnatal development and aging. *Neurobiol. Aging*, 4: 31–34.

Pepeu, G., Casamenti, F., Marconcini-Pepeu., I. and Scali, C. (1993a) The brain cholinergic system in aging mammals. *J. Reprod. Fertil., Suppl.*, 46: 155–162.

Pepeu, G., Casamenti, F., Scali, C. and Jeglinski, W. (1993b) Effect of serine phospholipids on memory and brain cholinergic mechanisms in aging rats. *Neurosci. Res. Commun.*, 13: S63–S66.

Sherman, K.A. and Friedman, E. (1990) Pre- and post-synaptic cholinergic dysfunction in aged rodent brain regions: new findings and an interpretative review. *Int. J. Dev. Neurosci.*, 8: 689–708.

Sims, N.R., Bowen, D.M., Smith, C.C.T., Flack, R.H.A., Davison, A.N., Snowden, J.S. and Neary, D. (1980) Glucose metabolism and acetylcholine synthesis in relation to neuronal activity in Alzheimer's disease. *Lancet*, 1: 333–335.

Smith, G. (1988) Animal models of Alzheimer's disease: experimental cholinergic denervation. *Brain Res. Rev.*, 13: 103–118.

Vannucchi, M.G. and Pepeu, G. (1987) Effect of phosphatidylserine on acetylcholine release and content in cortical slices from aging rats. *Neurobiol. Aging*, 8: 403–407.

Vannucchi, M.G., Casamenti, F. and Pepeu, G. (1990) Decrease of acetylcholine release from cortical slices in aged rats: investigations into its reversal by phosphatidylserine. *J. Neurochem.*, 55: 819–825.

Warburton, D.M. and Rusted, J.M. (1993) Cholinergic control of cognitive resources. *Neuropsychobiology*, 28: 43–46.

Williams, L.R. (1991) Exogenous nerve growth factor stimulates choline acetyltransferase activity in aging Fischer 344 male rats. *Neurobiol. Aging*, 12: 39–46.

F. Bloom (Editor)
Progress in Brain Research, Vol. 100

CHAPTER 10

Co-existence of chemical messengers in neurons

Guillermo Jaim-Etcheverry

Department of Cell Biology and Histology, School of Medicine, University of Buenos Aires, Paraguay 2155, 1121 Buenos Aires, Argentina

Introduction

While three decades ago the idea that a single neuron could store and release more than one active substance seemed farfetched, nowadays the co-existence of chemical messengers in central and peripheral nerve cells appears to be the rule rather than the exception.

During this period, the co-existence of putative transmitters in neurons has received serious consideration and experimental evidence has accumulated to substantiate this possibility. Hundreds of original reports, reviews and commentaries have been published on this topic and the mechanisms of neurotransmitter co-existence have been analyzed in numerous international meetings.

Work done in our laboratory has contributed to the evolution of this concept and is summarized here. Moreover, some general comments on the possible role of the mechanism of co-transmission are made. Detailed data on these problems may be found in the references mentioned in the text.

The co-existence of neurotransmitters in the vesicles of sympathetic fibers

In the late 1960s, at the Institute of Cell Biology directed by Eduardo De Robertis in Buenos Aires, together with Luis Zieher, we were investigating the specificity of a cytochemical reaction that was presumed to identify 5-hydroxytryptamine (serotonin) at the ultrastructural level. These studies were initiated with the idea of exploring the mechanisms of monoamine storage in sympathetic nerves. Thus, encouraged by the results obtained using rabbit blood platelets as a model (Jaim-Etcheverry and Zieher, 1968a), we rushed to try the reaction on the fibers that innervate the pineal gland of the rat. These nerves, studied in detail by Pellegrino de Iraldi et al. (1963) at the same laboratory some years earlier, have the peculiarity that, in addition to noradrenaline (NA), they physiologically contain serotonin taken up from the pinealocytes that produce the amine in high concentrations. In 1968, we could show that both NA and serotonin are stored in synaptic vesicles of pineal nerve fibers (Fig. 1) and we proposed that putative transmitters could co-exist within the dense-cored vesicles present in the nerve terminals (Jaim-Etcheverry and Zieher, 1968b).

The existence of cells storing more than one active substance was already known. Some endocrine cells, belonging to the APUD system described by Pearse (1969), had been shown to contain both a biogenic amine and a polypeptide hormone. Incidentally, with the cytochemical procedure that we were using, we could demonstrate at that time that serotonin was stored in the same granule that contained the peptide hormone, i.e. insulin in the guinea pig pancreas and calcitonin in the C cells of the sheep thyroid gland (Jaim-Etcheverry and Zieher, 1968c,d).

Our proposal that neurotransmitters could co-exist in the same vesicular storage organelles of neurons was initially received with a great scepticism that continued for many years thereafter. The reaction to these results when they were presented in a series of meetings in Europe during 1969 and in the United States in

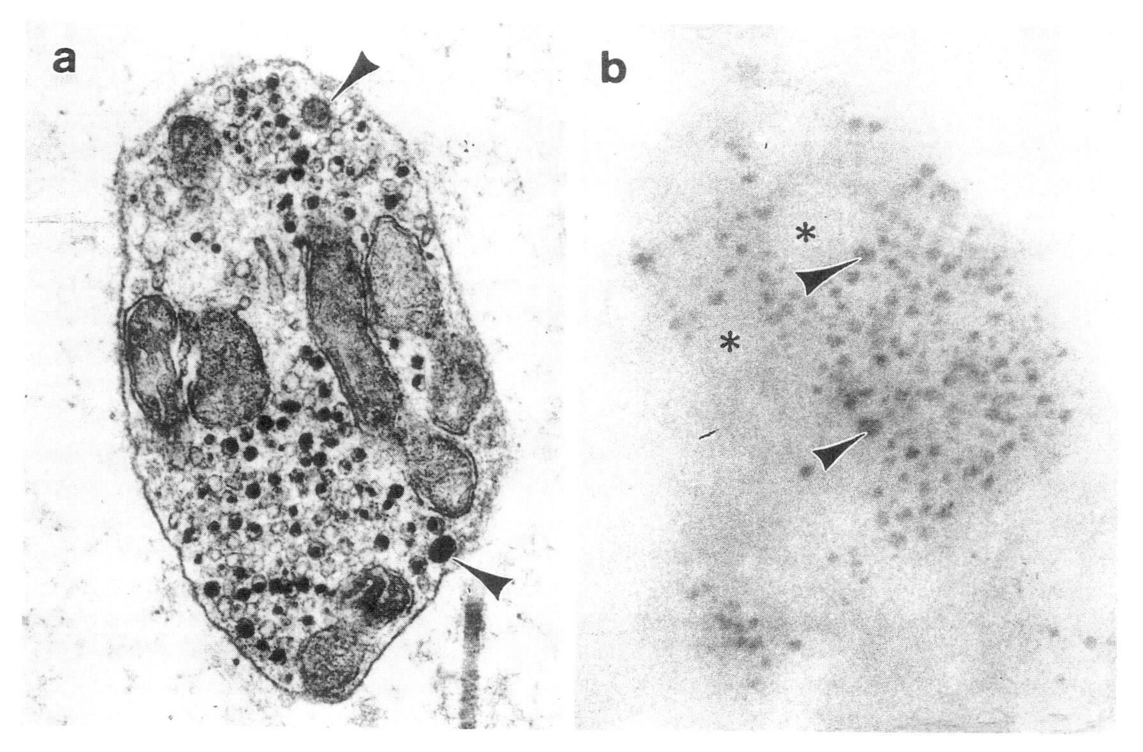

Fig. 1. Electron micrographs of adrenergic nerve terminals in the perivascular space of the pineal gland of normal rats. (a) The tissue has been conventionally processed with aldehyde fixation, followed by osmium tetroxide and lead staining of the thin section. Apart from mitochondria, small dense-cored vesicles are observed as well as some larger vesicles (arrrows) also containing a dense core (×50 000). (b) the tissue was processed with the formaldehyde-glutaraldehyde-dichromate sequence for the demonstration of serotonin storage sites. Small and large deposits corresponding to the cores observed in small and large (arrows) granular vesicles as well as the negative image of mitochondria (asterisk) are observed (×60 000) (from Jaim-Etcheverry and Zieher, 1982).

1972 was memorable. Some colleagues, patient enough to listen to peculiar theories coming from the South, conceded that we were probably dealing with a curiosity of nature.

Such a reaction was not unexpected since the co-existence hypothesis was challenging the accepted dogma of "one neuron, one transmitter". This was formulated by Eccles as "Dale's principle" on the basis of the statement made by Sir Henry Dale in the 1930s that a given class of nerve cells operates at all of its synapses using the same mechanism of transmission (see discussion in Eccles, 1986).

The mechanisms responsible for the co-storage of NA and serotonin

Apart from its significance for the interpretation of the process of peptide hormone storage in endocrine cells, our observations of the co-existence of two active molecules, a peptide and a monoamine in the same vesicle in endocrine cells and later also in peripheral sympathetic nerves (Jaim-Etcheverry and Zieher, 1969a), strengthened the possibility that NA and serotonin found in pineal nerves were in fact present in the same storage organelle. Such a mechanism was suggested by the distribution of the histochemically reactive sites corresponding to the cores of vesicles that occupied almost all the surface of the ending.

These leads prompted the initiation in 1968 of a series of studies aimed at the experimental analysis of the hypothesis of vesicular co-existence. Thus, the major thrust of our subsequent work was an attempt to demonstrate the undemonstrable: that a single nerve vesicle contains at the same time both NA and serotonin (for the detailed description of these studies, see Jaim-Etcheverry and Zieher, 1982).

The first question that we tried to answer was: are the vesicles in pineal sympathetic fibers unique in their ability to store serotonin? We could show that not only pineal fibers but also other sympathetic nerves such as those of the vas deferens, when studied in a condition that mimicked that found around the fibers innervating the pineal gland, have the ability to incorporate exogenous serotonin. The amine can gradually displace endogenous NA from the vesicles in a concentration-dependent manner and thus give a positive cytochemical reaction once it reaches a high intravesicular concentration (Jaim-Etcheverry and Zieher, 1969b; Zieher and Jaim-Etcheverry, 1971).

In 1969, octopamine was identified as a naturally occurring amine in mammalian adrenergic nerves and the possibility that it may serve as a "co-transmitter" was advanced (Molinoff and Axelrod, 1969). This role was similar to that proposed by us for serotonin in pineal nerves. Thus, the presence of octopamine within the vesicles of these nerves was considered feasible. They would contain the NA and octopamine that they synthesize, and serotonin that they take up. We reasoned that if the three amines were sharing storage space in the vesicles, the selective depletion of one of them would leave available intravesicular storage space and, as a consequence, the concentration of the other amines would rise. Figure 2 shows that by using compounds that deplete neuronal serotonin through two entirely different mechanisms, the concentration of NA and octopamine was markedly and selectively increased in the pineal gland. Apparently, as is schematically summarized in Fig. 3, due to the relative lack of specificity of the processes responsible for NA synthesis and reuptake, other molecules may be stored together with the neurotransmitter in the vesicles of adrenergic fibers (Jaim-Etcheverry and Zieher, 1971, 1975).

Consistent with the hypothesis that serotonin shares intravesicular storage space with NA was the finding that when serotonin is depleted from the vesicles, there is a rapid and transient enhancement of the activity of tyrosine hydroxylase due to the incorporation of NA to the vesicles and the release of the negative feedback control that cytoplasmic NA exerts on tyrosine hydroxylase activity (Rubio et al., 1977).

To propose the participation of neuronal serotonin in physiological mechanisms, it is important to demonstrate the release of the amine by nerve stimulation. We could show that when preganglionic nerves to both superior cervical ganglia of the rat were electrically stimulated, the reactive cores characteristic of pineal nerve vesicles almost totally disappear from the small vesicles but remain in the larger ones. Cores reacting cytochemically for serotonin as well as those giving a positive reaction for NA were depleted by

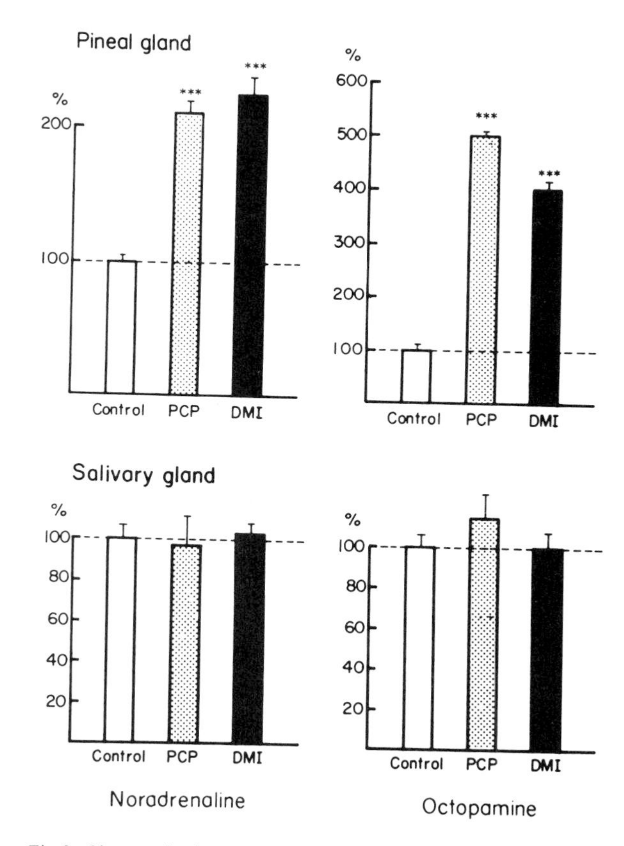

Fig.2. Changes in the content of NA and octopamine in the pineal and salivary glands of rats in which serotonin has been depleted from sympathetic nerves by two different mechanisms, i.e. by inhibiting its synthesis by the pinealocytes with *p*-chlorophenylalanine (PCP) or by blocking its uptake into the nerves with desmethylimipramine (DMI). Data are expressed as percent change from control values. $^{***}P < 0.001$ (from Jaim-Etcheverry and Zieher, 1982).

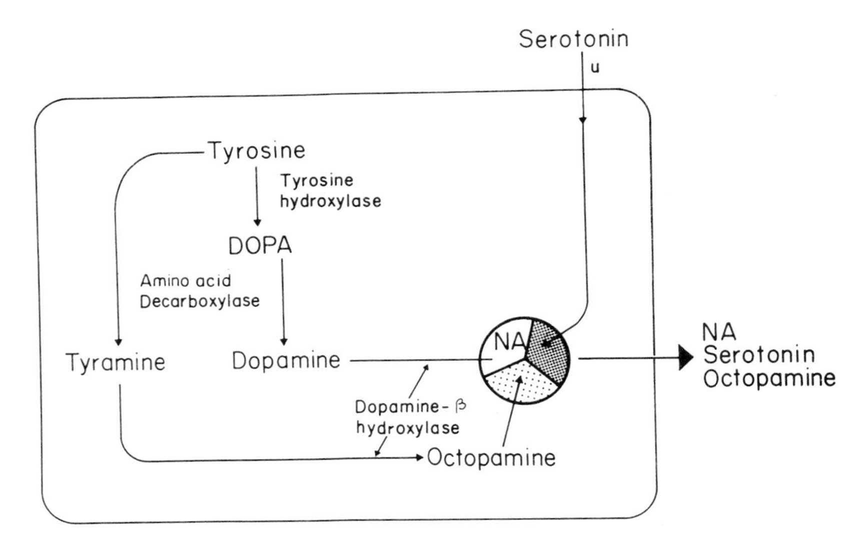

Fig. 3. Diagram showing the processes that can lead to the storage of several monoamines in the same vesicle in pineal adrenergic nerves. Whereas the lack of specificity of the reuptake process (u) is responsible for the accumulation of serotonin in the terminal, the lack of specificty of the enzymes responsible for NA synthesis results in the formation of octopamine. (from Jaim-Etcheverry and Zieher, 1982).

stimulation (Jaim-Etcheverry and Zieher, 1980). Thus, both the transmitter and the co-transmitter seem to be released by nerve impulses, a finding consistent with the hypothesis that they are stored within nerve vesicles.

On the basis of the evidence of these and other studies, we suggested that the presence of serotonin in pineal nerves could provide an efficient mechanism for the regulation of the control exerted by pineal sympathetic nerves on indole metabolism in the pineal parenchyma. The co-transmitter could act simply by modifying the amount of NA released by the nerve impulse, in this case during the 24-h cycle. This could be a special case of co-transmission that does not require the presence of the enzymatic machinery necessary for the synthesis of all the active molecules found in a given ending. This type of co-existence results from the peculiar anatomical and biochemical milieu surrounding a given terminal. Therefore, different endings of the same neuron would not necessarily contain similar active molecules. Moreover, the molecular signalling system of a given ending may change with time, depending on the activity of the surrounding neuronal or non-neuronal elements.

Neuropeptides and the phenomenon of co-existence

The study of transmitter co-existence, particularly in the CNS, poses technical difficulties due to the great heterogeneity of nervous tissue. Since the mid-1970s, the immunocytochemical approach so elegantly and thoroughly developed by Tomas Hökfelt and his group at the Karolinska Institutet in Sweden has provided many powerful insights into the mechanism of neurotransmitter co-existence. The exponential growth in the knowledge about the localization of neuropeptides both in the CNS and in the periphery thus gained, has greatly broadened our views on the possible co-existence in neurons of several molecules active in cellular communication. In this connection, the observation made by Hökfelt in 1977 that a somatostatin-like peptide co-exists with NA in the peripheral nervous system (Hökfelt et al., 1977) and the studies indicating that a substance P-like peptide co-exists with serotonin in the CNS (Hökfelt et al., 1978; Chan-Palay et al., 1978) were of particular significance.

This emerging morphological evidence favoring the mechanism of co-existence, led to a reformulation of

Dale's principle as stating that a neuron releases the same combination of transmitters from all of its terminals (Eccles, 1986). However, we were studying from the beginning a condition in which a set of terminals of a given neuronal population, that of the superior cervical ganglion, stored and released a mixture of substances different from those of terminals from the same neurons innervating other organs. This situation, resulting in our experimental model from the uptake of active substances from the vicinity of the nerve terminals, implies that the postsynaptic cell can infiltrate a "false" transmitter amongst the transmitter molecules of its afferent synapses and thus may control the extent to which it is affected by incoming impulses. A similar phenomenon to that described in the pineal may also occur in the CNS, especially where the serotonin neurons of the raphe are innervated by NA terminals (Bloom, 1974), but there is not yet such evidence probably because this situation is less amenable to experimental study.

There is another example of the heterogeneity of chemical messengers in axons of the same neuron. The invertebrate *Aplysia* was one of the first organisms in which transmitter co-existence was demonstrated (Brownstein et al., 1974; Osborne, 1981). Now there are data indicating that a single *Aplysia* neuron can store and release different neuropeptides from its individual endings (Sossin et al., 1990). This capacity not only depends on the uptake of one of the transmitters from the surrounding medium, as is the case in pineal sympathetic nerves, but rather is the result of the spatial segregation of the various peptides that are targeted to different axonal branches of the same neuron.

Possible physiological significance of transmitter co-existence

At present the phenomena of neurotransmitter co-existence and co-release, seem to provide neurons with important tools for neuronal communication and for its regulation. Different possibilities for co-existence have been described: (1) classical transmitter plus peptide(s); (2) more than one classical transmitter, either metabolically related or unrelated; and (3) more than one peptide, derived from a common prohormone or gene or from different prohormones. Virtually all types of neurons containing classical transmitters may contain one or more peptides. Recently, it has been suggested that, in addition to the biogenic amines and peptides, some neurons may also contain a fast excitatory amino acid as a third transmitter. In this case, the synapse could have the capacity to send fast (amino acid) , moderate (monoamine) and slow (peptide) signals (Nicholas et al., 1990; Hökfelt, 1991).

The particular combination of substances within well defined populations of neurons provides a way of chemically coding these groups. Subsets of neurons within the same nucleus or ganglion can be differentiated on the basis of their peculiar mixture of signalling molecules. The details of the physiological and clinical implications of the mechanism of transmitter co-existence have been extensively reviewed (see for references Hökfelt et al., 1986, 1988; Furness et al., 1989; Hökfelt, 1991).

Once the mechanism of co-existence was firmly established on anatomical grounds, there was a great interest in discovering its possible physiological significance. It has proven extremely difficult to evaluate the roles of these substances and to determine the ways in which more than one substance can participate in the transmission process. The morphological observations led to several questions, including: Are all the messengers present in neurons simultaneously released? Once liberated, are they active in the process of conveying significant signals? Does the anatomical co-existence imply that neurotransmission is plurichemical in nature?

One of the major problems under analysis has been the possible mode of action of multiple messengers. The neurons could release all their active molecules simultaneously and, in this case, the selectivity and specificity would be provided by the nature of the receptors and by their distribution. On the other hand, the selectivity could be presynaptic in nature. This would imply the storage of the messengers in different presynaptic loci. Moreover, both mechanisms may be operating at the same time.

Neuropeptides seem to be localized in the larger vesicles present in nerve terminals (diameter about 1000 Å) while classical transmitters are present in the more abundant smaller vesicles (500 Å in diameter) as well as in the larger ones (see Thureson-Klein and Klein, 1990). This differential localization has led to the speculation that, if the two types of vesicles could be selectively activated, the messenger substances could be differentially released. Such a release has been demonstrated to be dependent on the frequency of impulses and on the pattern of firing. Thus, at low firing rates, classical transmitters would be released from small vesicles at the synaptic cleft whereas at higher frequencies the large vesicles would release the peptides extrajunctionally (Lundberg et al., 1982).

Not much is known about the function of several messengers at the level of the single synapse. In most cases, the classical transmitters seem to be the important messenger but there are examples of a peptide conveying the primary message. There are many examples of the interaction between co-existing transmitters and peptides, these acting as auxiliary messengers with synergistic or antagonistic effects. These actions may be exerted either on the release of the classical transmitter or on its activity on its receptors. For example, in peripheral cholinergic nerves, vasoactive intestinal peptide (VIP) cooperates with acetylcholine (ACh) (Lundberg et al., 1982). Sympathetic fibers contain in many cases neuropeptide Y (NPY) in addition to NA . In some organs, NPY inhibits the release of NA presynaptically while in others both substances cooperate for producing vasoconstriction (Allen et al., 1982; Lundberg and Stjärne, 1984; Lundberg and Hökfelt, 1986) . There are many other situations both in the peripheral and in the central nervous system in which the possible significance of messenger co-existence have been thoroughly explored.

Apart from acting pre- and/or postsynaptically on the release of the classical transmitters, co-secreted peptides have been shown to interact with extracellular enzymes. Such is the case of the calcitonin gene related peptide (CGRP) which, when co-released with SP, seems to potentiate its action by inhibiting the enzyme responsible for SP degradation (see discussion in Hökfelt, 1991).

The interesting possibility has been recently raised that some of the messengers may participate in other forms of cell to cell interactions. They could exert long-term effects on their targets, both neurons or effector cells, acting as trophic factors related to maturation or to chemical differentiation.

The clinical significance of the mechanism of cotransmission in the pathology of the nervous system is being actively explored (see Hökfelt, 1991). For example, the detection of cholecystokinin in the mesencephalic dopaminergic neurons in the brain of schizophrenic patients treated with neuroleptics and the presence of galanin in cholinergic brain neurons, provide interesting examples of such a possibility.

Concluding remarks

Many questions remain unanswered and important details of the co-existence phenomenon are yet to be more clearly understood. Perhaps this mechanism only represents an evolutionary vestige although, as noted, it has been described in molluscs as well as in mammals. Moreover, there are now many indications of its potential role in the strategy that neurons use to communicate.

The alternative model that has emerged in the last decades derived from the study of multiple transmitter neurons is now accepted due to the evidence available for the modulatory interactions of co-transmitters, multiple post-synaptic receptors, multiple second messenger systems and the interactions between them (O'Donohue et al., 1985). During the last 25 years, we have witnessed the process by which this phenomenon has evolved from being an oddity of nature to becoming an established fact beyond discussion albeit its significance is not yet entirely clear. But there is now no doubt that several potentially active molecules coexist in both central and peripheral neurons. The available evidence suggest that their presence contributes to the subtlety and complexity that characterize the chemical signalling between neurons and their targets.

Acknowledgements

The original work reported, carried out together with Professor Luis Maria Zieher, was supported by the Consejo Nacional de Investigaciones Científicas y Técnicas and the Secretaría de Ciencia y Técnica, Argentina.

References

Allen, J., Tatemoto, K., Polak, J., Hughes, J. and Bloom, S. (1982) Two novel related peptides, neuropeptide Y (NPY) and peptide YY (PYY) inhibit the contraction of the electrically stimulated mouse vas deferens. *Neuropeptides*, 3: 71–77.

Bloom, F.E. (1974) Dynamics of synaptic modulation: perspectives for the future. In: F.O. Schmitt and F.G. Worden (Eds.), *The Neurosciences: Third Study Program*, MIT Press, Cambridge, MA, pp. 989–999.

Brownstein, M.J., Saavedra, J.M., Axelrod, J., Zeman, G.H. and Carpenter, D.O. (1974) Coexistence of several putative neurotransmitters in single identified neurons of Aplysia . *Proc. Natl. Acad. Sci. USA*, 7: 4662–4665.

Chan-Palay, V, Jonsson, G. and Palay, S.L. (1978) Serotonin and substance P coexist in neurons of the rat's central nervous system. *Proc. Natl. Acad. Sci. USA*, 75: 1582–1586.

Eccles, J. (1986) Chemical transmission and Dale's principle. In: T. Hökfelt, K. Fuxe and B. Pernow (Eds.) *Coexistence of Neuronal Messengers: A New Principle in Chemical Transmission. Progress in Brain Research,* Vol. 68, Elsevier, Amsterdam, pp. 3–13.

Furness, J.B, Morris, J.L., Gibbins, I.L. and Costa, M. (1989) Chemical coding of neurons and plurichemical transmission. *Annu. Rev. Pharmacol. Toxicol.*, 29: 289–306.

Hökfelt, T. (1991) Neuropeptides in perspective: the last ten years. *Neuron*, 7: 867–879.

Hökfelt, T., Elfvin, L.G., Elde, R., Schultzberg, M., Goldstein, M. and Luft, R. (1977) Occurrence of somatostatin-like immunoreactivity in some peripheral sympathetic noradrenergic neurons. *Proc. Natl Acad. Sci. USA*, 74: 3587–3591.

Hökfelt, T., Ljungdahl, A., Steinbusch, H., Verhofstad, A., Nilsson, G., Pernow, B. and Goldstein, M. (1978) Immunohistochemical evidence of substance P-like immunoreactivity in some 5-hydroxytryptamine-containing neurons in the rat central nervous system. *Neuroscience*, 3: 517–538.

Hökfelt, T., Fuxe, K. and Pernow, B. (Eds.) (1986) *Coexistence of Neuronal Messengers: A New Principle in Chemical Transmission.. Progress in Brain Research,* Vol. 68. Elsevier, Amsterdam, 411 pp.

Hökfelt, T., Meister, B., Melander, T., Schalling, M., Staines, W., Millhorn, D., Seroogy, K., Tsuruo, Y., Holets, V., Ceccatelli, S., Villar, M., Ju, G., Freedman, J., Olson, L., Lindh, B., Bartfai, T., Fisone, G., le Greves, P., Terenius, L., Post, C., Mollenholt, P., Dean, J. and Goldstein, M. (1988) Coexistence of multiple neuronal messengers: new aspects on chemical transmission. In: *Fidia Research Foundation Neuroscience Award Lectures*, pp. 2:61–113.

Jaim-Etcheverry, G. and Zieher, L.M. (1968a) Cytochemistry of 5-hydroxytryptamine at the electron microscope level. I. Study of the specificity of the reaction in isolated blood platelets. *J. Histochem. Cytochem.*, 16:162–171.

Jaim-Etcheverry, G. and Zieher, L.M. (1968b) Cytochemistry of 5-hydroxytryptamine at the electron microscope level. II. Localization in the autonomic nerves of the rat pineal gland. *Z. Zellforsch.*, 86: 393–400.

Jaim-Etcheverry, G. and Zieher, L.M. (1968c) Electron microscopic cytochemistry of 5-hydroxytryptamine (5-HT) in the beta cells of guinea pig endocrine pancreas. *Endocrinology*, 83: 917–923.

Jaim-Etcheverry, G. and Zieher, L.M. (1968d) Cytochemical localization of monoamine stores in sheep thyroid gland at the electron microscope level. *Experientia*, 24:593–595.

Jaim-Etcheverry, G. and Zieher, L.M. (1969a) Selective demonstration of a type of synaptic vesicle by phosphotungstic acid staining. *J. Cell Biol.*, 42: 855–860.

Jaim-Etcheverry, G. and Zieher, L.M. (1969b) Ultrastructural cytochemistry and pharmacology of 5-hydroxytryptamine in adrenergic nerve endings. I. Localization of exogenous 5-hydroxytryptamine in the autonomic nerves of the rat vas deferens. *J. Pharmacol. Exp. Ther.*, 166: 264–271.

Jaim-Etcheverry, G. and Zieher, L.M. (1971) Ultrastructural cytochemistry and pharmacology of 5-hydroxytryptamine in adrenergic nerve endings. III. Selective increase of norepinephrine in the rat pineal gland consecutive to depletion of neuronal 5-hydroxytryptamine. *J. Pharmacol. Exp. Ther.*, 178: 42–48.

Jaim-Etcheverry, G. and Zieher, L.M. (1975) Octopamine probably coexists with noradrenaline and serotonin in vesicles of pineal adrenergic nerves. *J. Neurochem.*, 25: 915–917.

Jaim-Etcheverry, G. and Zieher, L.M. (1980) Stimulation depletion of serotonin and noradrenaline from vesicles of sympathetic nerves in the pineal gland of the rat. *Cell Tissue Res.*, 207: 13–20.

Jaim-Etcheverry, G. and Zieher, L.M. (1982) Coexistence of monoamines in peripheral adrenergic neurons. In: A.C. Cuello (Ed.), *Co-Transmission*, Macmillan, London. pp. 189–206.

Lundberg, J. and Hökfelt, T. (1986) Multiple co-existence of peptides and classical transmitters in peripheral autonomic and sensory neurons – functional and pharmacological implications. In : T. Hökfelt, K. Fuxe and B. Pernow (Eds.) *Coexistence of Neuronal Messengers: A New Principle in Chemical Transmission. Progress in Brain Research,* Vol. 68, Elsevier, Amsterdam, pp. 241–262.

Lundberg, J. and Stjärne, L. (1984) Neuropeptide Y (NPY) depresses the secretion of ^{3}H-noradrenaline and the contractile response evoked by field stimulation in rat vas deferens. *Acta Physiol. Scand.*, 120: 477–479.

Lundberg, J.M., Hedlund, B., Ånggärd, A., Fahrenkrug, J., Hökfelt, T., Tatemoto, K. and Bartfai, T. (1982) Costorage of peptides and classical transmitters in neurons. In: S.R. Bloom, J.M. Polak and E. Lindenlaub (Eds.), *Systemic Role of Regulatory Peptides*. Schattauer, Stuttgart, pp. 93–119.

Molinoff, P. and Axelrod, J. (1969) Octopamine: normal occurrence in sympathetic nerves of rats. *Science*, 164: 428–429.

Nicholas, A., Cuello, A., Goldstein, M. and Hökfelt, T. (1990)

Glutamate-like immunoreactivity in medulla oblongata catecholamine/substance P neurons. *NeuroReport*, 1: 235–238.

O'Donohue, T.L., Millington, W.R., Handelmann, G.E., Contreras, P. and Chronwall, B.M. (1985) On the 50th anniversary of Dale's law: multiple neurotransmitter neurons. *Trends Pharmacol.*, 6: 305–308.

Osborne, N.N. (1981) Communication between neurones: current concepts. *Neurochem. Int.*, 3: 3–16.

Pearse, A.G.E. (1969) The cytochemistry and ultrastructure of polypeptide hormone producing cells of the APUD series and the embryologic, physiologic and pathologic implications of the concept. *J. Histochem Cytochem.*, 17: 303–313.

Pellegrino de Iraldi, A., Zieher, L.M. and De Robertis, E. (1963) 5-Hydroxytryptamine content and synthesis of normal and denervated pineal gland. *Life Sci.*, 1: 691–696.

Rubio, M.C., Jaim-Etcheverry, G. and Zieher, L.M. (1977) Tyrosine hydroxylase activity increases in the pineal gland after depletion of neuronal serotonin. N.S. *Arch. Pharmacol.*, 301: 75–78.

Sossin, W.S., Sweet-Cordero, A. and Scheller, R.H. (1990) Dale's hypothesis revisited: different neuropeptides derived from a common prohormone are targeted to different processes. *Proc. Natl. Acad. Sci. USA*, 87: 4845–4848.

Thureson-Klein, A. and Klein, R.L. (1990) Exocytosis from neuronal large dense-cored vesicles. *Int. Rev. Cytol.*, 121:67–126.

Zieher, L.M. and Jaim-Etcheverry, G. (1971) Ultrastructural cytochemistry and pharmacology of 5-hydroxytryptamine in adrenergic nerve endings. II. Accumulation of 5-hydroxtryptamine in nerve vesicles containing norepinephrine in rat vas deferens. *J. Pharmacol. Exp. Ther.*, 178: 30–41.

F. Bloom (Editor)
Progress in Brain Research, Vol. 100

CHAPTER 11

Cloning of kappa opioid receptors: functional significance and future directions

H. Akil and S.J. Watson

Mental Health Research Institute, Department of Psychiatry, University of Michigan Medical School, Ann Arbor, MI 48105, USA

Introduction

This review focuses on the molecular and anatomical study of kappa receptors, a class of opioid receptors which mediates unique and distinctive functions, including the modulation of drinking, eating, gut motility, temperature control, and various endocrine functions. Kappa agonists have been proposed to be potentially useful clinical compounds as neuroprotective agents, and as antinociceptive drugs with little drug abuse liability, although they have been reported to produce negative subjective effects. In some instances, the excessive activation of kappa receptors has been deemed harmful, and the use of specific kappa antagonists has been put forth as clinically useful (e.g. in trauma or spinal cord injury). A great deal of the complexity in the field appears to result from the existence of multiple kappa receptors which may mediate different functions. In order to understand the functions of this interesting class of receptors, it is important to first place them in the context of the entire opioid system to which they belong.

Endogenous opioids and their receptors

The study of opioid receptors predated the study of their endogenous ligands by many decades. This fact imparts some unique characteristics to the field. For example, unlike other receptors, opioid receptors are not defined as such on the basis of an endogenous ligand that they recognize, but on the basis of their interaction with certain pharmacological agents, such as heroin, morphine or their congeners. In particular, the antagonist naloxone has become the sine qua non for defining as "opiate" a binding site, a physiological function or a behavioral effect. Furthermore, it is possible for an endogenous opioid peptide to have non-opioid actions (e.g. non-naloxone reversible) and for an opioid binding site to have no identified endogenous ligand. While some of these features of the field may be attributable to historical accident, the nature of the opioid system does not simplify matters. Each receptor interacts with multiple endogenous ligands and each ligand with multiple opioid receptors. For these reasons, it is critical to define, as much as possible, the nature and number of "actors" in this rich and complex system, to identify the molecules involved on both sides of the synapse, and to describe their anatomical relationships and their functional interactions.

The 20 years which have elapsed since the discovery of the enkephalins have yielded a substantial body of knowledge regarding the endogenous opioids: their molecular biology, biogenesis, anatomical distribution, regulation and functions. We now know that three distinct genes encode unique opioid peptide precursors, termed pro-opiomelanocortin, pro-enkephalin and pro-dynorphin which give rise to a wide array of individual opioid and non-opioid peptides. During the same period, we have learned a great deal about the multiplicity of opioid receptors, and evolved many pharmacological tools (e.g. highly selective agonists and antagonists) to study them. Yet, we have no molecular or structural information about the receptors themselves. However, recently, two independent groups (Evans et al., 1992; Kieffer et al., 1992) cloned an opioid receptor of the delta type from a neuroblastoma × glioma cell line (NG-108). This has opened the door to the cloning of other opioid receptors including kappa, and to defining their structure, coupling to signaling pathways, anatomy and regulation.

Multiple opioid receptors

Opioid binding sites were first identified in the brain in the early 1970s. The issue of receptor multiplicity arose shortly thereafter. Martin et al. (1976) postulated the existence of three distinct opiate receptors: a morphine-preferring site termed *mu,* a ketocyclazocine-preferring site termed *kappa,* and an SKF 10, 047 preferring site termed *sigma.* Sigma was subsequently found not to mediate naloxone-reversible effects, and is no longer classified as an opiate site. Meanwhile, Kosterlitz, Hughes and their colleagues (Lord et al., 1977) proposed the existence of an enkephalin-preferring *delta* site. In addition to these three major classes (mu, delta and kappa), a number of receptor types have been proposed, often based on the use of bioassays from different species. The three major classes have been the subject of intense investigation. In the early 1980s, highly selective ligands became available for labeling the multiple receptor types, and the study of their biological functions was also facilitated by the recent syntheses of selective antagonists.

Relationships between multiple opioid peptides and receptors

As mentioned above, there is not a one-to-one correspondence between the endogenous ligands and their receptors, although some general associations do exist. The delta receptor(s) typically recognizes with high affinity the products of the proenkephalin precursor, particularly leucine- and methionine-enkephalin. However, these receptors also interact with excellent affinity with the POMC product β-endorphin 1-31, and with the prodynorphin peptide dynorphin A 1-8. The mu receptors also interact with products of each of the three opioid genes: β-endorphin, almost all proenkephalin products, and several products of prodynorphin, including dynorphin A. On the other hand, kappa receptors are thought to be somewhat more selective, interacting primarily with products of the prodynorphin family with high affinity and good selectivity. These pharmacological observations in vitro are confirmed by anatomical findings demonstrating that there is no exclusive anatomical relationship between a given opioid peptide family and a specific class of opioid receptors. For instance, prodynorphin products can be found in close proximity to kappa receptors in some sites and mu receptors in other sites.

Kappa receptor heterogeneity

The issue of multiplicity of kappa receptors is genuinely confusing. This is due to a number of reasons. The most fundamental one is the very definition of "kappa". While this receptor was originally identified by Martin (see Martin et al., 1976) on the basis of physiological effects of ethylketocyclazocine (EKC), this definition has been lost in the course of in vitro studies. It was evident from the outset that EKC and other drugs in its class (benzomorphans) interacted with other types of opioid receptors; many of these drugs (e.g. bremazocine) were identified as antagonists at the morphine site, and as agonists or partial agonists at this putative kappa site. Thus, one could not rely on these ligands to selectively label kappa sites in binding studies. Since relatively specific mu and delta ligands existed by the early 1980s, the strategy proposed by Kosterlitz et al. (1981) was to use the high affinity non-specific opiate bremazocine as the labeling compound and to block or suppress delta and mu binding with micromolar concentrations of their most selective ligands. This non-mu, non-delta component was termed kappa. It should be noted however, that such a definition assumes only three major classes of opioid receptors, and also assumes that the blocking drugs will suppress all subtypes of mu and delta receptors; in effect, this approach would cause all opioid sites heretofore not identified in brain to be designated as kappa. On the other hand, Goldstein and James (1984) defined kappa as the dynorphin receptor, due to various observations in bio-assays showing a high selectivity and potency of dynorphin A towards this site. Clearly, the two definitions, as the non-delta/non-mu site, versus the dynorphin A site, would not necessarily be expected to yield identical populations of receptors.

A second source of confusion is the nature of the labeling ligand. The original ligand commonly used was [^{3}H]EKC based on Martin's definition of kappa. However, [^{3}H]bremazocine was used shortly thereafter by numerous groups (Kosterlitz et al., 1981). Finally, as more specific kappa ligands were synthesized, particularly of the benzeneacetamide series, they were used to define kappa, e.g. [^{3}H]U-69 593 (Lahti et al., 1985). Each of these agents is unique in terms of its interactions with multiple types and subtypes of opioid receptors, and may therefore label different populations. A related issue is the exact conditions of the binding, particularly with regards to sodium ion concentration, and temperature, as both these variables are known to differentially affect the ligand binding characteristics of different subtypes of opioid receptors. A final issue is the regional and species differences, which are profound where kappa is concerned. Early studies detected little kappa binding in rat brain or spinal cord, and a great deal more binding in guinea pig brain, monkey and human brain. Different authors have examined kappa binding sites and their subtypes in guinea pig, rat, dog, mouse, calf and human tissue, without necessarily carrying out side-by-side comparisons. Thus, when a subtype is described, it is difficult to determine whether it can be generalized across species, or whether it is unique to that particular species and tissue.

Against this background, the issue of multiplicity of kappa binding sites can be briefly summarized. Reports of heterogeneity began in the early 1980s. For example, in 1981, Herz' laboratory (Pfeiffer et al., 1981) studied a binding preparation from human brain with suppressed mu and delta sites. They noted that dynorphin A was a highly potent ligand at a site that could be classified as kappa; yet this peptide could not displace 50% of the labeled population, which could be displaced by benzomorphans. This observation essentially contrasted the two definitions of kappa, as the dynorphin A site versus as the non-mu, non-delta site, presaging a great deal of subsequent work. At this point, there is a great deal of consensus about a kappa subtype termed kappa 1, which recognizes with high affinity members of the benzeneacetamide family, such as U-50 488 and U-69 593 synthesized by Upjohn. In addition, this site recognizes most of the products of the pro-dynorphin gene with high affinity, and is readily labeled by EKC, bremazocine and other benzomorphans, as well as by the non-specific antagonist, naloxone and the specific kappa antagonist nor-BNI. Some authors have suggested the existence of subtypes of kappa 1, based on affinity to prodynorphin products. While Dyn A bound both sites with equal affinity, Dyn B and alpha-neoendorphin recognizes the two sites with a 50-fold difference in potency.

The definition of kappa 2 and its subtypes is significantly more problematic. Chang et al. (1981) first suggested the existence of a benzomorphan binding site distinct from mu and delta. Many of the current classifications of the non-mu, non-delta, non-kappa 1 sites are consistent with the existence of such a site(s). Additionally, there is a proposed subdivision within the kappa 2 category, whereby the specific pharmacological profile of kappa 2 and its subtypes have been investigated, and significant differences shown in interactions with endogenous ligands. Taken together the various studies suggest the existence of 3–5 subtypes of benzomorphan binding sites in rat and guinea pig, with substantial species differences. Whether these classifications carry physiological implications remains to be determined. It should be recalled, however, that molecular cloning of other neurotransmitter receptor families has typically uncovered more subtypes than had been anticipated based on binding studies.

The anatomical distribution of kappa 1 versus kappa 2 sites in rat brain was shown to be different (Zukin et al., 1988). Thus, EKC labeling occurs in patches in the striatum, whereas U-69 593 labeling appears homogeneous over this region. EKC labels more lateral regions of the caudate-putamen-accumbens complex, whereas the kappa 1 sites appear to be somewhat more medial. Furthermore, there are a number of sites labeled by EKC but not by the Upjohn compound, such as certain hippocampal regions, habenula, interpeduncular nucleus, locus coeruleus and cerebellum.. However, not all investigators agree on these kappa 1 versus kappa 2 differences in rat, and different anatomical patterns are seen in guinea pig (cf. Mansour and Watson, 1993). These discrepancies are probably due to the various technical and species issues discussed above.

Coupling of kappa receptor to second messenger systems and ion channels

The signal transduction pathway of the kappa receptor(s) has been the subject of some controversy. By analogy to the mu and delta receptor transduction mechanisms, it was expected that kappa receptors would be coupled to G proteins and would produce inhibition of adenylyl cyclase. While some disagreement in the literature exists, there is sufficient evidence to suggest that this, indeed, may be the case. In addition, it has been proposed that kappa receptors may couple to the phospholipase C(PLC)-mediated cascade leading to the formation of IP3 and diacyl glycerol (DAG). The interaction with the PI turnover cascade is rather controversial with reports suggesting an increase and a decrease.

Werz and Macdonald (1982) were the first to show that kappa opioid receptors are coupled to calcium channels. This appeared to be in contrast to the mu and delta receptor coupling. These latter two receptors appeared to cause presynaptic inhibition by increasing potassium conductance (North, 1993), although calcium coupling of the mu receptor has since been also demonstrated. More recently, Macdonald and his coworkers have shown coupling of dynorphin receptors in dorsal root ganglion cultures to a large transient N-type calcium current.

It is clear that kappa receptors, like mu and delta, cause decreased neurotransmitter release via what is termed pre-synaptic inhibition (for review, see Mulder et al., 1984). This process is dependent on extracellular calcium, and given the fact that kappa opioids block calcium entry through the N-channel, it is typically thought that the two mechanisms are linked, i.e. that the inhibition of transmitter release is secondary to the calcium channel blockade, although a strict causal relation has not been established.

Cloning and structural characteristics of kappa receptors

The recent cloning of the mouse delta opioid receptor (Evans et al., 1992; Kieffer et al., 1992) has led to the rapid identification of kappa receptor clones. Bell, Reisine and their colleagues recognized that a mouse brain orphan receptor, previously classified as a member of the somatostatin receptor family, was indeed a member of the opioid family, with pharmacological characteristics consistent with a kappa 1 profile (Yasuda et al., 1993). Our own group, using the polymerase chain reaction (PCR), isolated and identified several opioid receptors from rat brain libraries. One of these clones was demonstrated to be the rat kappa 1 opioid receptor (Meng et al., 1993). The new sequence is most closely related to the mouse delta opioid receptor and to the somatostatin receptor family, with a 62% homology to mouse delta receptor and 49% homology to rat somatostatin receptor at the nucleic acid level. Computer hydrophobicity analysis of the encoded protein indicated that it has seven hydrophobic domains of 22–26 amino acid residues each, suggesting it may be a new member of the seven transmembrane domain G-protein coupled receptor family. Furthermore, its protein sequence possesses other features common to this family of receptors, e.g. two possible N-glycosylation sites in the N-terminal domain, and two putative palmotylation sites in the carboxy terminal region. In addition, it contains putative target sites for protein kinase C phosphorylation, Ca^{2+}/calmodulin dependent kinase, cAMP dependent protein kinase and casein kinase II.

The pharmacological profile exhibited a distinctive signature. The receptor expressed bound all kappa opioid receptor selective ligands tested with high affinity, including all the products of pro-dynorphin and all the arylacetamides (Upjohn compounds). The receptor also bound several non-selective opioid ligands such as bremazocine, EKC and naltrexone with high affinity. Furthermore, it showed high stereospecificity, with good affinity for (–)-naloxone and levorphanol and very poor binding affinity for their chiral isomers, (+)-naloxone and dextrorphan, respectively. It bound neither mu nor delta selective compounds with high affinity. Finally, it exhibited a strong requirement for the N-terminal tyrosine of dynorphin A, as demonstrated by a significant loss of affinity exhibited by Dyn 2-13. Based on this binding profile, and the high affinity to opioids of the arylacetamide family, we identified this clone as a kappa receptor, of the kappa 1 subtype.

To determine how this protein is functionally coupled to the second messenger system, we studied the ability of the expressed receptor (transfected into COS-1 cells) to inhibit adenylyl cyclase stimulation, and elevate IP_3 levels. The results showed inhibition of cAMP levels by selective kappa agonists, reversed by a selective antagonist. No effect on PI turnover could be detected. More recently, we have cloned the guinea pig kappa 1 receptor, have characterized its pharmacology and shown its ability to specifically alter calcium influx by using Fura-2 as a detector.

Kappa receptor mRNA distribution is widespread in the CNS with particularly high levels in the more ventral aspects of the basal ganglia, limbic structures and cortex. Within the nigrostriatal and mesolimbic dopamine systems, kappa receptor mRNA appears to be selectively distributed in the mesolimbic system with a localization primarily in the medial caudate-putamen and the medial aspect of the substantia nigra, as well as in the nucleus accumbens and ventral tegmental area. This distribution corresponds well to the distribution of kappa binding sites (both kappa 1 or total kappa), if one considers that some discrepancies are expected due to transport of the receptor proteins to sites distal from the cell bodies where mRNA synthesis occurs (Mansour et al., 1994).

Summary and future directions

The recent cloning of the kappa 1 receptor in mouse, rat, and guinea pig should allow us to answer many questions regarding the nature and unique functions of this receptor. The simplest question regards the existence of multiple members of this branch of the opioid family, i.e. the number of kappa subtypes within and across species. Of great interest at the structural level is how a single receptor molecule binds with great affinity and selectivity both a long peptide (Dyn A) and a compact opioid molecule. A related question is how the entire opioid receptor family achieves such exquisite selectivity between ligands (both endogenous and exogenous) in spite of an exceedingly high level of sequence identity amongst its members. Comparison of the sequences suggests a possible important role of the N-terminal and extracellular loops, as these regions are highly divergent across mu, delta and kappa. In contrast, the intracellular loops, especially loops 1 and 2 and the N-terminal segment of the carboxy-terminal domain, are extremely well conserved across opioid receptors (90%). This suggests very similar patterns of signal transduction. Yet, we know very well that at the functional level, these receptors mediate very distinctive and sometimes physiologically antagonistic functions. The relative importance of molecular events versus more complex neural events (e.g. circuits) in effecting these functions remains to be studied. Finally, the molecular mechanisms involved in desensitization and down regulation of kappa and other opioid receptors can now be elucidated, contributing to our understanding of mechanisms of tolerance and dependence at the whole animal level. Given that kappa receptors mediate very distinctive and clinically relevant functions, these recent breakthroughs promise an accelerated rate of research in the area, with both basic and clinical implications.

References

Chang, K.-J., Hazum, E. and Cuatrecacas, P. (1981) Novel opiate binding sites selective for benzomorphan drugs. *Proc. Natl. Acad. Sci. USA*, 78: 4141–4145.

Evans, C., Keith Jr, D., Morrison, H., Magendzo, K. and Edwards, R. (1992) Cloning of a delta opioid receptor by functional expression. *Science*, 258: 1952–1955.

Goldstein, A. and James, I. (1984) I. Site-directed alkylation of multiple opioid receptors II. Pharmacological selectivity. *Mol. Pharmacol.*, 25: 343–348.

Kieffer, B., Befort, K., Gaveriaux-Ruff, C. and Hirth, C. (1992) The delta opioid receptor: isolation of a cDNA by expression cloning and pharmacological characterization. *Proc. Natl. Acad. Sci. USA*, 89: 12048–12052.

Kosterlitz, H., Paterson and S., Robson, L. (1981) Characterization of the kappa-subtype of the opiate receptor in the guinea pig brain. *Br. J. Pharmacol.*, 73: 939-949.

Lahti, R., Mickelson, M., McCall, J. and Von Voigtlander, P. (1985) [^{3}H]U-69 593 a highly selective ligand for the opioid κreceptor. *Eur. J. Pharmacol.*, 109: 281.

Lord, J., Waterfield, A., Hughes, J. and Kosterlitz, H. (1977) Endogenous opioid peptides: multiple agonists and receptors. *Nature (London)*, 267: 495–499.

Mansour, A and Watson, S. (1993) Anatomical distribution of opioid receptors in mammalians: an overview. In: A. Herz (Ed.), *Opioids I*, Springer-Verlag, Berlin, pp. 79–105.

Mansour, A., Fox, C., Meng, F., Akil, H and Watson, S.J. (1994) Kappa 1 receptor mRNA distribution in the rat CNS: comparison to kappa receptor binding and prodynorphin mRNA. Mol. Cell. Neurosci., in press.

Martin, W., Eades, C., Thompson, J., Huppler, R. and Gilbert, P. (1976) The effects of morphine and nalorphine-like drugs in nondependent and morphine-dependent chronic spinal dog. *J. Pharmacol. Exp. Ther.*, 197: 517–532.

Meng, F., Xie, G., Thompson, R., Mansour, A., Goldstein, A., Watson, S. and Akil, H. (1993) Cloning and pharmacological characterization of a rat κ opioid receptor. *Proc. Natl. Acad. Sci. USA*, 90: 9954–9958.

Mulder, A., Frankhuyzen, A., Stoof, J., Wemer, J. and Schoffelmeer, A. (1984) Catecholamine receptors, opiate receptors and presynaptic modulation of transmitter release in brain. In: E. Usdin (Ed.), *Catecholamines, Neuropharmacology and the Central Nervous System – Theoretical Aspects*, Alan R. Liss, New York, p. 47.

North, R. (1993) Opioid actions on membrane ion channels. In: A. Herz (Ed.), *Opioids I*, Springer-Verlag, Berlin, pp. 773–797.

Pfeiffer, A., Pasi, A., Mehraein, P. and Herz, A. (1981) A subclassification of κ-sites in human brain by use of dynorphin 1–17. *Neuropeptides*, 2: 89–97.

Werz, M. and Macdonald, R. (1982) Opioid peptides decrease calcium-dependent action potential duration of mouse dorsal root ganglion neurons in cell culture. *Brain Res.*, 239: 315–321.

Yasuda, K., Raynor, K., Kong, H., Breder, C.D., Takeda, J., Reisine, T. and Bell, G.I. (1993) Cloning and functional comparison of κ and δ opioid receptors from mouse brain. *Proc. Natl. Acad. Sci. USA*, 90: 6736–6740.

Zukin, R., Eghbali, M., Olive, D., Unterwald, E., Tempel, A. (1988) Characterization and visualization of rat and guinea pig brain kappa opioid receptors: evidence for kappa 1 and kappa 2 opioid receptors. *Proc. Natl. Acad. Sci. USA*, 85: 4061–4065.

F. Bloom (Editor)
Progress in Brain Research, Vol. 100

CHAPTER 12

Vasoactive intestinal peptide and noradrenaline regulate energy metabolism in astrocytes: a physiological function in the control of local homeostasis within the CNS

Pierre J. Magistretti

Institut de Physiologie, Faculté de Médecine, Université de Lausanne, CH-1005 Lausanne, Switzerland

Introduction

The normal function of the central nervous system (CNS) depends on proper communication between billions of cells that are heterogeneous in structure and function. The cell types that constitute the CNS include: neurons, astrocytes (fibrous and protoplasmic), oligodendrocytes, microglial cells, ependymal cells, choroid plexus cells and cells of the vasculature, i.e. endothelial cells, pericytes and smooth muscle cells. Communication between neurons has been shown to be mediated by neurotransmitters. These molecules are released in a calcium-dependent manner from the depolarized axon terminal, interact with specific receptors on the postsynaptic membrane and, through elaborate transduction mechanisms, exert their effect(s) on cell function. Some variations on this theme, which greatly enrich the vocabulary of chemical neurotransmission have, however, been discovered in recent years. First, it has become clear that neurotransmitters released from neurons can also interact in a receptor-mediated manner with non-neuronal cells, such as for example astrocytes, choroid plexus cells or cells of the vasculature. Second, non-neuronal cells, in particular astrocytes and the microglia, can release molecules such as interleukins and prostanoids known to play a role in cell–cell communication in the immune system and in inflammatory processes, respectively (Bloom et al., 1993). To summarize, neurotransmitter-mediated neuron–neuron interaction is not the only communication line in the CNS, neurons also communicate with non-neuronal cells.

From the foregoing considerations, we have hypothesized, and in fact found experimental evidence indicating that neurotransmitters, released by neurons, can exert metabolic actions in non-neuronal cells of the CNS. This chapter focuses on the evidence indicating that the neurotransmitters vasoactive intestinal peptide (VIP) and noradrenaline (NA) act on astrocytes to regulate energy metabolism in the cerebral cortex.

Cytological substrate for interactions between neurons and non-neuronal cells

In the cerebral cortex, VIP is contained in a homogeneous population of radially oriented, bipolar interneurons (Magistretti and Morrison, 1988). Because their dendritic arborization diverges only minimally from the main axis of the cell, these intracortical neurons exert very localized input–output functions within radial cortical "columns" (Magistretti and Morrison, 1988) (Fig. 1). VIP neurons are distributed throughout the cortical mantle, with a slight rostro-caudal gradient. In addition, the density of VIP-containing neurons is such that the columnar ensembles that they define partially overlap, meaning that despite their radial nature, VIP neurons can "cover" the entire cerebral cortex (Magistretti and Morrison, 1988). The morphology of the neuronal circuits that contain NA is strikingly

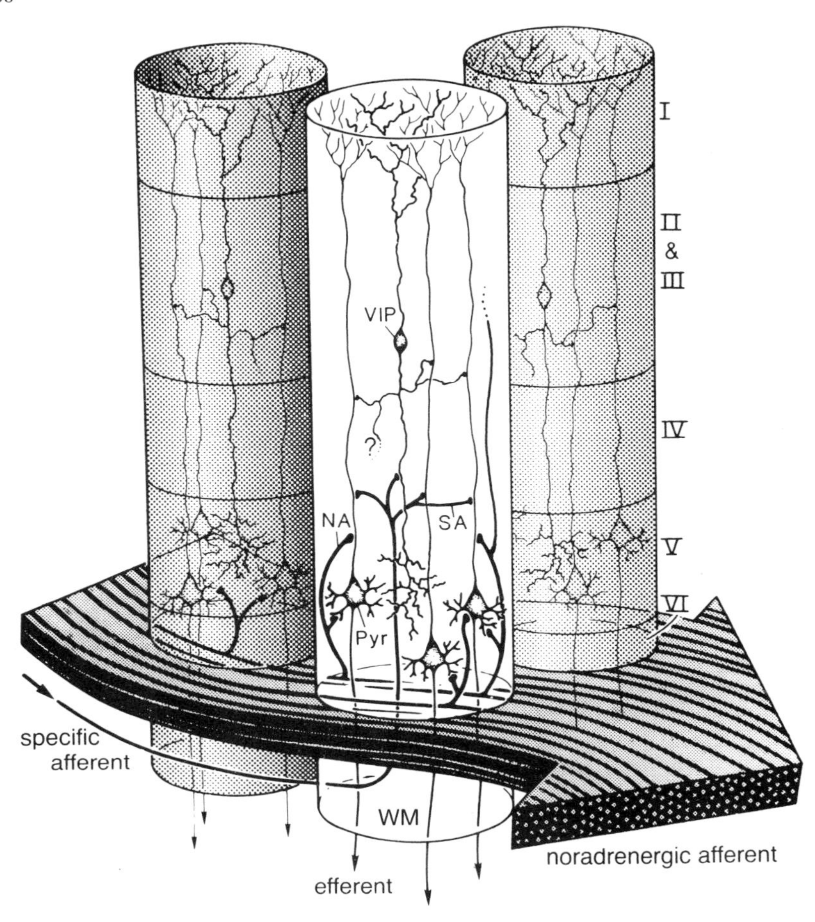

Fig. 1. Columnar organization of vasoactive intestinal peptide (VIP)-containing neurons. VIP, VIP-containing bipolar cells; NA, noradrenergic afferent; Pyr, pyramidal cells furnishing major afferent projections; SA, specific afferent (from the thalamus or from other cortical regions); WM, subcortical white matter. Cortical layers denoted by roman numerals (reproduced with permission from Magistretti and Morrison, 1988).

different from that of VIP intracortical neurons. Thus, noradrenergic axons originate in the locus coeruleus in the brainstem and enter the neocortex rostrally, adopting a tangential trajectory that spans the entire cortical mantle (Fig. 2). These characteristics allow the noradrenergic system to exert its actions globally and simultaneously across functionally distinct cortical areas (Magistretti and Morrison, 1988) (Figs. 1 and 2).

Receptors for VIP and NA have been characterized on astrocytes (Martin et al., 1992; Stone and Ariano, 1989). These receptors are functional, as they are coupled to second messenger systems, in particular the cAMP cascade (Martin et al., 1992; Stone and Ariano, 1989). Since synapses between neurons and astrocytes have not been described in the mammalian brain (Peters et al., 1991), the interaction between VIP- and NA-containing neurons should occur at extrasynaptic sites. In fact the co-existence of synaptic and extrasynaptic release of NA within the neocortex has received experimental support (Beaudet and Descarries, 1978). In the same brain area, the radially-oriented VIP-containing neurons (Figs. 1 and 2)) show an intense labeling of dendrites in immunohistochemical preparations both at the light and electron microscope (Hajòs et al., 1988; Magistretti and Morrison, 1988). In analogy with dopaminergic neurons in the pars reticulata of the substantia nigra and with amacrine cells in the retina, high neurotransmitter content in dendrites may indicate the occurrence of dendritic, possibly extrasynaptic, release (Niéoullon et al., 1977).

In a recent study (Martin et al., 1992) performed in purified preparations of mouse cerebral cortex consisting of primary astrocyte cultures, intraparenchymal microvessels and synaptosomal membranes, respectively, three VIP receptor subtypes, with differential cellular localization, were identified (Table I). The first subtype (VIP 1) is ubiquitous and of high affinity with K_d values of 3.3 nM (astrocytes), 1.4 nM (microvessels) and 4.9 nM (synaptosomes). Secretin does not interact with this site. The second receptor subtype (VIP 2) is exclusively present on synaptosomal membranes. It is a low affinity site, with a K_d of 42.8 nM. Secretin interacts with this site with an IC_{50} of 150 nM. The third subtype (VIP 3) is also of low affinity, with a K_d of 30.3 nM and is exclusively localized in microvessels. Secretin does not interact with this site. In addition to providing a classification for VIP receptor subtypes, these observations suggest that secretin may represent a useful tool to discriminate between neuronal and non-neuronal VIP binding sites.

Unequivocal evidence at the ultrastructural level indicates the intimate apposition, and in some cases the presence of true synapses, between neurotransmitter-containing profiles and intraparenchymal brain capillaries. In particular this has been demonstrated for VIP, acetylcholine, cholecystokinin and NA (Edvinsson et al., 1993). Figure 3, which has been redrawn from an original micrograph by Rennels and Nelson (1975), is a clear illustration of the cytological substrate for the interaction between (1) an axon terminal (i.e. the site of neurotransmitter release), (2) two elements of an intraparenchymal capillary, i.e. a pericyte and an endothelial cell and (3) an astrocyte. This cytological arrangement clearly provides the substrate for interactions between neurons and three types of non-neuronal cells. In addition, pharmacological studies have demonstrated the presence of receptors for NA and VIP, coupled to cAMP-generating systems in

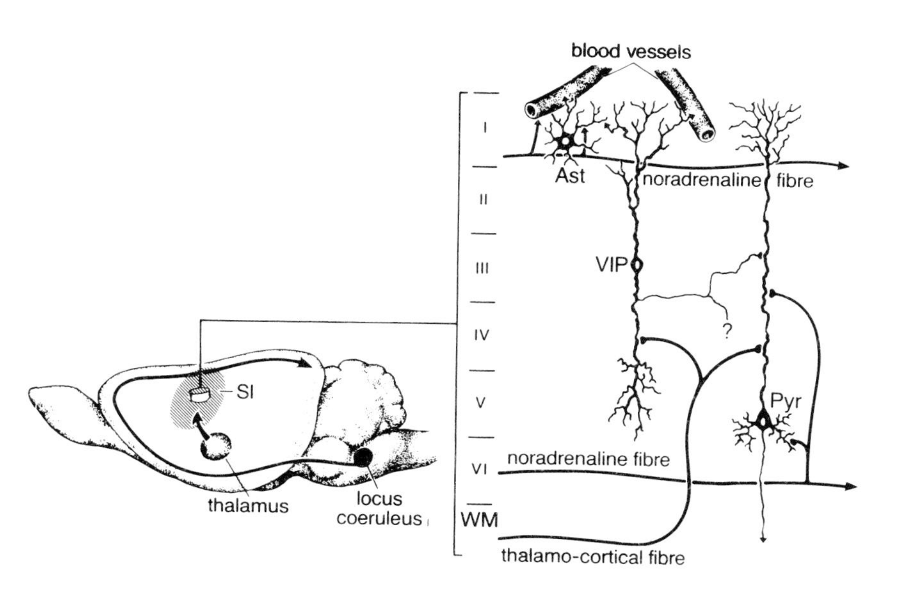

Fig. 2. Anatomical organization and putative targets of the noradrenaline (NA)- and vasoactive intestinal peptide (VIP)-containing neuronal circuits in rat cerebral cortex. Left: Noradrenergic fibers originate in locus coeruleus and project to the cerebral cortex, where they adopt a horizontal trajectory parallel to the pial surface. Right: VIP neurons are intrinsic to the cerebral cortex and are oriented vertically, perpendicular to the pial surface. Astrocytes (Ast), intraparenchymal blood vessels, and neurons such as certain pyramidal cells (Pyr) are potential target cells for VIP neurons. Roman numerals indicate cortical layers. VIP neurons can be activated by specific afferents (e.g., thalamocortical fibers). SI, primary sensory cortex; WM, white matter (reproduced with permission from Magistretti and Morrison, 1988).

TABLE I

Proposed classification for VIP receptor subtypes in the cerebral cortex

Subtype	Localization	K_d (nM)	Competition by secretin
VIP 1	Astrocytes	3.3	–
	Microvessels	1.4	–
	Neurons	4.9	–
VIP 2	Neurons	42.8	+
VIP 3	Microvessels	30.3	–

intraparenchymal microvessel preparations (Owman and Hardebo, 1986). Similar observations have been made in purified preparations of choroid plexus epithelial cells, where VIP and the β-adrenergic agonist isoproterenol stimulate cAMP formation (Crook and Prusiner, 1986). Furthermore, a very high density of VIP binding sites has been detected in the rat brain by autoradiography, in the subependymal layer at the level of the lateral ventricles (Martin et al., 1987). This observation is favorably correlated with the presence of VIP-immunoreactive neurons in the dentate gyrus possessing dendrites that extend into the ependyma of the lateral ventricles (Köhler, 1983).

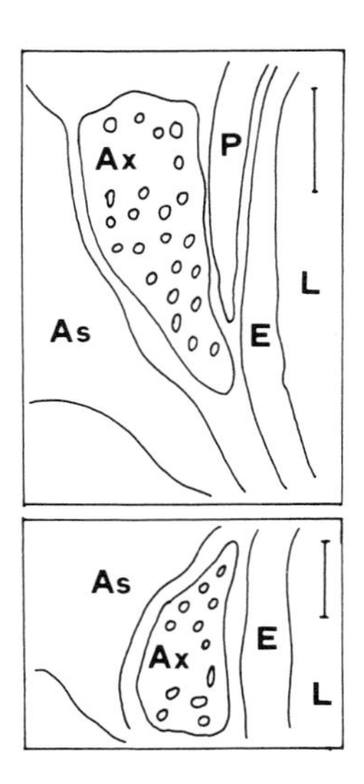

Fig. 3. Cytological substrate for interactions between (1) a neurotransmitter-containing axon terminal, (2) an astrocyte, (3) a pericyte and (4) an endothelial cell. Ax, axon terminal; As, astrocyte endfoot; P, pericyte process; E, endothelial cell; L, capillary lumen. Scale bar: 0.5 μm. (Redrawn and modified from an electron micrograph from Rennels and Nelson, 1975).

Regulation of glycogen levels by VIP and NA in astrocytes

Glycogen is the single largest energy reserve of the brain (Magistretti et al., 1993). It is predominantly localized in astrocytes, to the point where this cell type can be positively identified at the ultrastructural level by the presence of glycogen granules (for review see Magistretti et al., 1993).

VIP and NA readily promote a concentration-dependent glycogenolysis with EC_{50} values of 3 and 20 nM, respectively (Table II). The pharmacology of NA-induced glycogenolysis indicates both a β- and an α_1-adrenergic component (Sorg and Magistretti, 1991). Thus, both isoproterenol (β-adrenergic agonist) and methoxamine (α_1-adrenergic agonist) promote a concentration-dependent glycogenolysis, with EC_{50} values of 20 and 600 nM, respectively (Table II). A number of other neurotransmitters, for which the presence of receptors has been demonstrated on astrocytes (carbachol, glutamate, GABA) did not promote glycogenolysis (Sorg and Magistretti, 1991).

Peptides sharing sequence homologies with VIP have also been tested. As shown in Table II, PHI, secretin and the recently identified VIP-related peptide PACAP, are glycogenolytic, while the two structurally unrelated peptides somatostatin and neuropeptide Y (NPY) are without effect (Sorg and Magistretti, 1991). To further illustrate the tight regulation of glycogenolysis in astrocytes, adenosine and ATP are also glycogenolytic, with EC_{50} values of 0.8 and 1.3 μM, respectively (Table II). The action of VIP and NA is rapid, with initial rates of hydrolysis of 9.1 and 7.5 nmol/mg protein per min, respectively (Sorg and Magistretti, 1991). Interestingly, this value is close to the rate of ^{3}H-deoxyglucose uptake and phosphorylation by the same culture (Yu et al., 1993) and even by cerebral cortex in situ (Sokoloff et al., 1977). These observations indicate that the glycosyl units released by the neurotransmitter-evoked glycogenolysis can provide energy substrates that match the energy demands of cortical gray matter.

TABLE II

Glycogenolytic neurotransmitters in primary cultures of mouse cortical astrocytes

Substance	EC_{50} (nM)
VIP	3
PACAP	0.08
Secretin	0.5
PHI	6
Noradrenaline	20
– Isoproterenol (β)	20
– Methoxamine (α_1)	600
Adenosine	800
ATP	1300

VIP and NA, in addition to their glycogenolytic action discussed above (Sorg and Magistretti, 1991), which occurs within minutes, also induce a temporally delayed resynthesis of glycogen, resulting, within 9 h, in glycogen levels that are 6–10 times higher than those measured before application of either neurotransmitter (Sorg and Magistretti, 1992). The continued presence of the neurotransmitter is not necessary for this long-term effect since pulses as short as 1 min result in the doubling of glycogen levels 9 h later. The induction of glycogen resynthesis triggered by VIP or NA is dependent on protein synthesis, since both cycloheximide and actinomycin D abolish it entirely. These results indicate that the same neurotransmitter, e.g. VIP or NA, can elicit two actions with different time-courses. Thus, by increasing cAMP levels, VIP or NA simultaneously trigger a short-term effect, i.e. glycogenolysis, as well as a delayed one, i.e. transcriptionally regulated glycogen resynthesis. This longer-term effect ensures that sufficient substrate is available for the continued expression of the short-term action of VIP or NA (Fig. 4).

NA stimulates glucose uptake in astrocytes

Basal glucose uptake by astrocytes calculated from the specific activity of 2-[^{3}H]DG ranges between 3 and 9 nmol/mg protein per min (Yu et al., 1993), a value that compares very favourably with the glucose utilization of the grey matter as determined by the 2-[^{3}H]DG autoradiography technique in rodent cerebral cortex, assuming a protein content of 10% for brain tissue (Sokoloff et al., 1977). This observation would tend to suggest that glucose utilization in the cerebral cortex as measured by the 2-deoxyglucose technique may reflect, at least in part, glucose uptake by astrocytes. NA stimulates in a concentration-dependent manner 2-[^{3}H[DG uptake by astrocytes, with an EC_{50} of 1 μM (Yu et al., 1993). In contrast, VIP is without effect (Yu et al., 1993).

Given its morphological characteristics, the NA-containing neuronal system (see above and Figs. 1 and 2) would be ideally positioned to stimulate glucose uptake globally and simultaneously, throughout the cerebral cortex. In addition, this action, in parallel with the previously described glycogenolytic effect of NA would represent a coordinated regulatory mechanism to provide an adequate supply of metabolic substrates when energy demands of the active neuropil are increased.

Astrocyte end-feet surround intraparenchymal blood vessels (Peters et al., 1991) implying that at

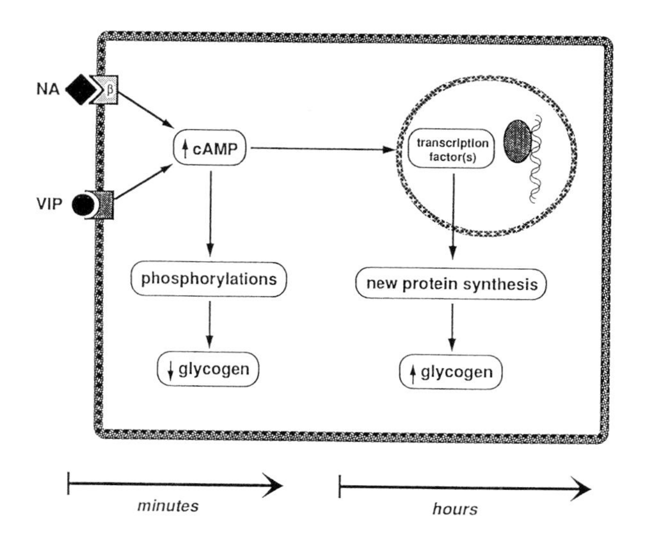

Fig. 4. Bidirectional effects of VIP and NA on glycogen in astrocytes. Short-term effect: within minutes after application, VIP or NA promote glycogenolysis. This effect is due to cAMP-dependent phosphorylation of pre-existing proteins. Long-term effect: within a few hours after application of VIP or NA, glycogen levels are increased 6–10 times above control levels. This effect is due to cAMP-dependent induction of new protein synthesis (reproduced with permission from Sorg and Magistretti, 1992).

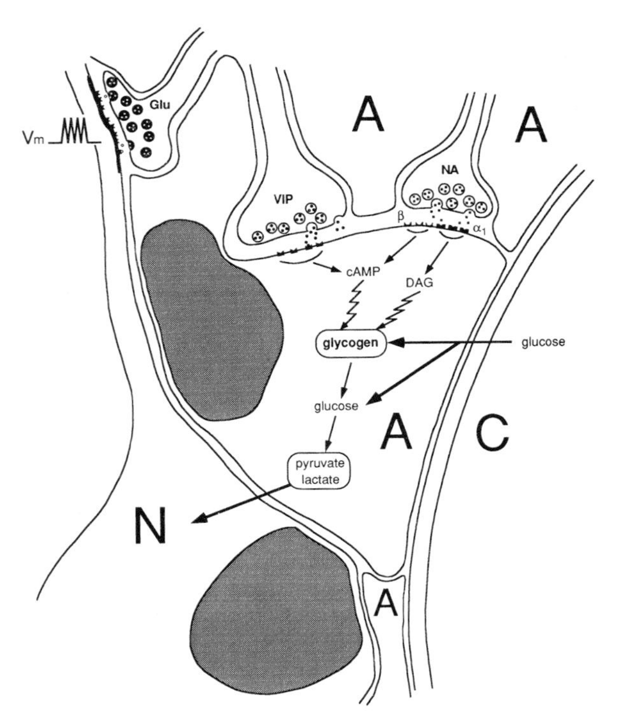

Fig. 5. Cytological substrate for the metabolic trafficking within the brain parenchyma. Glucose is taken up by astrocytic end-feet (A) which surround the capillaries (C). Lactate produced glycolytically from glucose, mobilized in part from glycogen, is released for its utilization by neurons (N). (see text for details). α, β, adrenergic receptor subtypes; cAMP, cyclic AMP; DAG, diacylglycerol; GLU, glutamate; NA, noradrenaline; VIP, vasoactive intestinal peptide; Vm, membrane potential.

least part of the glucose entering the brain parenchyma is taken up by astrocytes. It is therefore conceivable that either glucose crosses the astrocyte barrier unmetabolized or that it is first stored as glycogen and subsequently released by glycogenolytic neurotransmitters such as VIP or NA to provide a readily available metabolic substrate for neurons (Fig. 5).

Lactate is the major metabolic substrate released by astrocytes

In view of the foregoing, the question arises of the metabolic fate for the glucose taken up by astrocytes or for the glycosyl units mobilized from glycogen. Experimental evidence indicates that no glucose is released from astrocyte cultures, even when glucose is absent from the medium (Dringen et al., 1993a), consistent with the view that brain glucose-6-phosphatase activity is very low or not measurable (Sokoloff et al., 1977). It is therefore likely that energy substrates other than glucose are released by astrocytes and utilized by neurons. In vitro studies indicate that quantitatively, lactate is the main metabolic intermediate released by astrocytes at a rate of 15–30 nmol/mg protein per min (Dringen et al., 1993). This rate of release correlates well with the rate of glucose uptake by the grey matter (Sokoloff et al., 1977) or by astrocytes in culture (Yu et al., 1993), which range between 5 and 10 or 3 and 9 nmol/mg protein per min, respectively. Blockade of oxidative phosphorylation by azide or cyanide increases by 1.5–2-fold the release of lactate by astrocytes (Dringen et al., 1993) indicating that part of the glycosyl units mobilized from glycogen are oxidized by astrocytes rather than being exported as lactate. Other, quantitatively less important intermediates released by astrocytes are pyruvate (approximately 10 times less than lactate), α-ketoglutarate, citrate and malate (Shank et al., 1993). Fluxes of endogenous lactate between astrocytes and neurons have been quantified in vitro (Larrabee, 1992) and an avid lactate uptake has been demonstrated in neurons (Larrabee, 1992; Dringen et al., 1993b). It is also well established that lactate and pyruvate are adequate substrates for brain tissue in vitro (McIlwain and Bachelard, 1985; Schurr et al., 1988). In fact, synaptic activity can be maintained in cerebral cortical slices with only lactate or pyruvate as a substrate (McIlwain and Bachelard, 1985; Schurr et al., 1988). Thus, a metabolic compartmentation whereby glucose taken up by astrocytes is metabolized glycolytically to lactate or pyruvate (Fig. 5) which are then released in the extracellular space to be utilized by neurons, is consistent with the available biochemical and electrophysiological observations.

Observations reviewed here show that VIP and NA participate in the regulation of brain energy metabolism, by tightly regulating the glycogen content, and for NA, glucose uptake in astrocytes. In addition, cytological and pharmacological evidence strongly suggest that VIP- and NA-containing neurons interact with intraparenchymal blood vessels. The demonstration of homeostatic functions regulated by neurotransmitters contained in discrete neuronal circuits support the concept that such circuits may represent, within the brain, the counterpart of the autonomic nervous system which regulates, among other functions, blood flow, energy metabolism and local homeostasis in peripheral tissues.

Acknowledgements

This research is supported by a grant of Fonds National Suisse de la Recherche Scientifique (31-26427.89). The author wishes to thank Ms M. Emch for excellent secretarial help.

References

Beaudet, A. and Descarries, L. (1978) The monoamine innervation of rat cerebral cortex: synaptic and nonsynaptic axon terminals. *Neuroscience*, 3: 851–860.

Bloom, F.E., Campbell, I.L. and Mucke, L. (1993) Molecular and cellular mechanisms of neural-immune interactions. *Discussions Neurosci*, IX.

Crook, R.B. and Prusiner, S. (1986) Vasoactive intestinal peptide stimulates cyclic AMP metabolism in choroid plexus epithelial cells. *Brain Res.*, 384: 138–144

Dringen, R., Gebhardt, R. and Hamprecht, B. (1993a) Glycogen in astrocytes: possible function as lactate supply for neighboring cells. *Brain Res.*, 623: 208–214.

Dringen, R., Wiesinger, H. and Hamprecht, B. (1993b) Uptake of L-lactate by cultured rat brain nuerons. *Neurosci. Lett.*, 163: 5–7.

Edvinsson, L., MacKenzie, E.T. and McCulloch, J. (1993) *Cerebral Blood Flow and Metabolism*, Raven Press, New York.

Hajòs, F., Zilles, K., Schleicher, A. and Kalman, M. (1988) Types and spatial distribution of vasoactive intestinal polypeptide (VIP)-containing synapses in the rat visual cortex. *Anat. Embryol.*, 178: 207–217.

Köhler, C. (1983) A morphological analysis of vasoactive intestinal polypeptide (VIP)-like immunoreactive neurons in the area dentata of the rat brain. *J. Comp. Neurol.*, 221: 247–262.

Larrabee, M.G. (1992) Extracellular intermediates of glucose metabolism: fluxes of endogenous lactate and alanine through extracellular pools in embryonic sympathetic ganglia. *J. Neurochem.*, 59:1041–1052.

Magistretti, P.J. and Morrison, J.H. (1988) Noradrenaline- and vasoactive intestinal peptide-containing neuronal systems in neocortex: functional convergence with contrasting morphology. *Neuroscience*, 24: 367–378.

Magistretti, P.J., Sorg, O. and Martin, J.L. (1993) Regulation of glycogen metabolism in astrocytes: physiological, pharmacological, and pathological aspects. In S. Murphy (Ed.), *Astrocytes: Pharmacology and Function*, Academic Press, San Diego, pp. 240–243.

Martin, J.L., Dietl, M.M., Hof, P.R., Palacios, J.M. and Magistretti, P.J. (1987) Autoradiographic mapping of [monoe[^{125}I]iodo-Tyr^{10},MetO^{17}]-vasoactive intestinal peptide binding sites in the rat brain. *Neuroscience*, 23: 539–565.

Martin, J.L., Feinstein, D.L., Yu, N., Sorg, O., Rossier, C. and Magistretti, P.J. (1992) VIP receptors subtypes in mouse cerebral cortex: evidence for a differential localization in astrocytes, microvessels, and synaptosomal membranes. *Brain Res.*, 587: 1–12.

McIlwain, H. and Bachelard H.S. (1985) In *Biochemistry and the Central Nervous System*, Livingstone, Edinburgh, pp. 54–83.

Niéoullon, A., Cheramy, A. and Glowinski, J. (1977) Release of dopamine in vivo from cat substantia nigra. *Nature*, 266: 375–377.

Owman, C. and Hardebo, J.E. (1986) *Neural Regulation of Brain Circulation*, Elsevier, Amsterdam.

Peters, A., Palay, S.L. and Webster, H. de F. (1991) *The Fine Structure of the Nervous System: Neurons and their Supporting Cells*, W.B. Saunders, Philadelphia.

Rennels, M.L. and Nelson, E. (1975) Capillary innervation in the mammalian central nervous system: an electron microscopic demonstration. *Am. J. Anat.*, 144: 233–241.

Schurr, A., West, C.A. and Rigor, B.M. (1988) Lactate-supported synaptic function in the rat hippocampal slice preparation. *Science*, 240:1326–1328.

Shank, R.P., Leo, G.C. and Zielke, H.R. (1993) Cerebral metabolic compartmentation as revealed by nuclear magnetic resonance analysis of D-[1-^{13}C]glucose metabolism. *J. Neurochem.*, 61: 315–323.

Sokoloff, L., Reivich, M., Kennedy, C., Des Rosiers, M.H., Patlak, C.S., Pettigrew, K.D., Sakurada, O. and Shinohara, M. (1977) The [^{14}C]deoxyglucose method for the measurement of local cerebral glucose utilization: theory, procedure, and normal values in the conscious and anesthetized albino rat. *J. Neurochem.*, 28: 897–916.

Sorg, O. and Magistretti, P.J. (1991) Characterization of the glycogenolysis elicited by vasoactive intestinal peptide, noradrenaline and adenosine in primary cultures of mouse cerebral cortical astrocytes. *Brain Res.*, 563: 227–233.

Sorg, O. and Magistretti, P.J. (1992) Vasoactive intestinal peptide and noradrenaline exert long-term control on glycogen levels in astrocytes: blockade by protein synthesis inhibition. *J. Neurosci.*, 12: 4923–4931.

Stone, E.A. and Ariano, M.A. (1989) Are glial cells targets of the central noradrenergic system ? A review of the evidence. *Brain Res. Dev.*, 14: 297–309.

Yu, N., Martin, J.L., Stella, N. and Magistretti, P.J. (1993) Arachidonic acid stimulates glucose uptake in cerebral cortical astrocytes. *Proc. Natl. Acad. Sci. USA*, 90: 4042–4046.

SECTION III

Integrative Brain Research

F. Bloom (Editor)
Progress in Brain Research, Vol. 100

CHAPTER 13

Segmental and descending control of the synaptic effectiveness of muscle afferents*

Pablo Rudomin

Department of Physiology, Biophysics and Neurosciences, Centro de Investigación y de Estudios Avanzados, México D.F

Introduction

In 1957, Frank and Fuortes found that conditioning stimulation of group I afferents from flexor muscles depressed the monosynaptic EPSPs elicited in spinal motoneurons by stimulation of muscle spindle afferents (Ia afferents), without changing the membrane properties of the motoneurons or the time course of the Ia-EPSPs. At that time, models of motoneurons assigned practically no role to dendrites in the generation of monosynaptic responses from Ia afferents. That is, physiologically effective Ia synapses were assumed to be located predominantly near the soma (see Eccles, 1953, 1964), and it was therefore postulated that the inhibition resulted from interference with transmission of excitation from Ia afferents to motoneurons, i.e. to presynaptic inhibition.

Following the development of more realistic motoneuron models in which Ia synapses had a significant dendritic location (Rall, 1960; see also Smith et al., 1967), an alternative explanation to presynaptic inhibition was proposed. Namely, that the depression of the Ia EPSPs occurred at the motoneuron distal dendrites (Frank, 1959). Inhibitory conductances produced by the conditioning stimulus at the distal dendrites would not be detected at the motoneuron soma, but would be nevertheless able to depress Ia EPSPs generated in a close vicinity.

Subsequent work was addressed to determine the extent to which the depression of Ia EPSPs was indeed associated with changes in motoneuron membrane properties. The underlying assumption was that presynaptic and postsynaptic inhibition were mutually exclusive. In some cases Ia EPSP depression could not be fully explained by postsynaptic changes and presynaptic inhibition was assumed to be the main cause of EPSP depression (Eccles et al., 1962b; Eide et al., 1968; Cook and Cangiano, 1972; Rudomin et al., 1975b, 1991; Rudomin, 1992). In other cases, there seemed to be an appreciable postsynaptic inhibition and presynaptic inhibition was assumed to have a minor role (Granit et al., 1964; Carlen et al., 1980).

There is now good evidence that presynaptic inhibition is mediated by GABAergic interneurons that make axo-axonic synapses with the intraspinal terminals of the afferent fibers (see below). Electrophysiological studies have further indicated that these GABAergic interneurons also have monosynaptic inhibitory connections with motoneurons (Rudomin et al., 1987) and more recently, direct evidence for the existence of GABAergic boutons that make synapses with identified muscle spindle afferents as well as with the postsynaptic neurons has became available (Fyffe and Light, 1984; Maxwell et al., 1990). The functional implications of the co-existence of pre- and postsynaptic inhibition have been discussed elsewhere (Rudomin et al., 1987; Rudomin, 1990).

* This review is dedicated to K. Frank who died on February 25, 1993.

Sources of variability in synaptic transmission

Kuno (1964a,b) found that the monosynaptic potentials produced in a motoneuron by stimulation of a single Ia fiber showed considerable amplitude fluctuations including an appreciable number of failures. He also found that conditioning stimulation of group I fibers from flexors reduced the probabilities of occurrence of the largest EPSPs and increased the number of EPSP failures, without significantly changing the unitary EPSPs. These observations provided a more direct test on the possible existence of presynaptic inhibition. Failures in transmission could be due either to block of impulse conduction in the terminal arborizations of the afferent fibers (Henneman et al., 1984; Lüscher et al., 1990) or to failure in transmitter release (see Redman, 1990). Appraisal of these two possibilities has became an important issue because of their implications in the control of information transfer from Ia afferents to motoneurons.

One of the major problems involved in the analysis of the mechanisms that control the efficacy of synaptic transmission is that the signal to be analyzed occurs in the presence of appreciable noise, most of it of synaptic origin. During the last decade, deconvolution techniques have been used to separate the signal from the noise (Jack et al., 1981; Redman and Walmsley, 1983; see Redman, 1990 for a review). It was found that the amplitude of the noise-free EPSPs produced by stimulation of single Ia fibers varies in discrete steps, each of them with a finite probability. The discrete components appear separated by a constant amplitude interval, suggesting quantal steps in the generation of the single fiber EPSPs. Clements et al. (1987) have further shown that conditioning stimulation of group I fibers from flexors reduces the probabilities of occurrence of the large EPSP components and increases the probabilities of occurrence of the smallest components, without significantly changing the separation between them. Similar results have been obtained after administration of (–)-baclofen, a GABAb agonist (Edwards et al., 1989). These findings have been taken as evidence for the involvement of presynaptic inhibition in the depression of the synaptic potentials.

One assumption in the deconvolution procedure is that signal and noise interact linearly (see Redman, 1990). However, by comparing variance of the directly recorded EPSPs with the variance of the discrete components of the EPSPs obtained with deconvolution techniques, Solodkin et al. (1991) found that even in conditions of low synaptic noise, there is an appreciable non-linear interaction between the signal and noise. Furthermore, the deconvolution procedures require grouping of events of similar size into the same amplitude category (see Wong and Redman, 1980; Redman 1990). In practice, events with amplitudes differing more than 1.5 the standard deviation of the background noise are considered as belonging to a different amplitude category (Clements et al., 1987; Edwards et al., 1989; Kullman et al., 1989), which means that the separation between the discrete components of the noise-free EPSP may depend not only on the statistical properties of the transmitter release mechanisms associated with the generation of the EPSP, but also on the amount of background synaptic noise, even when there is little interaction between signal and noise (see also Solodkin et al., 1991; Clamann et al., 1991).

From the above considerations, it is clear that measurement of the noise-free components of monosynaptic EPSPs is not without problems, particularly in the presence of appreciable background synaptic noise, as it occurs under more physiological conditions. One of the challenges for future research is to seek for new approaches to have more reliable measurements of the signal in the presence of synaptic noise and to assess how the fluctuations of the signal are changed by segmental and descending inputs which are known to affect the synaptic effectiveness of the afferent fibers.

GABAergic origin of presynaptic inhibition

An important step in the understanding of the mechanisms involved in presynaptic inhibition was the discovery that stimulation of segmental and descending pathways depolarizes the intraspinal terminals of afferent fibers (Eccles et al., 1962a; Lundberg, 1964; Rudomin et al., 1983). The available evidence suggests quite strongly that this depolarization (primary afferent depolarization, or PAD) is produced by the activation of GABAergic interneurons that make axo-axonic synapses with the afferent fibers (for review see Schmidt, 1961; Burke and Rudomin, 1977; Davidoff and Hackmann, 1984; Rudomin, 1990, 1991).

It is now known that Ia afferent fibers have at least two different types of GABAergic receptors. Activation of GABAa receptors produces an outward movement of chloride ions and PAD in the afferent fibers. Presynaptic inhibition could result either from the depolarization itself, or from the associated conductance increase and block of conduction of action potentials in the terminal arborizations of the afferent fibers. Activation of GABAb receptors appears to reduce the inward calcium currents generated during the action potential and transmitter release (Gallagher et al., 1978; Curtis and Lodge, 1978; Rudomin et al., 1981; Price et al., 1984; Lev-Tov et al., 1988; Edwards et al., 1989; Peng and Frank, 1989a,b; Jiménez et al., 1991). It also reduces the frequency-dependent depression of monosynaptic EPSPs (Peshori et al., 1991; Pinco and Lev-Tov, 1993).

Quite recently, Sugita et al. (1992) have shown that in single neurons of the lateral amigdala and ventral tegmental area of the rat, GABAa and GABAb receptors are spatially separated and can be activated by independent sets of GABAergic interneurons. A similar segregation of GABAa and GABAb receptors appears to occur in afferent and descending inputs acting on preganglionic sympathetic neurons of newborn rats, which appear to have only GABAb receptors (Wu and Dun, 1992), as well as in the hippocampus where presynaptic inhibition occurs following activation of GABAb and μ-opioid receptors (Thompson et al., 1993).

A separate activation of GABAa and GABAb receptors in afferent fibers would allow, in principle, an independent control of impulse conduction at branch points or of transmitter release frequency behavior. This would increase the capability of the CNS to control information from sensory afferents in the spatial and temporal domain. Although at present time there is no direct evidence for this possibility, the observations of Stuart and Redman (1992) are relevant in this context. These investigators analyzed the effects of iontophoretic application of (–)-baclofen (a GABAb agonist) and of bicuculline and 2-OH-saclofen (GABAa and GABAb antagonists, respectively) on the presynaptic inhibition of Ia EPSPs produced by conditioning stimulation of group I afferents. They concluded that in the Ia afferent-motoneuron synapse, presynaptic inhibition is mediated primarily through the activation of GABAa receptors and that activation of GABAb receptors has a minor role in presynaptic inhibition. They also suggested that existing GABAb receptors in afferent fibers are extrasynaptic.

In contrast, Quevedo et al. (1992) have shown that intravenously applied (–)-baclofen abolishes practically all the PAD produced by stimulation of group I muscle and of cutaneous afferents and also the monosynaptic PAD that is produced following direct activation, by intraspinal microstimulation, of intermediate nucleus interneurons. They concluded that, in addition to the presence of GABAa and GABAb receptors in the afferent fibers, last-order PAD-mediating interneurons have GABAb autoreceptors. This implies that even if GABAb receptors in afferent fibers were extrasynaptic, (–)-baclofen, by acting on the GABAb interneuronal autoreceptors would depress the GABAa presynaptic inhibition.

Patterns of primary afferent depolarization

Studies on the PAD produced in single, functionally identified, muscle afferents have indicated that in the normal cat, Ia fibers (from muscle spindles) and Ib fibers (from tendon organs) have different PAD patterns (Jiménez et al., 1988). Namely, Ia fibers show PAD following stimulation of group I afferents from flexor muscles as well as by stimulation of the vestibular nucleus. Stimulation of cutaneous and joint afferents and of the motor cortex, bulbar reticular formation, red nucleus and raphe nucleus produces no PAD of Ia fibers, but inhibits the PAD produced in them by other inputs. In contrast, Ib fibers are depolarized by group I fibers and by all of the above descending inputs (Rudomin et al., 1983; Quevedo et al., 1993a) while cutaneous and joint afferents produce

PAD in a fraction of Ib fibers and inhibit the PAD in another fraction (Rudomin et al., 1986; Jiménez et al., 1988; Quevedo et al., 1993a).

In addition to substantial differences in the patterns of PAD of Ia and Ib fibers, it seems likely that separate last-order interneurons mediate the PAD of Ia and Ib fibers (Rudomin et al., 1983; Rudomin, 1990, 1991). This provides the structural basis for an independent control of the information conveyed by muscle spindles and tendon organs, that is, of muscle length and muscle tension, despite the marked convergence of these afferents on spinal interneurons (Jankowska et al., 1981).

More recently, Enríquez et al. (1991, 1992) recorded intrafiber PAD from functionally identified muscle spindle and tendon organ afferents. Two weeks to three months after crushing the medial gastrocnemius nerve, afferent fibers that became reconnected with receptors "in-parallel" (most likely from muscle spindles) showed increased PAD following stimulation of group I afferents from flexors. Stimulation of cutaneous afferents produced very little PAD, but was nevertheless able to inhibit the PAD produced by group I volleys. Unlike what has been found in normal animals, stimulation of the reticular formation produced a substantial PAD in most muscle spindle afferents. The PAD produced by stimulation of group I and cutaneous fibers in afferents reconnected with receptors "in series" (most likely muscle tendon organs) was also increased. However, in this case, there were no significant changes in the PAD elicited by stimulation of the reticular formation. Assuming that after nerve crush regenerating fibers followed the same path and became reconnected to their original muscle receptors (Barker et al., 1985, 1986), the above results would indicate that the changes in the PAD patterns observed after nerve crush were of central origin.

In humans with chronic spinal lesions, stimulation of cutaneous nerves produces a delayed presynaptic inhibition of Ia afferents (Roby-Brami and Busell, 1992), similar to that seen in the spinal cat after DOPA (Anden et al., 1967). It is therefore possible that the reticular-induced PAD in Ia fibers, regenerated after nerve crush, is also related to changes in the overall excitability of spinal neurons, and that previously inactive pathways became subsequently active.

In this context, it is interesting to mention that in the rat, impulses are conducted rostrally in afferent fibers but fail to penetrate long distances into the caudal branch of long range afferents. Seven to ten days after sectioning the dorsal roots in the area of suspected conduction failure, orthodromic conduction is restored. Failure and release of conduction appears to depend on the control of membrane potential in the primary afferents (Wall and McMahon, 1993).

Selective modulation of presynaptic inhibition

One recent and important advance in studies aimed at disclosing the functional role of presynaptic inhibition has been the development of a non-invasive paradigm to measure presynaptic inhibition (Hultborn et al., 1987a). The changes in the magnitude of Ia-induced heterosynaptic facilitation of H reflexes were used to estimate changes in presynaptic inhibition (see Rudomin et al. (1991) and Rudomin (1992) for additional tests on the validity of this technique). Hultborn et al. (1987a,b) showed that at the onset of a voluntary contraction in humans the background presynaptic inhibition of the Ia fibers arising from the contracting muscle is reduced, whereas the presynaptic inhibition of the Ia fibers innervating heteronymous muscles is increased. Such a differential control of the synaptic effectiveness of the muscle spindle afferents appears to be of supraspinal origin and may probably serve to increase motor contrast.

Further support for a differential control of presynaptic inhibition is provided by Burke et al. (1992) who examined the possibility that facilitation of transmission in the propriospinal-like system during voluntary contraction is due to a decrease in the background presynaptic inhibition of those muscle afferents connected with the propriospinal-like interneurons. They found no evidence of a decreased gating of the afferent input, regardless of whether this input was of muscular or cutaneous origin. In contrast, there was a clear reduction of presynaptic inhibition of the same set of muscle afferents synapsing with forearm motoneurons. In addition, Nielsen and Kagamihara (1992) and Nielsen et al. (1992) found that the changes in presynaptic inhibition that are observed at the onset of a voluntary contraction occur even during ischemic block of afferent input, so they are not due to changes in afferent feedback but are probably of descending origin.

More direct evidence on the selectivity of the cortical control exerted on the synaptic effectiveness of muscle afferents has been obtained by Eguíbar et al. (1993) who analyzed the effects of electrical stimulation of the motor cortex on the intraspinal threshold of two neighboring collaterals of the same muscle spindle or tendon organ afferent fiber in the cat spinal cord. They found that stimulation of the motor cortex was able to suppress the background PAD of muscle spindle afferents produced by group I conditioning volleys in one collateral, practically without affecting the PAD elicited by that same input in the other collateral. They also found that cortical stimulation could produce PAD in one collateral of a tendon organ afferent fiber and very little PAD in a nearby collateral.

These findings support the notion that the cerebral cortex is able to exert a highly selective control of the synaptic effectiveness of muscle afferent fibers. This selectivity is possible because individual collaterals of a single afferent fiber receive synapses from more than one last-order PAD mediating interneuron, and also because there is selectivity in the connections of the PAD-mediating interneurons. Some of these interneurons were found to be connected only with one of the two collaterals examined, while other interneurons were connected with both collaterals (Quevedo et al., 1993b).

Simultaneous measurements of PAD in pairs of single Ia or Ib fibers belonging to the same or to different muscles have further indicated that the most effective cortical regions affecting the PAD of Ia or Ib fibers of the same or of different muscles are distributed in discrete spots surrounded by less effective regions (Eguíbar et al., 1991, 1992). The spatial distribution of the most effective cortical spots affecting the PAD of a single Ia or Ib afferent may change with the amount of background PAD of that particular fiber. This stresses the contribution of segmental pathways in the assessment of cortical modulation of the synaptic effectiveness of the afferent fibers. The data also indicate that there is a partial overlap between the most active cortical spots affecting the PAD of single Ia and Ib fibers of the same or different origin. The existence of non-overlapping active regions can be taken as an expression of the selectivity of the cortical control on the synaptic efficacy of Ia and Ib fibers.

Concluding remarks

The information transmitted by the ensemble of Ia fibers will depend not only on the signals provided by the receptor organs, but also on the kind of signals introduced by the interneurons that make axo-axonic synapses with the afferent fibers. In the absence of any correlation in the activity of the PAD mediating interneurons, Ia afferents may function as independent channels. With increasing correlation, the redundancy in the line will also be increased (Rudomin and Madrid, 1972; Rudomin et al. 1975a; Rudomin, 1980). The nature of the redundant information introduced via the axo-axonic synapses has not been elucidated. However, considering the diversity of the segmental and descending inputs received by these interneurons (Jiménez et al., 1988; Rudomin et al., 1983, 1986), it seems unlikely that they convey information pertaining to a single functional parameter (i.e. on muscle length). A more attractive possibility is that by affecting the synaptic effectiveness of a substantial number of Ia fibers in a correlated manner, the interneurons mediating presynaptic inhibition act as a gating system that is able to switch-on or switch-off the information transferred along specific sets of Ia terminal arborizations (Rudomin et al., 1987).

Research in humans and in other vertebrates has now allowed disclosure of the selectivity of the supraspinal control of presynaptic inhibition. It is of particular interest that in humans, changes in presynaptic inhibition occur at the onset of a voluntary contraction and that no afferent feedback is required, suggesting that central structures are able to "select" the kind of sensory information that will be needed for the execution of a particular motor task. It seems likely that the descending commands involved in the selection of the sensory information are part of the cortical loop proposed by Asanuma (1989) to adjust the excitability of the motoneurons that will be involved in a particular movement. The descending control of sensory information may also be involved in the matching of the cortical representation of an intended movement with the movement itself (Georgopoulos, 1992), a situation that ensures that the executed movements are part of a survival strategy.

Acknowledgments

I thank Ismael Jiménez, Ph.D. and Jorge Quevedo, M.D., Manuel Enríquez, M.Sci., and José Ramón Eguíbar, M.D. for their enthusiastic collaboration throughout these years. This work was supported by United States Public Health Service Partly supported by NIH grant NS-09196 and grant 0319 N9197 from the Consejo Nacional de Ciencia y Tecnología, México.

References

Anden, N.E., Jukes, M.G.M. and Lundberg, A. (1967) The effect of DOPA on the spinal cord. I. Influence on transmission from primary afferents. *Acta Physiol. Scand.*, 67: 373–386.

Asanuma, H. (1989) Function of somesthetic input during voluntary movements. In: *The Motor Cortex*, Raven Press, New York, pp. 69–75.

Barker, D., Scott, J.J.A. and Stacey, M.J. (1985) Sensory reinnervation of cat peroneus brevis muscle spindles after nerve crush. *Brain Res.*, 333: 131–138.

Barker, D., Scott, J.J.A. and Stacey, M.J. (1986) Reinnervation and recovery of cat muscle receptors after long-term denervation. *Exp. Neurol.*, 94: 184–202.

Burke, R.E. and Rudomin, P. (1977) Spinal neurons and synapses. In: E.R. Kandel (Ed.), *Handbook of Physiology, Sect I. Vol. I. The Nervous System,* Am. Physiol. Soc., Bethesda, MD, pp. 877–944.

Burke, D., Gracies, J.M., Meunier, S. and Pierrot-Deseilligny, E. (1992) Changes in presynaptic inhibition of afferents to propriospinal-like neurons in man during voluntary contractions. *J. Physiol. (London),* 449: 673–687.

Carlen, P.L., Werman, R. and Yaari, Y. (1980) Post-synaptic conductance increase associated with presynaptic inhibition in cat lumbar motoneurones. *J. Physiol. (London)*, 298: 539–556.

Clamann, H.P., Rioult-Pedotti M.-S. and Lüscher, H.-R (1991) The influence of noise on quantal EPSP size obtained by deconvolution in spinal motoneurons of the cat. *J. Neurophysiol.*, 65: 67–75.

Clements, J.D., Forsythe, I.D. and Redman, S.J. (1987) Presynaptic inhibition of synaptic potentials evoked in cat spinal motoneurones by impulses in single group Ia axons. *J. Physiol. (London)*, 383: 153–169.

Cook, W.A. and Cangiano, A. (1972) Presynaptic and postsynaptic inhibition of spinal neurons. *J. Neurophysiol.*, 35: 389–403.

Curtis, D.R. and Lodge, D.R. (1978) GABA depolarization of spinal group I afferent terminals. In: R.W. Ryall and J.S. Kelly (Eds.), *Iontophoresis and Transmitter Mechanisms in the Mammalian Central Nervous System*, Elsevier, Amsterdam, pp. 258–260.

Davidoff, R.A. and Hackman, J.C. (1984) Spinal Inhibition. In: R.A. Davidoff (Ed.), *Handbook of the Spinal Cord.* Dekker, New York, pp. 385–459.

Eccles, J.C. (1953) *The Neurophysiological Basis of Mind,* Clarendon Press, Oxford.

Eccles, J.C. (1964) *The Physiology of Synapses.* Academic Press, New York.

Eccles, J.C., Magni, F. and Willis, W.D. (1962a) Depolarization of central terminals of group I afferent fibres from muscle. *J. Physiol. (London)*, 160: 62–93.

Eccles, J.C., Schmidt, R.F. and Willis, W.D. (1962b) Presynaptic inhibition of the spinal monosynaptic reflex pathway. *J. Physiol. (London)*, 161: 282–297.

Edwards, F.R., Harrison, P.J., Jack J.B. and Kullman, D.M. (1989) Reduction by baclofen of monosynaptic EPSPs in lumbosacral motoneurones of the anesthetized cat. *J. Physiol. (London)*, 416: 539–556.

Eguíbar, J.R., Quevedo, J., Jiménez, I. and Rudomin, P. (1991) Selective modulation of the PAD of single Ia and Ib afferents produced by surface stimulation of the motor cortex in the cat. *Soc. Neurosci. Abstr.*, 17: 1024.

Eguíbar, J.R., Quevedo, J., Jiménez, I. and Rudomin, P. (1992) Selective connectivity of last-order interneurons mediating PAD of group I fibers according to the muscle of origin. *Soc. Neurosci. Abstr.*, 18: 524.

Eguíbar, J.R., Quevedo, J., Jiménez, I. and Rudomin, P. (1993) Differential control exerted by the motor cortex on the synaptic effectiveness of two intraspinal branches of the same group I afferent fiber. *Soc. Neurosci. Abstr.*, 19: in press.

Eide, E., Jurna, I. and Lundberg, A. (1968) Conductance measurements from motoneurons during presynaptic inhibition. In: C. Von Euler, A. Skoglund and U. Soderberg (Eds.), *Structure and Function of Inhibitory Neuronal Mechanisms,* Pergamon Press, New York, pp. 215–219.

Enríquez, M., Hernández, O. Jiménez, I. and Rudomin, P. (1991) Is the PAD evoked in Ia fibers related to their responses to stretch? *Soc. Neurosci. Abstr.*, 17: 1024.

Enríquez, M., Jiménez, I. and Rudomin, P. (1992). PAD patterns of regenerated group I afferents after peripheral nerve crush in the cat. *Soc. Neurosc. Abstr.*, 18: 514.

Frank, K. (1959) Basic mechanisms of synaptic transmission in the central nervous system. *Inst. Radio Eng. Trans. Med. Electron.*, 6: 85–88.

Frank, K. and Fuortes, M.G.F. (1957) Presynaptic and postsynaptic inhibition of monosynaptic reflexes. *Fed. Proc.*, 16: 39–40.

Fyffe, R.E.W. and Light, A.R. (1984) The ultrastructure oo group Ia afferent fibre synapses in the lumbosacral spinal cord of the cat. *Brain Res.*, 300: 201–209.

Gallagher, J.P., Higashi, H. and Nishi, S. (1978) Characterization and ionic basis of GABA-induced depolarizations recorded *in vitro* from cat primary afferent neurones. *J. Physiol. (London)*, 275: 263–282.

Georgopoulos, A.P. (1993) Cortical representation of intended movements. In P. Rudomin, M.A. Arbib, F. Cervantes-Pérez and R. Romo (Eds.), *Neuroscience: From Neural Networks to Artificial Intelligence,* Springer-Verlag, Berlin, pp. 398–412.

Granit, R., Kellerth, J.O. and Williams, T.D. (1964) Intracellular aspects of stimulating motoneurones by muscle stretch. *J. Physiol. (London)*, 174: 435–452.

Henneman, E., Lüscher, H.R. and Mathis, J. (1984) Simultaneously active and inactive synapses of single Ia fibres on cat spinal motoneurones. *J. Physiol. (London)*, 352: 147–161.

Hultborn, H., Meunier, S., Morin, C. and Pierrot-Deseilligny, E. (1987a) Assessing changes in presynaptic inhibition of Ia fibres: a study in man and the cat. *J. Physiol. (London)*, 389: 729–756.

Hultborn, H., Meunier, S., Pierrot-Deseilligny E. and Shindo, M. (1987b) Changes in presynaptic inhibition of Ia fibres at the onset of voluntary contraction in man. *J. Physiol. (London)*, 389: 757–772.

Jack, J.J.B., Redman, S.J. and Wong, K. (1981) The components of synaptic potentials evoked in cat spinal motoneurones by impulses in single group Ia afferents. *J. Physiol. (London)*, 321: 65–96.

Jankowska, E., Johannisson, T. and Lipski, J. (1981) Common interneurones in reflex pathways of ankle extensors in cat. *J. Physiol. (London)*, 310: 381–402.

Jiménez, I., Rudomin, P. and Solodkin, M. (1988) PAD patterns of physiologically identified afferent fibers from the medial gastrocnemius muscle. *Exp. Brain. Res.*, 71: 643–657.

Jiménez, I., Rudomin, P. and Enríquez, M. (1991). Differential effects of (-)-baclofen on Ia and descending monosynaptic EPSPs. *Exp. Brain Res.*, 85: 103–113.

Kullman, D.M., Martin, R.L. and Redman, S.J. (1989) Reduction by general anaesthetics of group Ia excitatory postsynaptic potentials and currents in the cat spinal cord. *J. Physiol. (London)*, 412: 277–296.

Kuno, M. (1964a) Quantal components of excitatory synaptic potentials in spinal motoneurons. *J. Physiol. (London)*, 175: 81–89.

Kuno, M. (1964b) Mechanism of facilitation and depression of the excitatory synaptic potential in spinal motoneurones. *J. Physiol. (London)*, 175: 100–112.

Lev-Tov, A., Meyers, D.E.R. and Burke, R.E. (1988) Activation of $GABA_b$ receptors in the intact mammalian spinal cord mimics the effects of reduced presynaptic Ca^{++} influx. *Proc. Natl. Acad. Sci. USA.*, 85: 5330–5333.

Lundberg, A. (1964) Supraspinal control of transmission in reflex pathways to motoneurons and primary afferents. In: J.C. Eccles and J.P. Schade (Eds.), *Physiology of Spinal Neurons,* Elsevier, Amsterdam, pp. 197–219.

Lüscher, H.R. (1990) Transmission failure and its relief in the spinal monosynaptic arc. In: M.D. Binder and L.M. Mendell (Eds.), *The Segmental Motor System,* Oxford University Press, New York, pp. 328–348.

Maxwell, D.J., Christie, W.M., Short, A.D. and Brown, A.G. (1990) Direct observations of synapses between GABA-immunoreactive boutons and muscle afferent terminals in lamina VI of the cat's spinal cord. *Brain Res.*, 530: 215–222.

Nielsen, J. and Kagamihara, Y. (1992) The regulation of disynaptic reciprocal Ia inhibition during co-contraction of antagonistic muscles in man. *J Physiol (London),* 456: 373–391.

Nielsen, J., Kagamihara, Y., Crone, C. and Hultborn, H. (1992) Central facilitation of Ia inhibition during tonic ankle dorsiflexion revealed after blockade of peripheral feedback. *Exp. Brain. Res.*, 88: 651–656.

Peng, Y.Y. and Frank, E. (1989a) Activation of GABA-a receptors causes presynaptic and postsynaptic inhibition at synapses between muscle spindle afferents and motoneurons in the spinal cord of bullfrogs. *J. Neurosci.*, 9: 1516–1522.

Peng, Y.Y. and Frank, E. (1989b) Activation of GABA-b receptors causes presynaptic inhibition at synapses between muscle spindle afferents and motoneurons in the spinal cord of bullfrogs. *J. Neurosci.*, 9: 1502–1515.

Peshori, K.R., Collins III, W.F. and Mendell, L.M. (1991) Change in EPSP amplitude modulation during high frequency stimulation is correlated with changes in EPSP amplitude. A baclofen study. *Soc. Neurosci. Abstr.*, 17: 647.

Pinco, M. and Lev-Tov, A. (1993) Synaptic excitation of alpha motoneurons by dorsal root afferents in the neonatal rat spinal cord. *J. Neurophysiol.*, 70: 406–417.

Price, G.W., Wilkin, G.P., Turnbul, M.J. and Bowery, N.G. (1984) Are baclofen-sensitive $GABA_b$ receptors present on primary afferent terminals of the spinal cord?. *Nature*, 307: 71–72.

Quevedo, J., Eguíbar, J.R., Jiménez, I. and Rudomin, P. (1992) Differential action of (–)-baclofen on primary afferent depolarization produced by segmental and descending inputs. *Exp. Brain. Res.*, 91: 29–45.

Quevedo, J., Eguíbar, J.R., Jiménez, I., Schmidt, R.F. and Rudomin, P. (1993a) Primary afferent depolarization of muscle afferents elicited by stimulation of joint afferents in cats with intact neuraxis and during reversible spinalization. *J Neurophysiol.*, in press.

Quevedo, J., Eguíbar, J.R., Jiménez, I. and Rudomin, P. (1993b) Connectivity patterns of single last-order PAD mediating interneurons with two branches of the same group I fiber. *Soc. Neurosci. Abstr.*, 19: in press.

Rall, W. (1960) Membrane potential transients and membrane time constants. *Exp. Neurol.*, 2: 503–532.

Redman, S.J. (1990) Quantal analysis of synaptic potentials in neurons of the central nervous system. *Physiol. Rev.*, 70: 165–198.

Redman, S.J. and Walmsley, B. (1983) Amplitude fluctuations in synaptic potentials evoked in cat spinal motoneurones at identified group Ia synapses. *J. Physiol. (London)*, 343: 135–145.

Roby-Brami, A. and Bussel, A. (1992) Inhibitory effects on flexor reflexes in patients with a complete spinal cord lesion. *Exp. Brain. Res.*, 90: 201–208.

Rudomin, P. (1980) Information Processing at Synapses in the vertebrate spinal cord: presynaptic control of information transfer in monosynaptic pathways. In: H.M. Pinsker and W.D. Willis (Eds.), *Information Processing in the Nervous System,* Raven Press, New York., pp. 125–155.

Rudomin, P. (1990) Presynaptic control of synaptic effectiveness of muscle spindle and tendon organ afferents in the mammalian spinal cord. In: M.D. Binder and L.M. Mendell (Eds.), *The*

Segmental Motor System, Oxford University Press, New York, pp. 349–380.

Rudomin, P (1991) Presynaptic inhibition of muscle spindle and tendon organ afferents in mammalian spinal cord. *Trends Neurosci.*, 13: 499–505.

Rudomin, P. (1992) Validation of the changes in heterosynaptic facilitation of monosynaptic responses of spinal motoneurons as a test for presynaptic inhibition. In: L. Jami, E. Pierrot-Deseilligny and D. Zytnicki (Eds.), *Muscle Afferents and Spinal Control of Movement,* Pergamon Press, Oxford, pp. 439–455.

Rudomin, P. and Madrid, J. (1972) Changes in correlation between monosynaptic responses of single motoneurons and in information transmission produced by conditioning volleys to cutaneous nerves. *J. Neurophysiol.*, 35: 44–54.

Rudomin, P., Burke, R.E., Núñez, R., Madrid, J. and Dutton, H, (1975a) Control by presynaptic correlation: A mechanism affecting information transmission from Ia fibers to motoneurons. *J. Neurophysiol.,* 38: 267–284.

Rudomin, P., Núñez, R. and Madrid, J. (1975b) Modulation of Synaptic effectiveness of Ia and descending fibers in the cat spinal cord. *J. Neurophysiol.,* 38, 1181–1195.

Rudomin, P., Engberg, I. and Jiménez, I. (1981) Mechanisms involved in presynaptic depolarization of group I and rubrospinal fibers in cat spinal cord. *J. Neurophysiol.*, 46: 532–548.

Rudomin, P., Jiménez, I., Solodkin, M. and Dueñas, S. (1983) Sites of action of segmental and descending control of transmission on pathways mediating PAD of Ia and Ib afferent fibers in the cat spinal cord. *J Neurophysiol.*, 50: 743.

Rudomin, P., Solodkin, M. and Jiménez, I. (1986) Response patterns of group Ia and Ib fibers to cutaneous and descending inputs in the cat spinal cord. *J. Neurophysiol.*, 56: 987–1006.

Rudomin, P., Solodkin, M. and Jiménez, I. (1987) Synaptic potentials of primary afferent fibers and motoneurons evoked by single intermediate nucleus interneurons in the cat spinal cord. *J. Neurophysiol.*, 57: 1288–1313.

Rudomin, P., Jiménez, I. and Enríquez, M. (1991). Effects of stimulation of group I afferents on heterosynaptic facilitation of monosynaptic reflexes produced by Ia and descending inputs: a test for presynaptic inhibition. *Exp. Brain Res.*, 85: 93–102.

Schmidt, R.F. (1971) Presynaptic inhibition in the vertebrate central nervous system. *Ergebn. Physiol.*, 63: 20–101.

Smith, T.G., Wuerker, R.B. and Frank, K. (1967) Membrane impedance changes during synaptic transmission in cat spinal motoneurons. *J. Neurophysiol.,* 30: 1072–1096.

Solodkin, M., Jiménez, I., Collins III, W.F., Mendell, L.M. and Rudomin, P. (1991) Interaction of baseline synaptic noise and Ia EPSPs: evidence for appreciable negative-correlation under physiological conditions. *J. Neurophysiol.*, 65: 927–945.

Stuart, G.J. and Redman S.J. (1992) The role of GABAa and GABAb receptors in presynaptic inhibition of Ia EPSPs in cat spinal motoneurones. *J. Physiol. (London),* 447: 675–692.

Sugita, S., Johnson, S.W. and North, R.A. (1992) Synaptic inputs to GABAa and GABAb receptors originate from discrete afferents neurons. *Neurosci. Lett.,* 134: 207–211.

Thompson, S.M., Capogna, M. and Scanziani, M. (1993) Presynaptic Inhibition in the hippocampus. *Trends Neurosci.,* 16: 222–227.

Wall, P.D. and McMahon, S.B. (1993) Long range afferents in rat spinal cord. III. Failure of impulse transmission in axons and relief of the failure following rhizotomy of dorsal roots. *Philos. Trans. R. Soc. London B.,* in press.

Wong, K. and Redman, S.J. (1980) The recovery of a random variable from a noisy record with application to the study of fluctuations of synaptic potentials. *J. Neurosci. Methods*, 2: 389–409.

Wu, S.Y. and Dun, N.J. (1992) Presynaptic GABAb receptor activation attenuates synaptic transmission to rat sympathetic preganglionic neurons in vitro. *Brain. Res.,* 572: 94–102.

F. Bloom (Editor)
Progress in Brain Research, Vol. 100

CHAPTER 14

Noradrenergic control of cerebello-vestibular functions: modulation, adaptation, compensation

O. Pompeiano

Dipartimento di Fisiologia e Biochimica, Via S. Zeno 31, 56127 Pisa, Italy

Introduction

The central nervous system receives noradrenergic (NA) afferents from the locus coeruleus (LC). Although small in size, this dorsal pontine structure gives rise to widespread projections ending within cerebrocortical and subcortical structures, cerebellum, brain stem and spinal cord (cf. Foote et al., 1983). Several lines of evidence indicate that the NA LC neurons control high functions, such as attention, orientation, sleep-waking cycle, learning, memory, and also intervene in development, regeneration and plasticity of the brain (cf. Barnes and Pompeiano, 1991). Moreover, there is evidence that the number of NA LC neurons decreases in humans with age and that a more severe neuron loss in the LC occurs in the senile dementias of the Alzheimer type and Parkinson's disease (Chan-Palay and Asan, 1989).

In an attempt to identify the sensory inputs that could contribute to the low and regular discharge rate of the LC neurons in the animal at rest, we have recently recorded the activity of these neurons in decerebrate cats and found that one of the main sources of tonic activation of the LC originated from macular gravity and neck receptors. In particular, a large proportion of LC complex neurons, including those projecting to the spinal cord, responded not only to static but also to dynamic stimulation of both types of receptors (cf. Pompeiano et al., 1991b). Moreover, these NA neurons contributed to the control of posture as well as to the gain regulation of the vestibulospinal (VSR) and cervicospinal reflexes (cf. Pompeiano et al., 1991a).

The finding that the LC neurons not projecting to the spinal cord are under the control of macular labyrinth and neck receptors raises the question of the possible role that the corresponding inputs exert in the NA regulation of high brain functions. On the other hand, the demonstration that LC neurons, including those projecting to the spinal cord, intervene in the control of postural responses to gravity may be relevant in order to understand the equilibrium disturbances that occur with age, as well as the compensatory mechanisms that appear after labyrinthine lesion or functional inactivation of macular receptors during exposure to microgravity.

The NA influences on posture and VSRs utilize not only coeruleospinal but also coeruleocerebellar projections, which act on Purkinje (P) cell dendrites in the molecular layer and to a lesser extent also on P cell body and superficial granular cell layer through different types of adrenoceptors (cf. Foote et al., 1983; Barnes and Pompeiano, 1991). In the present report, we summarize the results of recent experiments suggesting that the NA signals acting on the cerebellar anterior vermis not only exert a short-term modulatory influence on the VSR, but may also intervene in long-term plastic changes which are at the basis of vestibular adaptation and compensation.

Noradrenergic influences on the cerebellar cortex: modulation of the VSR gain

In decerebrate cats, slow rotation about the longitudinal axis of the whole animal (roll tilt at 0.026–0.15 Hz, ±10°) leading to stimulation of labyrinth receptors, produces a contraction of limb extensors during ipsilateral tilt and a relaxation during contralateral tilt (Lindsay et al., 1976; Schor and Miller, 1981) (Fig. 1*A*). These effects, related to animal position, depend particularly upon activation or inactivation of macular utricular receptors and can be attributed, in part at least, to parallel changes in the firing rate of lateral vestibulospinal (VS) neurons (cf. Pompeiano, 1990), exerting an excitatory influence on ipsilateral limb extensor motoneurons (Lund and Pompeiano, 1968).

We have first shown that the P cells located in the paramedial zone B of the cerebellar anterior vermis, which projects to the lateral vestibular nucleus (Corvaja and Pompeiano, 1979), where they exert an inhibitory influence, respond to roll tilt of the animal with a predominant pattern of simple spike discharge opposite in sign with respect to that of the VS neurons (Denoth et al., 1979). The conclusion of this study, i.e. that the cerebellar vermis contributes positively to the VSR gain, was proved by the fact that microinjection into zone B of the cerebellar anterior vermis of the GABA-A agonist muscimol or the GABA-B agonist baclofen (0.25 μl in 2–16 μg/μl saline), leading to local inactivation of the P cells, reversibly decreased the amplitude of the VSR recorded from the forelimb extensor triceps brachii during animal tilt (Andre et al., 1992, 1994).

We then investigated the role that the NA afferent input to the cerebellar cortex exerts on the VSRs in decerebrate cats (cf. Andre et al., 1991a–c). Unilateral microinjection into zone B of the cerebellar anterior vermis of small doses of α_1-, α_2- or β-adrenergic agonists (metoxamine, clonidine or isoproterenol, respectively, 0.25 μl at the concentration of 2–16 μg/μl saline) increased the gain of the multiunit electromyogram (EMG) responses recorded from the triceps brachii of both sides during animal tilt (Fig. 2), while opposite results were obtained by injecting the corresponding antagonists (prazosin, yohimbine or propranolol, 0.25 μl at 8–16 μg/μl saline). These effects appeared 5–10 min after the injection, reached peak values after 20 min and disappeared within 2 h.

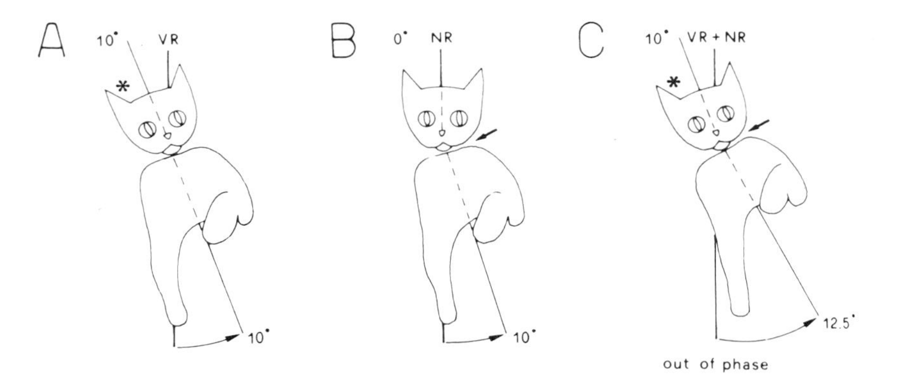

Fig. 1. Schematic representation of different head and/or body displacements, leading to individual or combined stimulation of labyrinth and neck receptors. *A*, Vestibular stimulation: 10° rotation about the longitudinal axis of the whole animal (to the right side, asterisk) produced selective stimulation of labyrinth receptors leading to contraction of the side-down (right) and relaxation of the side-up (left) forelimb extensors. *B*, Neck stimulation: 10° rotation about the longitudinal axis of the body (to the left side, arrow), while maintaining the horizontal position of the head, produced selective stimulation of neck receptors leading to relaxation of the side-down (left) and contraction of the side-up (right) forelimb extensors. *C*, out of phase neck-vestibular stimulation: 10° tilt of the head to the right side (asterisk), leading to stimulation of labyrinth receptors as in *A*, was associated with 12.5° rotation of the body. This produced a 2.5° body-to-head rotation to the left (arrow), which was thus out of phase with respect to head displacement. The resulting neck input increased the postural asymmetry elicited by the pure labyrinth signal (compare with *A*). From Andre et al., 1993.

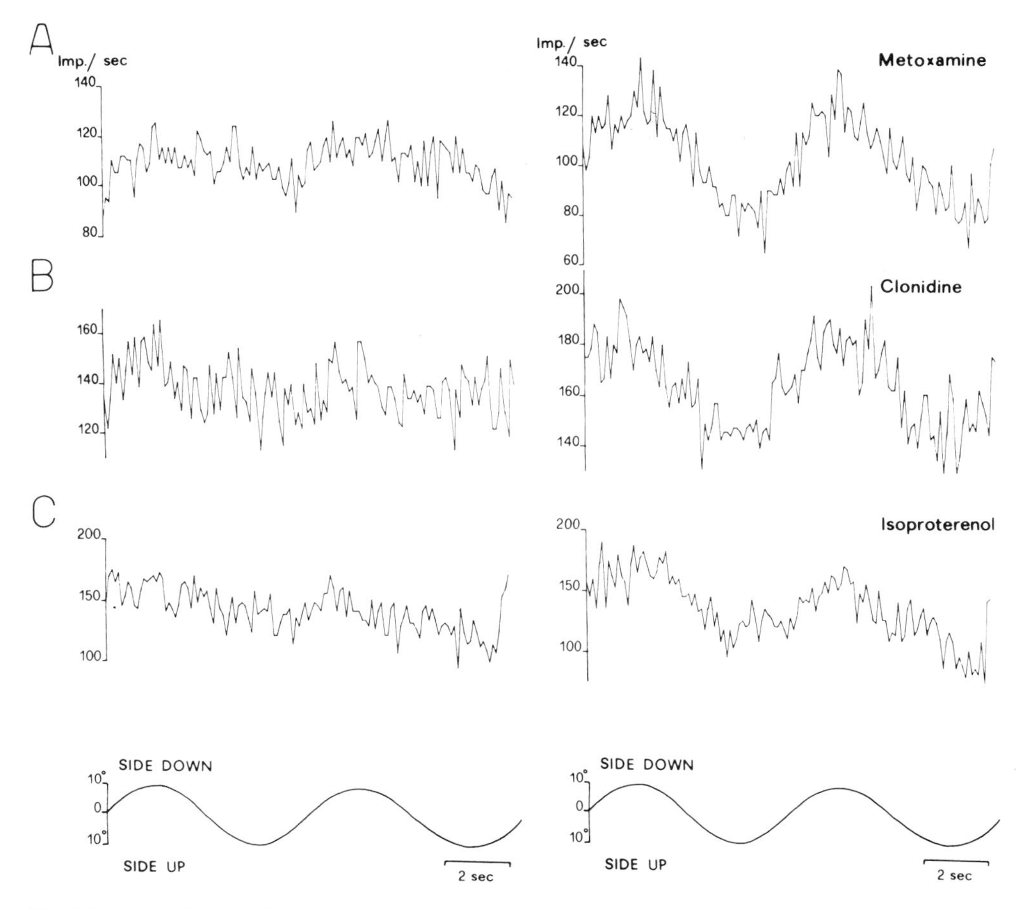

Fig. 2. Increase of the vestibulospinal reflex (VSR) gain elicited by recording the electromyogam (EMG) responses of the triceps brachii to animal tilt after microinjection of an α1- (*A*), α2- (*B*) and β- (*C*) adrenergic agonist into the ipsilateral vermal cortex of the cerebellar anterior lobe (culmen). Precollicular, decerebrate cats. Sequential pulse density histograms (SPDHs) showing the averaged multiunit responses of the triceps brachii of one side to roll tilt of the whole animal at 0.15 Hz, ±10°. Each record is the average of six sweeps, containing the responses to two successive cycles of rotation (128 bins, 0.1 s bin width). The lower traces indicate the animal displacement. The traces on the left side were taken before, while those on the right side after individual or multiple injections of 0.25 μl of metoxamine (4 μg/μl), clonidine (4 μg/μl) and isoproterenol (16 μg/μl) solutions. The response gain increased on average from 0.61 to 2.08 impulses/s per degree in *A*, from 0.60 to 1.94 impulses/s per degree in *B* and from 1.24 to 2.72 impulses/s per degree in *C*. From Andre et al., 1991a,b.

Since activation of VS neurons during ipsilateral tilt depends not only upon an increased discharge of the excitatory labyrinthine afferents, but also on a reduced discharge of the overlying inhibitory P cells, we postulated that the NA system potentiates the gain of the VSR by increasing the amplitude of modulation of the P cells to given parameters of labyrinth stimulation. This conclusion was supported by the results of experiments showing that iontophoretic application of norepinephrine (NE), while depressing the spontaneous activity of P cells, enhanced their responses to both excitatory (mossy fibers and climbing fibers) and inhibitory (basket and stellate cells) inputs, as well as to the corresponding excitatory (glutamate, aspartate) and inhibitory (GABA) transmitters (cf. Woodward et al., 1991). This modulatory action of NE on P cells would act not only by increasing the signal-to-noise ratio of the evoked versus spontaneous activity, but also to gate the efficacy of subliminal synaptic input conveyed by the classical afferent systems, thus improving information transfer within local circuits (cf. Woodward et al., 1991). Second messengers have been implicated in these effects (cf. Waterhouse et al., 1991).

Noradrenergic influences on the cerebellar cortex: adaptation of the VSR gain

Roll tilt of the whole animal increases the contraction not only of the side-down limb extensors but also of the side-up neck extensors (cf. Schor and Miller, 1981), thus stabilizing the position of the head and body in space. In the free-moving condition, this vestibulo-collic reflex generates a proprioceptive neck input, which acts synergistically with the labyrinth input to maintain the postural adjustments during animal tilt. In fact, a contraction of limb extensors can be elicited not only during side-down animal tilt (Fig. 1*A*), but also during side-up neck rotation (Fig. 1*B*) (cf. Lindsay et al., 1976). In decerebrate cats, in which the proprioceptive neck input does not occur due to fixation of the head at the stereotaxic equipment, the VSRs are barely compensatory. We decided, therefore, to investigate whether, in this preparation, a sustained sinusoidal roll tilt of the whole animal performed selectively or associated with appropriate neck rotation produced an adaptive increase in gain of the VSR (Andre et al., 1993) and, if so, whether the NA afferent input to the cerebellar anterior vermis intervened in the regulation of these adaptive changes of the VSR gain (Pompeiano et al., 1992, 1994).

All the experiments started with baseline measurements of VSRs, obtained by recording intermittently groups of averaged multiunit responses of the triceps brachii to 12 cycles of animal tilt at 0.15 Hz, ±10°, performed at regular intervals of 8–10 min for at least 0.5–1 h. After this control period was over, two groups of adaptive experiments were performed. In the first group, a sustained roll tilt of the whole animal at 0.15 Hz, ±10° was applied continuously for 3 h, while the resulting VSRs were recorded on-line from the triceps brachii of one or both sides. In a second group of experiments, a sustained roll tilt of the head at 0.15 Hz, ±10° was associated with a synchronous rotation of the body at 0.15 Hz, ±12.5°. This produced an additional neck input due to 2.5° of out of phase body-to-head displacement (Fig. 1*C*, see arrow), which exerted a synergistic influence on the postural changes affecting the limb extensors during labyrinth stimulation. This strategy was applied almost continuously for 3 h, being interrupted only every 10–15 min to record the EMG responses to pure labyrinth stimulation. At the end of this adaptive period, the baseline measurements of the pure VSR elicited by animal tilt (at 0.15 Hz, ±10°) were obtained intermittently every 8–10 min for about 1 hour, as during the control period.

The 3-h period of continuous tilt of the whole animal represented a very poor means to induce an adaptive increase in gain of the VSR, which occurred in only one-third of the experiments. However, the gain of the VSR progressively increased in all the experiments submitted to a 3-h period of sustained out of phase head and body rotations, and remained almost unmodified during the first hour of post-adaptation. Figure 3 illustrates the time course of the average increase in gain of the VSR obtained in one of these experiments during the out of phase neck-vestibular stimulation, the dashed line representing the average initial value obtained in the baseline measurements of the VSR before the adaptation of this reflex was started. Moreover, Fig. 4 (dots) illustrates the average curve of adaptation of the VSR recorded from the triceps brachii of one or both sides (11 muscles) in eight experiments. In these instances, the difference between the mean value obtained at the end of the third hour of stimulation and the baseline value was statistically significant ($P < 0.001$, paired *t*-test). This adaptive change in gain increased by increasing the efficacy of neck rotation. In no instance, however, did the gain of the pure VSR change if tested intermittently in non-adaptation experiments.

The possibility that the neuronal changes, which are at the basis of the VSR adaptation, occurred within the zone B of the cerebellar anterior vermis is indirectly supported by the fact that the corresponding P cells not only showed a mossy fiber discharge to animal tilt as reported in the previous section (Denoth et al., 1979), but also responded with a mossy fiber and a climbing fiber discharge to the neck signals (Denoth et al., 1979, 1980). In particular, the proprioceptive neck input evoked during sustained neck-vestibular stimu-

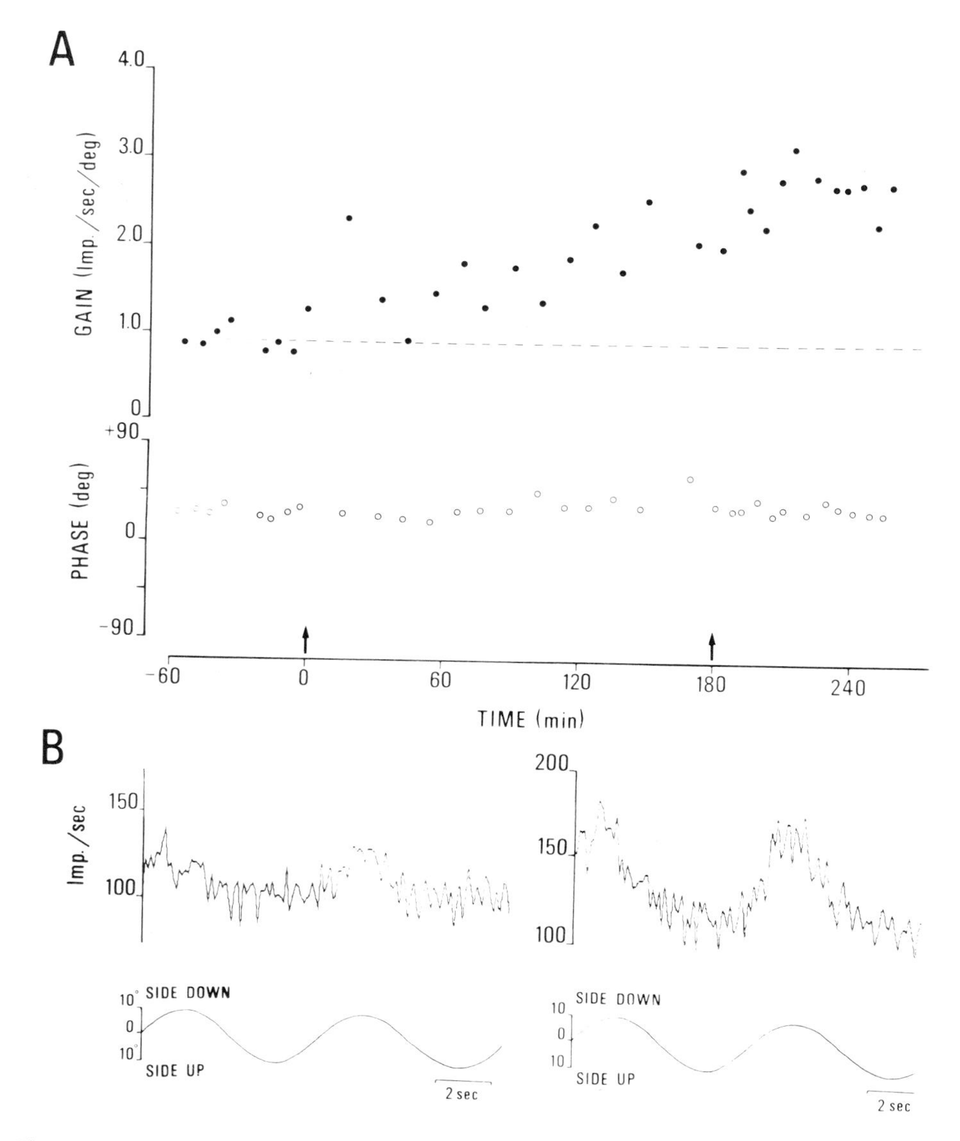

Fig. 3. Adaptive changes of the VSR gain elicited by recording the EMG responses of the triceps brachii to animal tilt during out of phase neck-vestibular stimulation. Precollicular, decerebrate cat. *A*, Changes in gain (upper diagram) and phase angle (lower diagram) of the VSR were obtained by averaging at regular intervals the multiunit responses of the left triceps brachii to roll tilt of the whole animal at 0.15 Hz, ±10° before, during (between 0 and 180 min) and after the adaptive period elicited by a sustained combination of 10° head and 12.5° body rotation at 0.15 Hz. The two arrows indicate the duration of these out of phase head-body rotations, which were briefly interrupted every 13 min to record the VSR. Each symbol represents the averaged response (AR) to 12 cycles of animal tilt at the parameters indicated above. *B*, SPDHs showing AR of the left triceps brachii to animal tilt at 0.15 Hz, ±10° recorded before (left trace) and 50 min after the 3-h period of adaptation (right trace). The response gain increased from 1.01 impulses/s per degree (phase lead of +42.4°) to 2.54 impulses/s per degree (phase lead of +33.3°). From Andre et al., 1993.

lation could, through climbing fibers, not only increase the proportion of P cells showing an out of phase modulatory response to labyrinth stimulation, but also give rise to an efficient process of adaptation, whatever the mechanism at the basis of the adaptive process may be. This model is similar to that proposed to explain the adaptive increase in gain of the vestibulo-ocular reflex (VOR) which occurs during con-

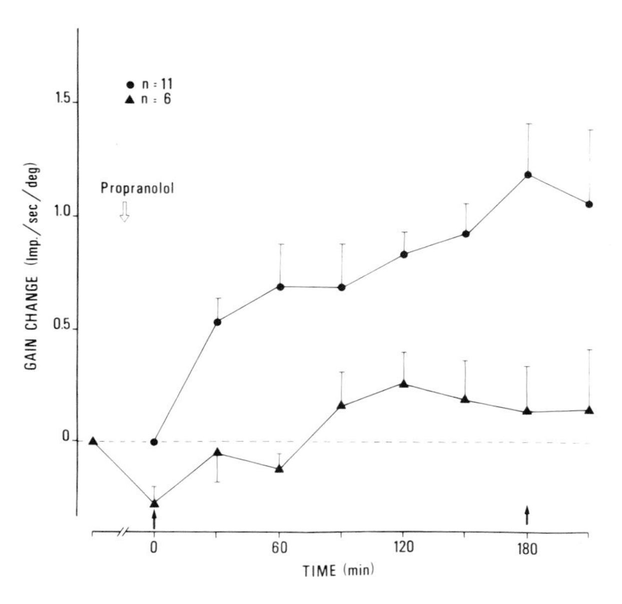

Fig. 4. Effects of unilateral microinjection into the cerebellar anterior vermis of the β-adrenergic antagonist propranolol on adaptive changes of the VSR gain elicited by recording the EMG responses of the triceps brachii to animal tilt during out of phase neck-vestibular stimulation (between arrows). Precollicular decerebrate cats, showing the adaptive changes in gain of the VSR evoked in the triceps brachii of one or both sides by a sustained combination of 10° head rotation and 12.5° body rotation at 0.15 Hz. These adaptive changes were recorded either in the control condition (dots, 8 experiments, 11 periods of adaptation) or after microinjection of propranolol (0.25 μl, 8 μg/μl saline) into zone B of the cerebellar anterior vermis (triangles, 3 experiments, 6 periods of adaptation). Propranolol injection (arrow) was usually performed 1 h prior to the beginning of the adaptive stimulation. The VSRs were elicited during roll tilt of the whole animal at 0.15 Hz, ±10°, and evaluated with respect to the control value (indicated by 0 in the ordinate). In particular, groups of AR to 12 cycles of animal tilt were recorded every 10–15 min after brief interruption of the sustained head-body rotation, as well as during the post-adaptation period. The curves represent the average of the adaptive changes in gain of the VSR obtained in all the experiments during 30-min periods (each symbol indicates mean ± SE values of the responses). From Andre et al., 1993 and Pompeiano et al., 1994.

tinuous sinusoidal rotation of the animal, associated with an out of phase moving screen leading to a synergistic visual-vestibular interaction (cf. Ito, 1982). In these instances, the visual input conveyed through climbing fibers to the specific corticocerebellar area which controls the VOR gain, i.e. the flocculus (see van Neerven et al., 1989 for references), produced a long-term depression of the mossy fiber responses of the P cells to labyrinth stimulation, due to desensitization of the excitatory synapses made by the parallel fibers on the P cell dendrites (Ito, 1990). Experiments are required to find out whether similar cellular mechanisms can be induced by the neck input in the cerebellar anterior vermis during adaptation of the VSR gain.

After the demonstration that adaptive changes affected the gain of the VSR, we investigated whether these changes could be modified by microinjection of β-adrenergic agents into zone B of the cerebellar anterior vermis of decerebrate cats (Pompeiano et al., 1992, 1994). When a 3-h period of continuous roll tilt of the animal was ineffective, microinjection into the cerebellar anterior vermis of the β-adrenergic agonist isoproterenol (0.25 μl at 8 μg/μl saline) performed 30–60 min prior to a second period of sustained stimulation produced not only a small amplitude early increase in the VSR gain, due to a direct modulatory influence of the β-agonist on the P cells activity (cf. Andre et al., 1991b), but also a prominent and delayed increase which reached a peak at the end of the third hour of stimulation and persisted later, thus being attributed to adaptation of the VSR. On the other hand, inactivation of the NA system following intravermal injection of the β-adrenergic antagonists, propranolol or sotalol (0.25 μl at 8 μg/μl saline), either suppressed the adaptive increase in gain following a sustained roll tilt of the animal, or prevented the occurrence of the adaptive increase in the VSR gain which always appeared in the absence of any intravermal injection during a sustained out of phase head and body rotations.

Figure 4 (triangles) illustrates the average changes in gain of the VSR recorded from the triceps brachii of both sides (six muscles) in three adaptation experiments after microinjection of propranolol into zone B of the cerebellar anterior vermis. In these instances, the difference between the baseline value and that obtained at the end of the adaptation period was not statistically significant (paired *t*-test). Moreover, by using the multifactor analysis of variance, a test which allows the comparison of variables in different groups of animals, it appeared that the changes in gain of the

VSR were severely depressed after microinjection into the cerebellar vermis of the β-adrenergic blocker propranolol with respect to those obtained in the absence of intravermal injections ($P < 0.0001$, MANOVA). Similar results were also obtained in other experiments after intravermal injections of the β-adrenergic antagonist sotalol at the same dose used for propranolol. These findings closely resemble those obtained in previous experiments showing that microinjection into the flocculus of rabbits of isoproterenol increased, while sotalol reduced the adaptation of the VOR to retinal slip (van Neerven et al., 1990).

The modalities by which the NA system contributes to a more efficient process of adaptation of these reflexes are still unknown. It has been postulated that a complex chain of events involving excitatory amino acid receptors and certain second messenger systems at the P cell level underlie the long-term depression that is at the basis of the VOR adaptation (Ito, 1990; Ito and Karachot, 1991). The NA system could potentiate these adaptive processes by acting through β-adrenoceptors.

It is of interest that trans-synaptic signals may elicit not only rapid responses in neurons, ranging from milliseconds (e.g. opening of ligand-gated channels) to seconds and minutes (e.g. second messenger-mediated events), but also slower, long-term responses that are mediated by changes in gene expression (cf. Morgan and Curran, 1991 for references). These genes fall into two general classes: the immediate early genes (IEGs), whose transcription is activated rapidly and transiently within minutes of stimulation and the late response genes, whose expression is induced (or repressed) more slowly, over hours, via a mechanism that is generally dependent on new protein synthesis (cf. Sheng and Greenberg, 1990 for references). It has been demonstrated that many IEGs encode transcription factors that are likely to control the expression of late response genes. The protein products of the latter are then thought to mediate more specific long-term neuronal responses. IEGs may thus function as "third messenger" molecules in signal transduction mechanisms that convert specific stimuli into long-term gene expression of target proteins.

Among the numerous IEGs so far identified, the best known is c-*fos*. While c-*fos* transcription, which can be mediated by several second messenger systems, is rapidly induced within minutes of extracellular stimulation, the related Fos proteins, synthesized following mRNA expression, can be detected for several hours (cf. Sheng and Greenberg, 1990; Morgan and Curran, 1991 for references).

There is also evidence that a Fos-like protein binds to a protein product of the Jun family induced by another IEG, the c-*jun*, to form a heterodimeric transcription factor complex. The resulting Fos/Jun complex moves to the cell nucleus to bind genomic DNA at promoter regions, called AP-1 sites (cf. Doucet et al., 1990). These "activator protein" sites would then control the expression of "downstream" genes that are relevant for long-term responses of neurons to trans-synaptic signals (cf. Curran et al., 1990; Doucet et al., 1990).

With respect to the adaptive experiments reported in this section, we can hypothesize that induction of c-*fos* and/or other IEGs represents a mechanism by which impulses elicited by sustained neck-vestibular stimulation are transduced into long-term biochemical changes that are required for adaptation of the VSR gain. The NA system could then potentiate the expression of these early genes. This hypothesis receives some indirect support by the evidence that: (1) the NA LC neurons respond to the same modalities of labyrinth and neck stimulation used in our adaptive stimulation (Pompeiano et al., 1991b); (2) an increase in NE release may induce an increase in the c-*fos* expression in some target regions, as shown in the cerebral cortex of rats (Bing et al., 1992). It appeared also that this effect was predominantly mediated through β-adrenoceptors. Experiments are required to investigate whether the central NA system driven by the labyrinth and neck signals may, through β-adrenoceptors, modify the expression of the c-*fos* and/or other IEGs and related proteins in the cerebellar anterior vermis and possibly also in other target systems, which could then contribute to the plastic changes underlying adaptation (cf. Ito et al., 1993, for structures involved in VOR adaptation).

Noradrenergic influences on vestibular compensation

Unilateral labyrinthectomy (UL) produces phenomena of deficit, characterized by a decreased postural activity in the ipsilateral limbs and an increased activity in the contralateral limbs, tilting of the head and rolling movements towards the lesioned side and also horizontal nystagmus with the fast phase directed to the intact side. These findings depend upon a reduced discharge in the ipsilateral and an increased discharge in the contralateral vestibular neurons (cf. Smith and Curthoys, 1989).

There are only few data about the possible role of the NA LC system in vestibular compensation. In particular, it appears that intracisternal injection of the NA agonist clonidine in compensated frogs produced decompensation of roll head tilt, whereas the NA antagonist phentolamine caused overcompensation (observation by Abeln and Flohr reported in Flohr and Lüneburg, 1985). Experiments were also performed in unanesthetized cats in which chronically implanted guide tubes allowed microinjection of NA agents in different brain stem structures (see methods in Tononi et al., 1989). In unilaterally labyrinthectomized and compensated animals, microinjection into the dorsal pontine tegmentum of the β-adrenergic antagonist propranolol, which suppressed the inhibitory influence exerted by the NA LC neurons on cholinoceptive pontine reticular structures (d'Ascanio et al., 1989; Pompeiano et al., 1991a), led to the reappearance of the postural and motor deficits induced by the UL (unpublished observations).

Kaufman et al. (1992, 1993) have recently studied the effects of UL on the induction of Fos protein in the brain of Long Evans rats. In these experiments, the animals were killed 24 h and 2 weeks after a lesion performed by injection of sodium arsanilate (5 mg in 50 μl of saline) through the tympanic membrane into the middle ear of one side. The behavioral signs of this lesion appeared slowly within several hours of the arsanilate injections and by 24 h were characterized by a significant head tilt, neck deviation and circling towards the lesioned side. At this time, a bilateral Fos labeling was found in the medial and inferior vestibular nuclei and the praepositus hypoglossi. However, a greater number of labeled neurons were found in the vestibular nuclei of the ipsilateral side and in the praepositus hypoglossi of the contralateral side. Within the inferior olive, the contralateral β-subnucleus exhibited strong labeling, while the dorsomedial cell column had slight labeling in some rats only. Moreover, ipsilateral labeling was found in the cerebellar uvula and nodulus. Within the midbrain, the dorsolateral periaqueductal gray showed Fos expression that was greater on the ipsilateral side, and there was a bilateral expression in the interstitial nucleus of Cajal and the Darkschewitsch nucleus. These effects were almost gone in rats killed 2 weeks after UL, i.e. when compensation of the vestibular syndrome had already occurred.

Experiments are in progress in our laboratory to investigate: (1) whether the pattern of Fos protein induction following a unilateral *surgical* lesion of the labyrinth closely corresponds to that described by Kaufman et al. (1992, 1993) after a unilateral *chemical* lesion; (2) whether the increase in the Fos protein induction revealed by immunohistochemistry is also associated with a parallel increase in the expression of the c-*fos* mRNA, as detected by using in situ hybridization; finally, (3) whether the time course of the c-*fos* mRNA and Fos protein induction parallels the development of the postural and motor deficits, which occur soon after the surgically induced labyrinthine lesion and largely disappear 24–48 h after this lesion.

Attempts should also be made to find out whether the NA LC system facilitates the c-*fos* expression after UL. This hypothesis is supported by the fact that LC neurons of one side respond to labyrinth stimulation with a predominant response pattern characterized by an increased discharge during contralateral tilt (Pompeiano et al., 1991b), a stimulus which leads to a postural asymmetry similar to that elicited by ipsilateral labyrinthectomy (Lindsay et al., 1976; Schor and Miller, 1981). Since NA efferents project to vestibular nuclei (Schuerger and Balaban, 1993), cerebellar cortex (Foote et al., 1983; Andre et al., 1991a–c) and inferior olive (Powers et al., 1990), asymmetric changes in unit discharge of LC neurons following UL could contribute at least in part to the asymmetric changes in c-*fos* expression observed in these target systems.

The possibility that asymmetric changes in activity of LC neurons determine parallel changes of gene expression in several target systems could be of great significance for the developing brain. It has been asserted that the vestibular system exerts a powerful influence on the development of the brain and behavior (Ornitz, 1993), and that asymmetric prenatal development of the labyrinths, attributed to the position of the human fetus during the final trimester, is probably at the origin of cerebral lateralization (Previc, 1991). The demonstration that changes in the head position leading to macular stimulation produce asymmetric changes in unit discharge in the LC (Pompeiano et al., 1991b) and that this structure may, through ipsilateral ascending projections (Foote et al., 1983), increase the c-*fos* expression in the cerebral cortex (Bing et al., 1992), thus exerting a potential long-term regulation on high brain functions, gives support to these hypotheses.

Acknowledgements

Most of the experiments reported in this review were made in collaboration with Drs. P. Andre, P. d'Ascanio and D. Manzoni. The research was supported by the National Institute of Neurological and Communicative Disorders and Stroke Research grant NS 07685-25 and by grants of the Ministero dell'Università e della Ricerca Scientifica e Tecnologica, and the Agenzia Spaziale Italiana (ASI 91 RS-77 and 92 RS-123), Roma, Italy.

References

Andre, P., d'Ascanio, P., Gennari, A., Pirodda, A. and Pompeiano, O. (1991a) Microinjections of α_1- and α_2-noradrenergic substances in the cerebellar vermis of decerebrate cats affect the gain of vestibulospinal reflexes. *Arch. Ital. Biol.*, 129: 113-160.

Andre, P., d'Ascanio, P., Manzoni, D. and Pompeiano, O. (1991b) Microinjections of β-noradrenergic substances in the cerebellar vermis of decerebrate cats modify the gain of vestibulospinal reflexes. *Arch. Ital. Biol.*, 129: 161–197.

Andre, P., d'Ascanio, P. and Pompeiano, O. (1991c) Noradrenergic agents into the cerebellar anterior vermis modify the gain of the vestibulospinal reflexes in the cat. In C.D. Barnes and O. Pompeiano (Eds.), *Neurobiology of the Locus Coeruleus, Progress in Brain Research*, Vol. 88, Elsevier, Amsterdam, pp. 463–484.

Andre, P., d'Ascanio, P., Manzoni, D. and Pompeiano, O. (1992) Depression of the vestibulospinal reflex by intravermal microinjection of GABA-A and GABA-B agonists in decerebrate cats. *Pflügers Arch.*, 420: R159, n. 51.

Andre, P., d'Ascanio, P., Manzoni, D. and Pompeiano, O. (1993) Adaptive modification of the cat's vestibulospinal reflex during sustained vestibular and neck stimulation. *Pflügers Arch.*, 425: 469–481.

Andre, P., d'Ascanio, P., Manzoni, D. and Pompeiano, O. (1994) Depression of the vestibulospinal reflex by intravermal microinjection of $GABA_A$ and $GABA_B$ agonists in the decerebrate cat. *J. Vestib. Res.*, 4: in press.

Barnes, C.D. and Pompeiano, O. (Eds.) (1991) *Neurobiology of the Locus Coeruleus, Progress in Brain Research*, Vol. 88, Elsevier, Amsterdam, pp. XIV–642.

Bing, G., Stone, E.A., Zhang, Y. and Filer, D. (1992) Immunohistochemical studies of noradrenergic-induced expression of c-fos in the rat CNS. *Brain Res.*, 592: 57–62.

Chan-Palay, V. and Asan, E. (1989) Alterations in catecholamine neurons of the locus coeruleus in senile dementia of the Alzheimer type and Parkinson's disease with and without dementia and depression. *J. Comp. Neurol.*, 287: 373–392.

Corvaja, N. and Pompeiano, O. (1979) Identification of cerebellar corticovestibular neurons retrogradely labeled with horseradish peroxidase. *Neuroscience*, 4: 507–515.

Curran, T., Sonnenberg, J.L., MacGregor, P. and Morgan, J.I. (1990) Transcription factors on the brain-Fos, Jun and the Ap-1 binding site. *Neurotoxicity of Excitatory Amino Acids*, 4: 175–184.

d'Ascanio, P., Horn, E., Pompeiano, O. and Stampacchia, G. (1989) Injections of a β-adrenergic antagonist in pontine reticular structures modify the gain of vestibulospinal reflexes in decerebrate cats. *Arch. Ital. Biol.*, 127: 275–303.

Denoth, F., Magherini, P.C., Pompeiano, O. and Stanojević, M. (1979) Responses of Purkinje cells of the cerebellar vermis to neck and macular vestibular inputs. *Pflügers Arch.*, 381: 87–98.

Denoth, F., Magherini, P.C., Pompeiano, O. and Stanojević, M. (1980) Responses of Purkinje cells of the cerebellar vermis to sinusoidal rotation of neck. *J. Neurophysiol.*, 43: 46–59.

Doucet, J.P., Squinto, S.P. and Basan, N.G. (1990) Fos-jun and the primary genomic response in the nervous system: possible physiological role and pathophysiological significance. In: Bazan, N.G. (Ed.), *Molecular Neurobiology*, Humana Press, Clifton, NJ, pp. 27–55.

Flohr, H. and Lüneburg, U. (1985) Neurotransmitter and neuromodulator systems involved in vestibular compensation. In: A. Berthoz and G. Melvill Jones (Eds.), *Adaptive Mechanisms in Gaze Control: Facts and Theories*. Elsevier, Amsterdam, pp. 269–277.

Foote, S.L., Bloom, F.E. and Aston-Jones, G. (1983) Nucleus locus coeruleus: new evidence of anatomical and physiological specificity. *Physiol. Rev.*, 63: 844–914.

Ito, M. (1982) Cerebellar control of the vestibulo-ocular reflex-around the flocculus hypothesis. *Anun. Rev. Neurosci.*, 5: 275–296.

Ito, M. (1990) Long-term depression in the cerebellum. *Semin. Neurosci.*, 2: 381–390.

Ito, M. and Karachot, L. (1991) Messengers mediating long-term desensitization in cerebellar Purkinje cells. *NeuroReport*, 1: 129–132.

Ito, M., Lisberger, S.G. and Sejnowski, T.J. (1993) Cerebellar flocculus hypothesis. *Nature*, 363: 24–25.

Kaufman, G.D., Anderson, J.H. and Beitz, A.J. (1992) Brainstem Fos expression following acute unilateral labyrinthectomy in the rat. *NeuroReport*, 3: 829–832.

Kaufman, G.D., Anderson, J.H. and Beitz, A.J. (1993) Otolith-brain stem connectivity: evidence for differential neural activation by vestibular hair cells based on quantification of Fos expression in unilateral labyrinthectomized rats. *J. Neurophysiol.*, 70: 117–127.

Lindsay, K.W., Roberts, T.D.M. and Rosenberg, J.R. (1976) Asymmetric tonic labyrinth reflexes and their interaction with neck reflexes in the decerebrate cat. *J. Physiol. (London)*, 261: 583–601.

Lund, S. and Pompeiano, O. (1968) Monosynaptic excitation of alpha-motoneurons from supraspinal structures in the cat. *Acta Physiol. Scand.*, 73: 1–21.

Morgan, J.I. and Curran, T. (1991) Stimulus-transcription coupling in the nervous system: involvement of inducibile proto-oncogenes *fos* and *jun*. *Annu. Rev. Neurosci.*, 14: 421–451.

Ornitz, E.M. (1983) Normal and pathological maturation of vestibular function in the human child. In R. Romand (Ed.), *Development of Auditory and Vestibular Systems*, Academic Press, San Diego, CA, pp. 479–536.

Pompeiano, O. (1990) Excitatory and inhibitory mechanisms involved in the dynamic control of posture during the vestibulospinal reflexes. In: L. Deecke, J.C. Eccles and V.B. Mountcastle (Eds.), *From Neuron to Action*, Springer-Verlag, Berlin, pp. 107–123.

Pompeiano, O., Horn, E. and d'Ascanio, P. (1991a) Locus coeruleus and dorsal pontine reticular influences on the gain of vestibulospinal reflexes. In: C.D. Barnes and O. Pompeiano (Eds.), *Neurobiology of the Locus Coeruleus, Progress in Brain Research*, Vol. 88, Elsevier, Amsterdam, pp. 435–462.

Pompeiano, O., Manzoni, D. and Barnes, C.D. (1991b) Responses of locus coeruleus neurons to labyrinth and neck stimulation. In: C.D. Barnes and O. Pompeiano (Eds.), *Neurobiology of the Locus Coeruleus, Progress in Brain Research*, Vol. 88. Elsevier, Amsterdam, pp. 411–434.

Pompeiano, O., Andre, P., d'Ascanio, P. and Manzoni, D. (1992) Local injections of β-noradrenergic substances in the cerebellar anterior vermis of cats affect adaptation of the vestibulospinal reflex (VSR) gain. *Soc. Neurosci. Abstr.*, 18(1): 508, n. 215.6.

Pompeiano, O., Manzoni, D., d'Ascanio, P. and Andre, P. (1994) Injections of β-noradrenergic substances in the cerebellar anterior vermis of cats affect adaptation of the vestibulospinal reflex gain. *Arch. Ital. Biol.*, 132: 117–145.

Powers, R.E., O'Connor, D.T. and Price, D.L. (1990) Noradrenergic innervation of human inferior olivary complex. *Brain Res.*, 523: 151–155.

Previc, F.H. (1991) A general theory concerning the prenatal origins of cerebral lateralization in humans. *Psychol. Rev.*, 98: 299–334.

Schor, R.H. and Miller, A.D. (1981) Vestibular reflexes in neck and forelimb muscles evoked by roll tilt. *J. Neurophysiol.*, 46: 167–178.

Schuerger, R.J. and Balaban, C.D. (1993) Immunohistochemical demonstration of regionally selective projections from locus coeruleus to the vestibular nuclei in rats. *Exp. Brain Res.*, 92: 351–359.

Sheng, M. and Greenberg, M.E. (1990) The regulation and function of c-*fos* and other immediate early genes in the nervous system. *Neuron*, 4: 477–485.

Smith, P.F. and Curthoys, I.S. (1989) Mechanisms of recovery following unilateral labyrinthectomy: a review. *Brain Res. Rev.*, 14: 155–180.

Tononi, G., Pompeiano, M. and Pompeiano, O. (1989) Modulation of desynchronized sleep through microinjection of β-adrenergic agonists and antagonists in the dorsal pontine tegmentum of the cat. *Pflügers Arch.*, 415: 142–149.

van Neerven, J., Pompeiano O. and Collewijn, H. (1989) Depression of the vestibulo-ocular and optokinetic responses by intrafloccular microinjection of GABA-A and GABA-B agonists in the rabbit. *Arch. Ital. Biol.*, 127: 243–263.

van Neerven, J., Pompeiano, O., Collewijn, H. and van der Steen, J. (1990) Injections of β-noradrenergic substances in the flocculus of rabbit affect adaptation of the VOR gain. *Exp. Brain Res.*, 79: 249–260.

Waterhouse, B.D., Sessler, F.M., Liu, W. and Lin, C.-S. (1991) Second messenger-mediated actions of norepinephrine on target neurons in central circuits: a new perspective on intracellular mechanisms and functional consequences. In: C.D. Barnes and Pompeiano O. (Eds.), *Neurobiology of the Locus Coeruleus, Progress in Brain Research*, Vol. 88, Elsevier, Amsterdam, pp. 351–362.

Woodward, D.J., Moises, H.C., Waterhouse, B.D., Yeh, H.H. and Cheun, J.E. (1991) The cerebellar norepinephrine system: inhibition, modulation, and gating. In: C.D. Barnes and O. Pompeiano (Eds.), *Neurobiology of the Locus Coeruleus, Progress in Brain Research*, Vol. 88. Elsevier, Amsterdam, pp. 331–341.

F. Bloom (Editor)
Progress in Brain Research, Vol. 100

CHAPTER 15

Chemical transmission in the brain: homeostatic regulation and its functional implications

Michael J. Zigmond

Department of Neuroscience, University of Pittsburgh, 570 Crawford Hall, Pittsburgh, PA 15260, USA

Homeostasis of neuronal function

The concept of homeostasis was first applied to organismal physiology and later to cell biology and biochemistry. Over the past 25 years it has become clear that this phenomenon also can be detected in many of the interactions between neurons and their targets, both effectors and other neurons. One of the earliest suggestions of homeostatic regulation of neuronal communication came from the work of Arvid Carlsson who observed that chlorpromazine and haloperidol, which produced a reserpine-like akinesia in rats, also caused an increase in the concentration of its *o*-methylated metabolite, 3-methyoxytyramine (Carlsson and Lindqvist, 1963). Carlsson had already shown that reserpine acted primarily by depleting the brain of dopamine (DA), thus the increase in DA turnover produced by these "neuroleptics" appeared to be a paradox.

Carlsson subsequently proposed that (1) chlorpromazine blocked DA receptors, (2) this action was responsible for the behavioral deficits observed, and (3) the blockade triggered an increase in DA synthesis and release by removing feedback inhibition normally exerted by DA. His innovative hypothesis proved to be correct. In the years to follow it was shown that neuroleptic drugs do act as DA receptor antagonists (Creese et al., 1978), and that a wide variety of negative feedback loops existed which can account for the drug-induced increase in DA metabolism. These feedback loops can be categorized in several ways: local versus multi-synaptic, synthesis-modulating versus release-modulating, and rapid versus slow (Fig. 1).

Local versus multi-synaptic loops

Some of the regulation of DA synthesis and release occurs at the level of the terminal. For example, once DA is released into the synapse, it can influence dopaminergic activity by acting on receptors located on the terminal from which it was released. Slightly longer loops may also exist. For example, some evidence suggests a local DA-glutamate-DA negative feedback loop through which DA can inhibit glutamate release, thus reducing an inhibitory influence that glutamate can exert on the synthesis and release of DA (Cheramy et al., 1990; but see also Keefe et al., 1993). Although we have labeled these influences as *synaptic homeostasis* (Zigmond and Stricker, 1985), this may be a misnomer as non-synaptic communication may be involved (see below). Furthermore, additional feedback loops also exist, including a neuronal projection from neostriatum back to the level of the cell soma (Bunney, 1979), and input resulting from the release of DA from dendrites (Groves et al., 1975)

Synthesis modulating versus release modulating feedback loops

Although Carlsson focussed on DA release, DA synthesis is also subject to homeostatic regulation; indeed, under normal circumstances, the two processes

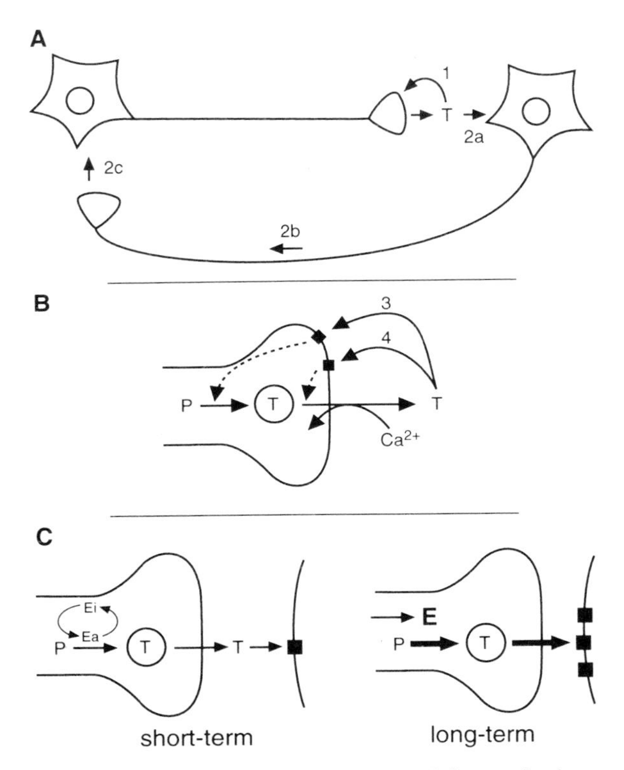

Fig. 1. Categories of feedback loops that can influence the dynamics of transmitter (T) release: (*A*) Local (1) versus multi-synaptic loops (2); (*B*) synthesis-modulating (3) versus release-modulating (4) influences; and (*C*) rapid (e.g. activation of enzyme activity from Ei to Ea, left panel) versus slow (e.g. induction of enzyme levels and synthesis of new receptors, right panel).

are coupled. Thus, whereas regulation of release presumably maintains some constancy in the availability of extracellular transmitter, regulation of synthesis ensures that intracellular transmitter stores are constant even if release is not. Both of these processes can be subject to local as well as long-loop regulation. For example, the synthesis of DA in the neostriatum is increased when impulse flow is elevated (Murrin and Roth, 1976) and when antagonist is applied to the presynaptic terminal (Nybäck and Sedvall, 1968).

Rapid versus slow feedback loops

The Carlsson experiments involved changes that occurred over minutes. However, we now know that the regulation of DA synthesis as well as the transduction of DA into a postsynaptic response via specific receptor proteins are also regulated via processes that develop over several days or weeks. For example, whereas tyrosine hydroxylase, the rate limiting enzyme in catecholamine biosynthesis, can be modulated by phosphorylation of the tyrosine hydroxylase molecule, it is also subject to genomic regulation via induction of new tyrosine hydroxylase protein (see Zigmond and Stricker, 1985).

Homeostatic regulation of neuronal interactions: a widespread phenomenon

Homeostasis is evident in the functioning of a variety of chemically defined neurons, including some that utilize norepinephrine, serotonin, and acetylcholine, both in the peripheral nervous system and in the brain (Zigmond and Stricker, 1985). The homeostatic property of these neuronal systems may have implications for many aspects of neurobiology (Fig. 2). Some examples: the capacity of a neuron to adjust its rate of release to the concentration of transmitter in the adjacent extracellular fluid may permit a few neurons present early in development to exert a disproportionately greater influence on brain function (Fig. 2*A*). Furthermore, the coupling of synthesis to release should enable a neuron to function over a broad range of firing rates without running out of transmitter or generating an excess (Fig. 2*B*). And, the availability of both short- and long-term processes for increasing the capacity for transmitter biosynthesis should permit neurons to respond rapidly to changing demands, and to make more permanent adjustments when the increased demand becomes long-lasting, thereby facilitating a rapid response to still greater challenges (Fig. 2*C*).

Functional response to brain damage

Over the past two decades, my colleagues and I have focussed on the implications of homeostasis of cell–cell interactions for the response to neuronal injury, in particular that associated with Parkinson's disease (for review see Zigmond and Stricker, 1989; Zigmond et al., 1993). We use an animal model of Parkinson's disease in which rats are treated with the neurotoxin 6-

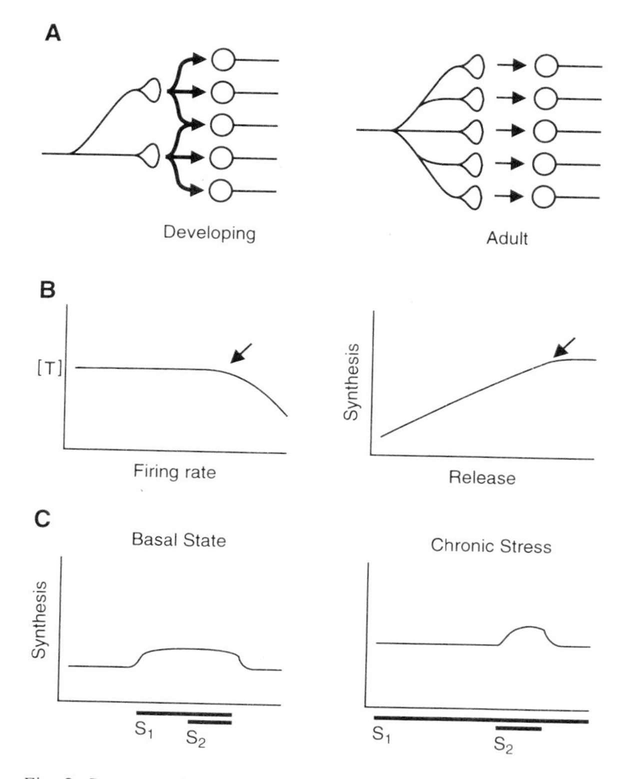

Fig. 2. Some possible implications of the homeostatic properties of neuronal systems: (*A*) during development a small number of initial "pioneer" neurons (left panel) might release transmitter at a higher rate so as to influence a wider field of target neurons than would be predicted by the density of innervation; (*B*) coupling synthesis to release should enable a neuron to function over a broad range of firing rates without running out of transmitter [T] or generating an xcess (until limits of regulation are exceeded, see arrow); and (*C*) short-term processes for increasing the capacity for transmitter biosynthesis (left panel) may permit neurons to respond rapidly to changing demands (S_1) but preclude an additional response to still further demands imposed simultaneously (S_2), whereas long-term processes (right panel) may allow more permanent adjustments when the increased demand becomes long-lasting (S_1), thereby facilitating a rapid response to still greater challenges (S_2).

hydroxydopamine (6-OHDA) in order to produce a selective loss of DA neurons within the brain. Most initial studies of centrally administered 6-OHDA did not detect gross neurological deficits despite the loss of up to 90% of striatal DA. This was curious since previous psychopharmacological studies indicated that disruption of dopaminergic transmission caused a marked reduction in behavioral output, whereas dopaminergic agonists caused behavioral activation. It was soon observed that more extensive bilateral damage to dopaminergic neurons of the nigrostriatal pathway did cause considerable behavioral dysfunction, including impairments of sensorimotor integration, akinesia, and aphasia. Moreover, more subtle behavioral testing indicated functional abnormalities even in partially lesioned animals showing no gross behavioral deficits: when these animals were exposed to a variety of physiological and environmental challenges they exhibited signs of akinesia.

Severely lesioned animals with gross initial neurological deficits often gradually recovered if maintained for a period by intragastric feedings. Why was this? And why were such large lesions required to produce initial deficits? These questions intrigue us because of their possible implications for neuroplasticity, and they also draw our attention to the parallel between this animal model and patients with Parkinson's disease, a disorder involving the degeneration of DA neurons of the nigrostriatal bundle which does not produce obvious clinical deficits until the loss of DA

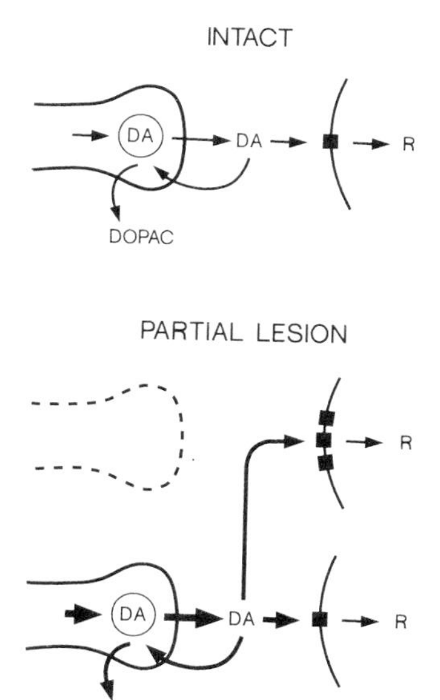

Fig. 3. A model for compensation to partial injury of a projection of dopamine (DA)-containing neurons. It is proposed that the field of influence of those neurons that remain after the lesion is increased due to an increase in the synthesis and release of DA, a decrease in the uptake of DA resulting from the degeneration of neighboring neurons, and an increase in the number of receptors present on denervated targets.

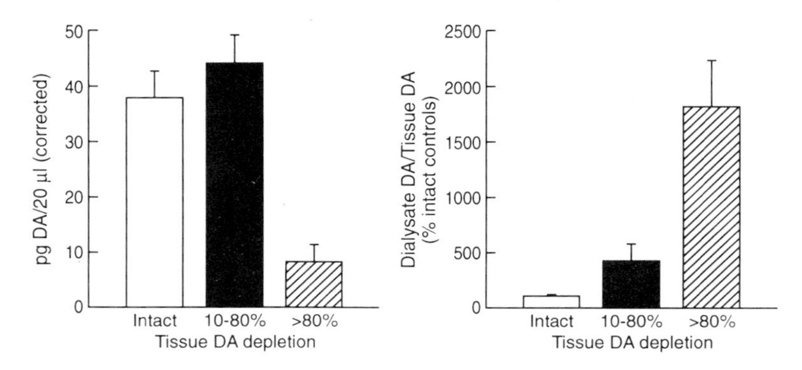

Fig. 4. Extracellular DA concentration in the neostriatum after partial destruction of DA-containing neurons (from Abercrombie et al., 1990). Left panel: The concentration of DA in extracellular fluid as measured by in vivo microdialysis versus the loss of DA in neostriatal tissue, a measure of the extent of terminal degeneration. Right panel: Extracellular DA expressed relative to tissue DA. These results suggest that in lesioned animals the DA terminals that remain are contributing a larger amount of DA to the extracellular pool than in control animals.

neurons is almost complete (Zigmond and Stricker, 1989).

We hypothesize that the capacity to withstand extensive neuronal loss within the dopaminergic nigrostriatal bundle as well as in other neuronal systems derives from homeostatic processes that regulate neuronal interactions. Specifically, we propose that after damage to the dopaminergic system, the remaining DA neurons compensate by increasing their capacity to synthesize and release transmitter (Fig. 3), and this appears to indeed be the case (Zigmond et al., 1993). For example, *in vivo* measurements using microdialysis indicate that the extracellular concentration of DA in striatum remains normal until 80% of the dopaminergic afferents to striatum have been destroyed (Fig. 4) (Abercrombie et al., 1990) and this is accompanied by indicators of increased DA turnover. A detailed analysis of the electrophysiological properties of the residual DA cells do not suggest any changes that might account for this observation: The average firing rate of the cells, the firing pattern, and the apparent number of active cells are within the normal range except under circumstances in which the loss of DA neurons is almost complete.

In contrast to the apparent absence of electrophysiological changes within residual DA neurons, neurochemical evidence suggests that the net amount of DA released per terminal in response to a given set of impulses is increased after the lesion (Fig. 5) (Stachowiak et al., 1987). This results from an increase in transmitter synthesis and release, and a decrease in transmitter deactivation by high affinity uptake. Several additional observations at both the behavioral and cellular levels support the hypothesis that these changes are compensatory, i.e. they maintain dopaminergic control over striatal function until the loss of DA neurons is extreme (Fig. 6) (Nisenbaum et al., 1986; MacKenzie et al., 1989).

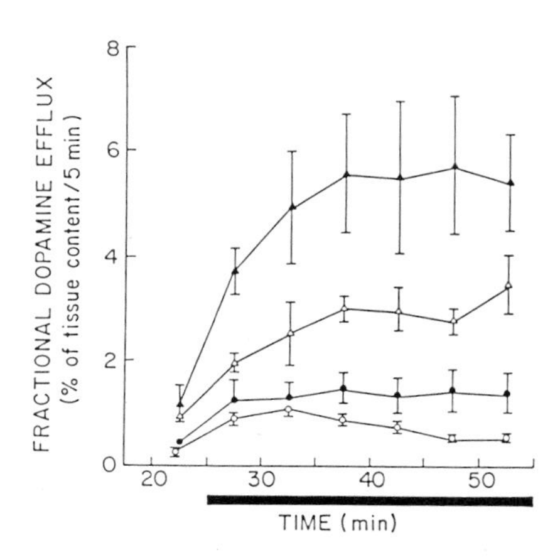

Fig. 5. Fractional DA overflow from neostriatal slices prepared from control (open circles) and rats lesioned with 6-hydroxydopamine so as to produce various levels of DA depletion (50–80%, closed circles; 80–96%, open triangles; and >96%, closed triangles). Slices were stimulated at 2 Hz for 30 min (bar) and endogenous DA overflow was expressed as a percent of the DA present in the slices at the outset of the experiment (from Stachowiak et al., 1987).

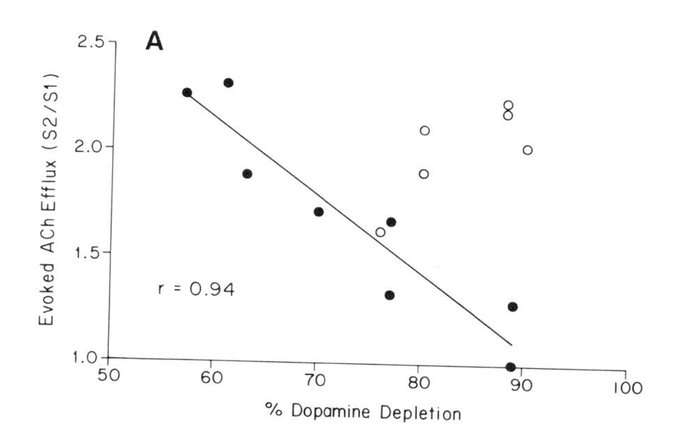

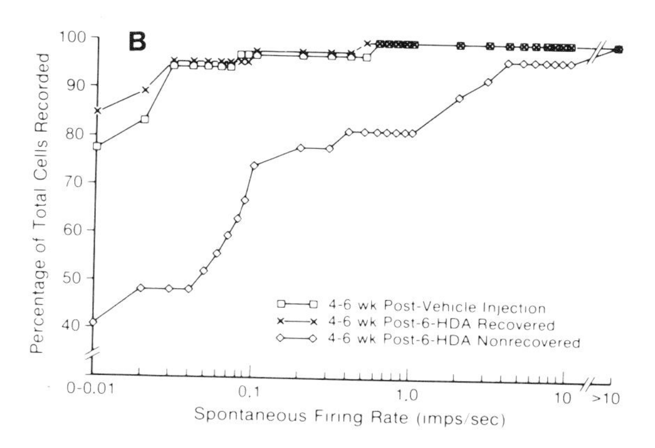

Fig. 6. Evidence that after partial lesions the apparent increase in the availability of DA from residual neurons serves a compensatory role. (*A*) Neostriatal slices were stimulated at 8 Hz for 1 min with standard buffer (S1) and in the presence of the DA receptor antagonist, sulpiride (10 μM; S2), and an estimate of the amount of ACh release was determined in each case. The ratio S2/S1 provides an index of the capacity of endogenous DA to exert an inhibitory influence over ACh release. Shown are results obtained 3 days (closed circles) and 2 months (open circles) after 6-OHDA-induced lesions. Decreases in S2/S1 below the control value of 2.0–2.5 indicate a loss of dopaminergic control (from MacKenzie et al., 1989). (*B*) The cumulative frequency distribution of spontaneous firing rate of a group of neurons in the medial neostriatum of control rats (squares) and rats treated 4–6 weeks earlier with 6-OHDA that have either recovered (crosses) or still show significant behavioral deficits (diamonds). The presence of a large number of neurons firing at relatively high frequencies in the last of these groups suggests the absence of an adequate supply of DA (from Nisenbaum et al., 1986).

Neuroregulatory deficits: lesions without neuropathology?

Consideration of the normal biochemical plasticity of catecholaminergic neurons suggests the existence of a family of disorders that result from neuroregulatory abnormalities caused by a failure of "synaptic homeostasis" (Fig. 7). For example, as discussed above, compensatory hyperactivity appears to allow a small number of DA neurons to perform the functions normally subserved by the full complement of these neurons. If so, it seems reasonable to further propose that deficits in the ability to invoke such compensations may exist in circumstances such as senescence, leading to accelerated rates of behavioral decline (Fig. 7*A*). Furthermore, might such conditions as affective disorders result from an inability of monoaminergic neurons to compensate properly for changes in demand, leading to too much or too little transmitter (Fig. 7*B,C*)? If so, perhaps much of pharmacotherapy actually represents attempts to restore homeostatic mechanisms.

It should be noted that if such neuroregulatory deficits occur in certain disease states, the attendant pathology would not be apparent during any conventional neuropathological examination, either upon imaging the brain or during postmortem analysis. Instead, such disorders would appear to have no organic basis and thus a new generation of tests would be needed to detect neurobiological abnormalities. For example, a deficiency in the autoregulation of transmitter release might require a functional assay, such as an examination of changes in transmitter release after autoreceptor blockade, or an analysis of the presence of the regulatory receptor protein itself.

Some additional implications

Neurochemistry versus electrophysiology

An important implication of the existence of local controls over transmitter release is that the influence of a neuron on its target cannot be fully estimated by monitoring its electrophysiological activity at the soma. Instead, local influences exerted at the terminal region can be expected to modify the impact of action potentials on transmitter release. Indeed, in the extreme case, local influences might actually *trigger* re-

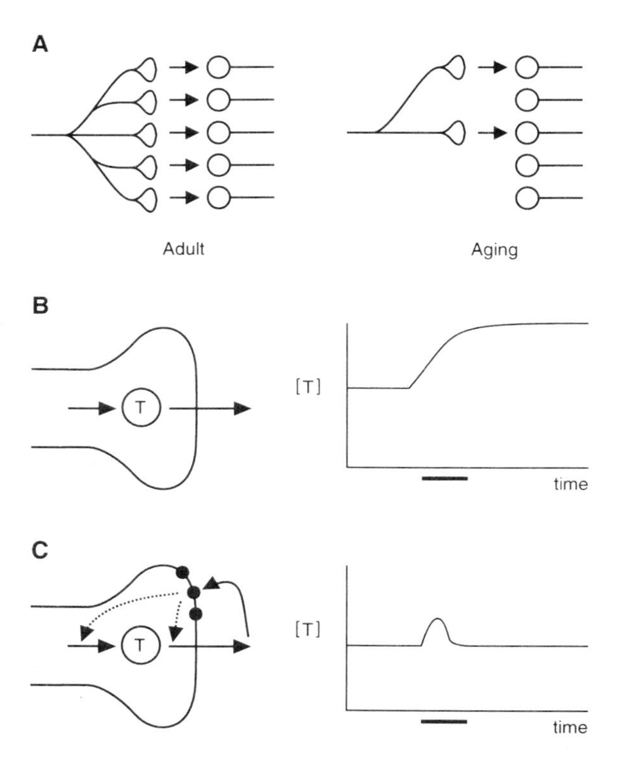

Fig. 7. Models for dysfunction resulting from a breakdown in synaptic homeostasis: (*A*) absence of compensation to the loss of neurons as might occur during senescence (compare with the adaptation postulated to occur during development, shown in Fig. 2*A*); (*B*) reduced compensation of transmitter synthesis and release during a period of increased activity (mania?) compared to the normal case (dashed line); and (*C*) overcompensation of transmitter synthesis and release during a period of increased activity (anhedonia?).

lease in the absence of impulses initiated at the cell soma. There are many examples of this phenomenon occurring at a pharmacological level and there is reason to believe that it can happen under physiological (or at least pathophysiological) conditions (Chesselet, 1984; Abercrombie and Zigmond, 1990). One value of such a scheme would be the conversion of a general system to a more specific one by allowing a cell with multiple terminal branches to deliver different amount of transmitter to different sites.

Neurochemistry versus morphology

A corollary of local regulation of transmitter release is that some influences between neurons can occur in the absence of conventional synapses. This is because many of the terminals which have receptors on them do not have any axo-axonic synapses. Thus, input to those receptors, if it exists, must occur through a route that is not detectable by conventional morphological criteria. This is hardly a new concept. *Non-synaptic, neurohumoral, paracrine,* and *volume transmission* are all terms that have been applied to interactions between neurons and their targets (Beaudet and Descarries, 1978; Cuello, 1983; Schmitt, 1984; Vizi, 1984; Fuxe and Agnati, 1991). Evidence for such interactions are well documented in the autonomic nervous system and invertebrates. Moreover, application of these concepts to the mammalian CNS has been proposed by many authors on the basis of a considerable body of evidence, including the apparent ability of neurons to influence distant targets (Vizi, 1984; Fuxe and Agnati, 1991), and differences between the distribution of some transmitters and their receptors (Herkenham, 1988) (but see also Bloom, 1991). Of course, if both types of transmission co-exist, the functions subserved by each must of necessity be quite different. Whereas classical synaptic transmission may transfer information from cell to cell with high fidelity and considerable privacy, non-synaptic transmission would be incapable of either of these tasks (Fig. 8).

Neurochemistry versus neurochemistry

No one methodological approach can yield a complete picture; neurochemistry is no exception. What do we measure when we examine "release"? It is not the amount of transmitter released into a synapse, as even the best collection procedure lacks the necessary temporal and the spatial resolution for such a determination. More importantly, virtually every synapse has mechanisms to reduce the spread of transmitter from the synaptic cleft; enzymes like acetylcholine esterase to rapidly degrade the transmitter, and proteins like those in the plasma membranes of monoaminergic neurons which return transmitter into the terminal. These processes presumably sharpen the communication between cells, extending the privacy of axonal conduction to synaptic transmission. It does not follow that *all* transmitter found in extracellular fluid is

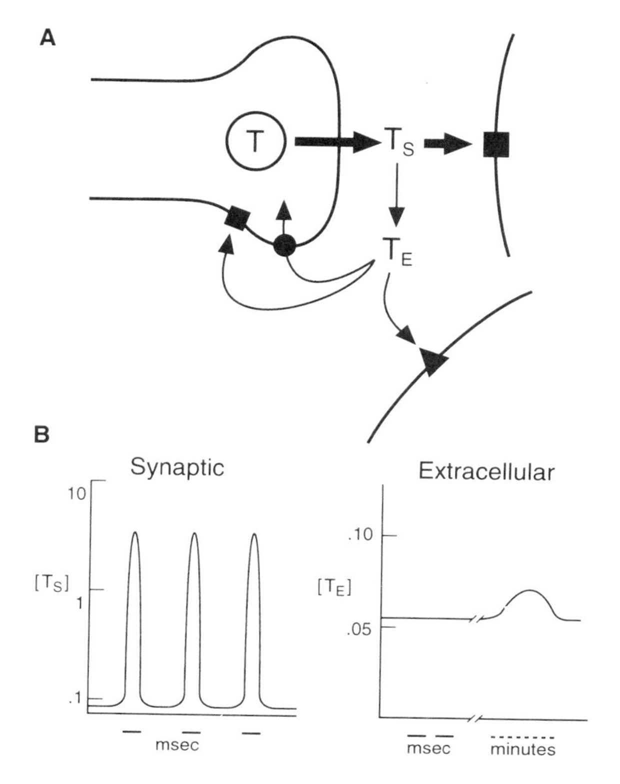

Fig. 8. Some distinctions between synaptic and extrasynaptic communication. Bottom, left: at a resting state the concentration of transmitter in a synaptic space (T_S) would be expected to be very low and to increase rapidly but transiently to a high level in response to an action potential (bars). Bottom, right: in contrast, in the resting state the concentration of transmitter in the surrounding extracellular fluid (T_E) also would be expected to be low, but no significant response to a single action potential would be expected and even extended trains of action potentials should only lead to small and slow changes in transmitter level. Ordinate shows arbitrary units of transmitter concentration. (Note difference in scale.)

merely a faint echo of the biologically significant event. As noted above, cells may communicate with each other through chemical signals that take other than synaptic routes and thus the extracellular pool of a transmitter may be biologically active. Nonetheless, it is clear that a great deal of information is lost by current methods for monitoring transmitter efflux (Finlay and Zigmond, 1994).

Directions for the future

Much has been learned in the 40 years that have elapsed since Carlsson hypothesized about feedback modulation of neuronal activity. Yet, as always, answers lead to questions:

1. Is transmitter release *really* regulated and if so why? Does this not subvert the objective of chemical transmission, which is to respond faithfully to the electrical impulses that are received from the soma?
2. Do electrophysiological and anatomical methods for evaluating cell–cell interactions fail to disclose critical information about these events? Is transmitter release at the terminal not always reflected by the electrophysiological activity measured in the soma? Do neurons have a field of influence beyond those seen in electronmicrographs? If so what type of information might be transmitted in this non-conventional way?
3. Do deficits of homeostasis occur at the level of chemical communication and do they underlie some of the many psychiatric disorders (depression, anxiety) and neurological disorders (dystonia, Tourettes Syndrome) for which no clear neuropathology is evident? If so, can such regulatory dysfunctions be detected and repaired?

Answers to such questions will require the applications of multiple approaches to the study of neuronal interactions, as well as an increasing dialogue between basic and clinical investigators.

Acknowledgements

Many colleagues have contributed to the research and concepts summarized in this review. Special thanks go to Elizabeth D. Abercrombie, Anthony A. Grace, Janet M. Finlay, and Edward M. Stricker. Thanks also to Beth A. Vojta for her assistance in preparing the manuscript and Blaine Walker for his help in preparing the figures. This work was supported in part by U.S. Public Health Service Grant NS19608, MH43947, MH29670, MH45156 and MH00058.

References

Abercrombie, E.D., and Zigmond, M.J. (1990) Striatal dopamine release: In vivo evidence for local initiation. *Ann. N. Y. Acad. Sci.*, 604: 575–578.

Abercrombie, E.D., Bonatz, A.E. and Zigmond, M.J. (1990) Effects of L-DOPA on extracellular dopamine in striatum of normal and 6-hydroxydopamine-treated rats. *Brain Res.*,. 525: 36–44.

Beaudet, A. and Descarries, L. (1978) The monoamine innervation of rat cerebral cortex: synaptic and non synaptic axon terminals. *Neuroscience,* 3: 851–860.

Bloom, F.E. (1991) An integrative view of information handling in the CNS. In: K. Fuxe and L.F. Agnati (Eds.), *Advances in Neuroscience, Volume 1: Transmission in the Brain: Novel Mechanisms for Neural Transmission*, Raven Press, New York, pp. 11–23.

Bunney, B.S. (1979) The electrophysiological pharmacology of midbrain dopaminergic systems. In: A.S. Horn, J. Korf and B.H.C. Westerink (Eds.), *The Neurobiology of Dopamine*, Academic Press, New York, pp. 417–452.

Carlsson, A. and Lindqvist, J. (1963) Effect of chlorpromazine and haloperidol on formation of 3-methoxytyramine and normetanephrine in mouse brain. *Acta Pharmacol. Toxicol.* 20: 140–144.

Carlsson, A., Kehr, W., Linqvist, M., Magnusson, T., and Atack, C.V. (1972) Regulation of monoamine metabolism in the central nervous system. *Pharmacol. Rev.*, 24: 371–384.

Cheramy, A., Barbeito, L., Godeheu, G., Desce, J.M., Pittaluga, A., Galli, T., Artaud, F., and Glowinski, J. (1990) Respective contributions of neuronal activity and presynaptic mechanisms in the control of the in vivo release of dopamine. *J. Neural. Trans.*, 29: 183–193.

Chesselet, M.-F. (1984) Presynaptic regulation of neurotransmitter release in the brain: facts and hypothesis. *Neuroscience*, 12: 347–375.

Creese, I., Burt, D.R. and Snyder, S.H. (1978) Biochemical actions of neuroleptic drugs: focus on the dopamine receptor. In: L. Iversen, S.D. Iversen, and S.H. Snyder (Eds.), *Handbook of Psychopharmacology, Volume 10: Neuroleptics and Schizophrenia*, Plenum Press, New York.

Cuello, A.C. (1983) Nonclassical neuronal communications. *Fed. Proc.*, 42: 2912–2922.

Finlay, J. and Zigmond, M.J. (1994) A critical analysis of neurochemical methods for monitoring transmitter dynamics in brain. In: F.E. Bloom and D. Kupfer (Eds.), *Psychopharmacology: Fourth Generation of Progress*, Raven Press, New York, in press.

Fuxe, K. and Agnati, L.F. (1991) Two principal modes of electrochemical communication in the brain: Volume versus wiring transmission. In: K. Fuxe and L.F. Agnati (Eds.), *Advances in Neuroscience, Volume 1: Transmission in the Brain: Novel Mechanisms for Neural Transmission*, Raven Press, New York, pp. 11–23.

Groves, P.M., Wilson, C.J., Young, S.J. and Rebec, G.Y. (1975) Self-inhibition by dopaminergic neurons. *Science,* 190: 522–529.

Herkenham, M. (1987) Mismatches between neurotransmitter and receptor localizations in brain: observations and implications. *Neuroscience*, 23: 1–38.

Keefe, K.A., Zigmond, M.J. and Abercrombie, E.D. (1993) In vivo regulation of extracellular dopamine in the neostriatum: influence of impulse activity and local excitatory amino acids. *J. Neural Transm.*, 91: 223–240.

MacKenzie, R.G., Stachowiak, M. and Zigmond, M.J. (1989) Dopaminergic inhibition of striatal acetylcholine release after 6-hydroxydopamine. *Eur. J. Pharmacol.*, 168: 43–52.

Murrin, R.H. and Roth, R.H. (1976) Dopaminergic neurons: effects of electrical stimulation on dopamine biosynthesis. *Mol. Pharmacol.*, 12: 463–475.

Nisenbaum, E.S., Stricker, E.M., Zigmond, M.J. and Berger, T.W. (1986) Long-term effects of dopamine-depleting brain lesions on spontaneous activity of Type II striatal neurons: relation to behavioral recovery. *Brain Res.*, 398: 221–230.

Nybäck, H. and Sedvall, G. (1968) Effect of chlorpromazine on accumulation and disappearance of catecholamines formed from tyrosine-14C in brain. *J. Pharmacol. Exp. Ther.*, 162: 294–301.

Schmitt, F.O. (1984) Molecular regulators of brain function: a new view. *Neuroscience.*, 13: 991–1001.

Stachowiak, M.K., Keller, R.W. Jr., Stricker, E.M., and Zigmond, M.J. (1987) Increased dopamine efflux from striatal slices during development and after nigrostriatal bundle damage. *J. Neurosci.*, 7: 1648–1654.

Vizi, E.S. (1984) *Non-Synaptic Interactions Between Neurons: Modulation of Neurochemical Transmission.*, Wiley, New York.

Zigmond, M.J. and Stricker, E.M. (1985) Adaptive properties of mono-aminergic neurons. In: A. Lajtha (Ed.), *Handbook of Neurochemistry*, Vol. 9, Plenum Press, New York, pp. 87–102.

Zigmond, M.J. and Stricker, E.M. (1989) Animal models of Parkinsonism using selective neurotoxins: Clinical and Basic Implications. In: J.R. Smythies and R.J. Bradley (Eds.), *International Review of Neurobiology*, Vol. 31, Academic Press, New York, pp. 1–79.

Zigmond, M.J., Abercrombie, E.D., Berger, T.W., Grace, A.A. and Stricker, E.M. (1993) Compensatory responses to partial loss of dopaminergic neurons: studies with 6-hydroxydopamine. In: J.S. Schneider and M. Gupta (Eds.), *Current Concepts in Parkinson's Disease Research*, Hogrefe & Huber, Toronto, pp. 99–140.

F. Bloom (Editor)
Progress in Brain Research, Vol. 100

CHAPTER 16

Conservation of basic synaptic circuits that mediate GABA inhibition in the subcortical visual system

R. Ranney Mize

Department of Anatomy and the Neuroscience Center, Louisiana State University Medical Center, 1901 Perdido Street, New Orleans, LA 70112, USA

Introduction

Inhibition is fundamental to visual function in the subcortical visual system (SVS). Inhibitory mechanisms are thought to be involved in both spatial and temporal aspects of visual processing. In the spatial domain, inhibition is thought to contribute to center-surround antagonism, to movement and directional selectivity, to binocular inhibition, and to spatial frequency tuning of the receptive fields of single neurons. In the temporal domain, inhibition is thought to play a role in shaping the transient responses of single cells and in controlling levels of adaptation under different illumination conditions. Inhibitory processing is thus involved in a multitude of visual functions (reviewed in Mize et al., 1992).

The inhibitory neurotransmitter, gamma-aminobutyric acid (GABA), is the chief and often the sole neurotransmitter generating intrinsic inhibition in the SVS. Accordingly, GABA inhibition has been invoked to explain each of the above functions. In this chapter, the anatomical organization and physiological processes that underlie GABAergic inhibition in the SVS are reviewed, focusing upon the cell types and synaptic circuitry that have been identified in the SVS. What is known of the physiological actions of GABA in these regions is described briefly. There are three basic types of GABA containing synaptic profile in the SVS. These are relatively homogeneous and form similar synaptic circuits in most of the structures examined. GABA neurons and circuits may therefore be common to a variety of structures and species and represent a highly conserved feature of brain development.

Heterogeneity of SVS function and GABA cell types

The cell types, receptive field characteristics and global functions of the SVS are quite heterogeneous. The lateral geniculate nucleus (LGN) is involved principally in the state-dependent gating of signal transmission from the retina to visual cortex. Retinorecipient nuclei of the pretectum also vary in function. The nucleus of the optic tract (NOT) helps to control the stabilization of gaze and optokinetic nystagmus. The olivary pretectal nucleus (OPN) controls the pupillary light reflex. The principal site of retinal termination in the midbrain, the superior colliculus, is involved in visual attention, detection of moving objects in the visual field and in the generation of saccadic eye movements. SVS functions are therefore remarkably diverse. (for review, see Sherman and Spear, 1982).

The receptive field properties of cells in these regions is equally diverse. Receptive fields in the LGN have a center-surround organization, and may be on or off center and have X, Y, or W receptive fields. The proportion of these cells varies in different regions of the cat LGN. Both X and Y cells are commonly found in the A layers of the dLGN while the parvocellular C layers contain exclusively W cells. The medial interlaminar nucleus (MIN) contains primarily Y cells, while the ventral LGN contains mostly W cells. Cells in the NOT are directionally selective and contribute to optokinetic afternystagmus. OPN cells respond to changes in ambient illumination. Cells within the superior colliculus receive primarily W input from the retina with some Y input located in deeper laminae. Many are directionally selective with silent inhibitory surrounds. Others in the deep layers are multimodal and/or respond in association with saccades (for review, see Sherman and Spear, 1982).

The cells that contain GABA in these SVS structures are correspondingly heterogeneous. The principal local circuit interneuron in the A layers of the dLGN is a type 3 neuron with a small cell body and an exquisite tangle of thin dendrites that form an asymmetric vertically oriented field. These dendrites have complex appendages that contain pleomorphic synaptic vesicles. These cells represent about 25% of all neurons in the A layers (Guillery, 1969; Hamos et al., 1985; Uhlrich and Cucchiaro, 1992). Other GABA containing interneurons in the dLGN are interlaminar cells with larger cell bodies, somatic spines, and stout dendrites that cross laminar boundaries and may not receive retinal input (Montero, 1989). Neurons above the dLGN within the perigeniculate nucleus (PGN) and the thalamic reticular nucleus (TRN) often have horizontally distributed dendrites and axons that dip into the dLGN. Small cells containing GABA are also found throughout the parvocellular C laminae, MIN, and the vLGN (for review, see Uhlrich and Cucchiaro, 1992).

The superior colliculus also has a variety of GABA containing neurons that are about 25–40% of the total population. At least three GABAergic cell types have been described in cat: horizontal cells have oblong, fusiform cell bodies and two stout horizontally oriented presynaptic dendrites that course parallel to the pial surface. They receive limited numbers of synapses, most from visual cortex. Granule 1 type cells are small cells, often with vertically oriented dendrites, that have moderate synaptic input and a definitive axon. Granule 2 neurons also receive moderate synaptic input and have both somatic and dendritic spines that contain synaptic vesicles (Mize, 1992).

The GABA cell types of the OPN and NOT are less well characterized but consist of both small and medium sized neurons with variable morphologies. They may include both interneurons and projection cells (cf. Campbell and Lieberman, 1985; Van der Want et al., 1992).

Homogeneity of GABA synaptic ultrastructure

Despite the heterogeneous and distinctive cell types found to contain GABA, the ultrastructure of synapses established by these cells is surprisingly homogeneous within each of the retinorecipient zones studied. Two and often three GABAergic synaptic types have been identified consistently in the dorsal and ventral lateral geniculate nuclei, medial intralaminar nucleus, two retinorecipient nuclei of the pretectum and the superior colliculus.

The P1 presynaptic dendrite

The best studied of the GABAergic synaptic profiles is a presynaptic dendrite that is one component of the retinal triad. Retinal triads consist of a retinal synaptic terminal (called an R or RT or RLP), a conventional dendrite (D) from a relay cell, and an intermediate synaptic element (referred to as a P or PSD or F2 profile) that is postsynaptic to the RT but presynaptic to the D (Fig. 1). In the dLGN, the retinal triad is surrounded by a glial sheath, which anatomically and functionally segregates it within a structure called a glomerulus. The triad also receives input from acetylcholine fibers from the brainstem and GABA containing fibers from the perigeniculate nucleus (Fig. 1) (Sillito, 1992).

Retinal terminals in these triads contain large, round synaptic vesicles and mitochondria with a pale matrix and broken up cristae (Figs. 2–4). In the A layers of the dLGN, the RT is often the central element with a scallop shape produced by indentations from dendrites that surround it (Fig. 2*A*). The intermediate presynaptic dendrite, called an F2 in cat LGN (Guillery, 1969), contains a pleomorphic population of round, ovoid, and occasionally flattened synaptic vesicles that are almost always loosely scattered

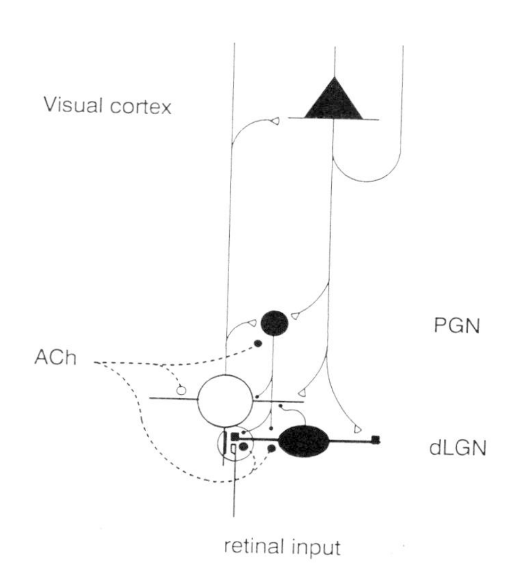

Fig. 1. Synaptic circuitry of the dorsal lateral geniculate nucleus (LGN), illustrating the relationship between retinal afferents, relay cell dendrites and GABA containing interneurons. The units of the retinal triad are outlined by the circle, and include a retinal terminal (open rectangle), a presynaptic dendrite (black square), and a relay neuron (black line), as well as an inhibitory synapse from acetylcholine (ACh) containing synapses from the brainstem. A perigeniculate (PGN) neuron is also shown providing feed backward inhibition to the dLGN (from Sillito, 1992).

through the profile (Fig 2*A*). The triad is a triplicate synapse in which the RT makes synaptic contact with both the D and F2 profiles, while the F2 also contacts the D profile, thus forming a triangular synaptic arrangement (the "triad", Fig. 1).

F2s in dLGN are invariably inhibitory. They are labeled by antibodies to GABA and GAD (glutamic acid decarboxylase, the synthetic enzyme for GABA), contain at least a few distinctively flattened vesicles, and establish symmetric synaptic contacts, all typical features of inhibitory synapses. In the LGN, these profiles can be traced back to their dendrites of origin and are found to be the dendritic spine-like appendages of one class of dLGN interneuron (Hamos et al., 1985).

Synaptic triads between RTs, presynaptic dendrites, and Ds are also found in every other retinorecipient zone of the SVS studied. Presynaptic dendrites receiving retinal input and containing sparse accumulations of pleomorphic synaptic vesicles are found in MIN (Fig. 2*B*), vLGN (Fig. 2*C*,*D*), parvocellular C laminae of the dLGN (Fig. 2*E*), the nucleus of the optic tract (Figs. 3*A*), the olivary pretectal nucleus (Figs. 3*B*,*C*), and the superior colliculus (Fig. 4*A*).

The intermediate elements of these triads, which for consistency are subsequently called P1 profiles, are labeled by antibodies to GABA and establish symmetric synaptic contacts, usually with D profiles. In the cat and monkey superior colliculus, the P1s have also been shown to be the spine-like appendages of GABA containing granule cells (Mize, 1992) and may be spine-like dendritic appendages in the OPN of rat as well (Campbell and Lieberman, 1985).

Although the basic triad synapse and the P1 profile have been identified in every SVS structure examined (Figs. 2–4), their frequency and morphology do differ for different relay cell types and in different nuclei. Triads are frequently associated with X cells in the dLGN, much less frequently with Y cells (where they are estimated to comprise only 10% of the input) (see Uhlrich and Cucchiaro, 1992). Triads are common in MIN, OPN and SC, less common in vLGN, the parvocellular C layers of dLGN, and the NOT. In the vLGN and parvocellular C layers, RTs form synaptic contacts with only about one-third the number of F2 profiles that are contacted by RTs in the dLGN and MIN (Mize and Horner, 1984; Mize et al., 1986). Only about 5–10% of RTs contacting W cells in the parvo C layers form triadic synapses (Raczkowski et al., 1988), a percentage much smaller than that for X cells in the dLGN (see Uhlrich and Cucchiaro, 1992).

The size of triadic profiles also varies in different regions of the SVS. In SC, for example, retinal terminals are often small and P1s are accordingly of smaller diameter than those seen in the dLGN (Fig. 4*A*). The size of RTs and P1s in both the vLGN and the parvo C layers of dLGN is often less than half that of those in the dLGN and MIN (Figs. 2*C*–*E* versus Figs. 2*A*,*B*) (Mize and Horner, 1984; Mize et al., 1986).

Despite these differences in frequency and size, the basic triadic arrangement of the synaptic complex is apparent throughout much of the SVS. In addition, the morphology of the P1 is quite similar in all structures examined. It contains a mixed, pleomorphic population of synaptic vesicles that have a characteristic sparse distribution throughout the profile (Figs. 2–4).

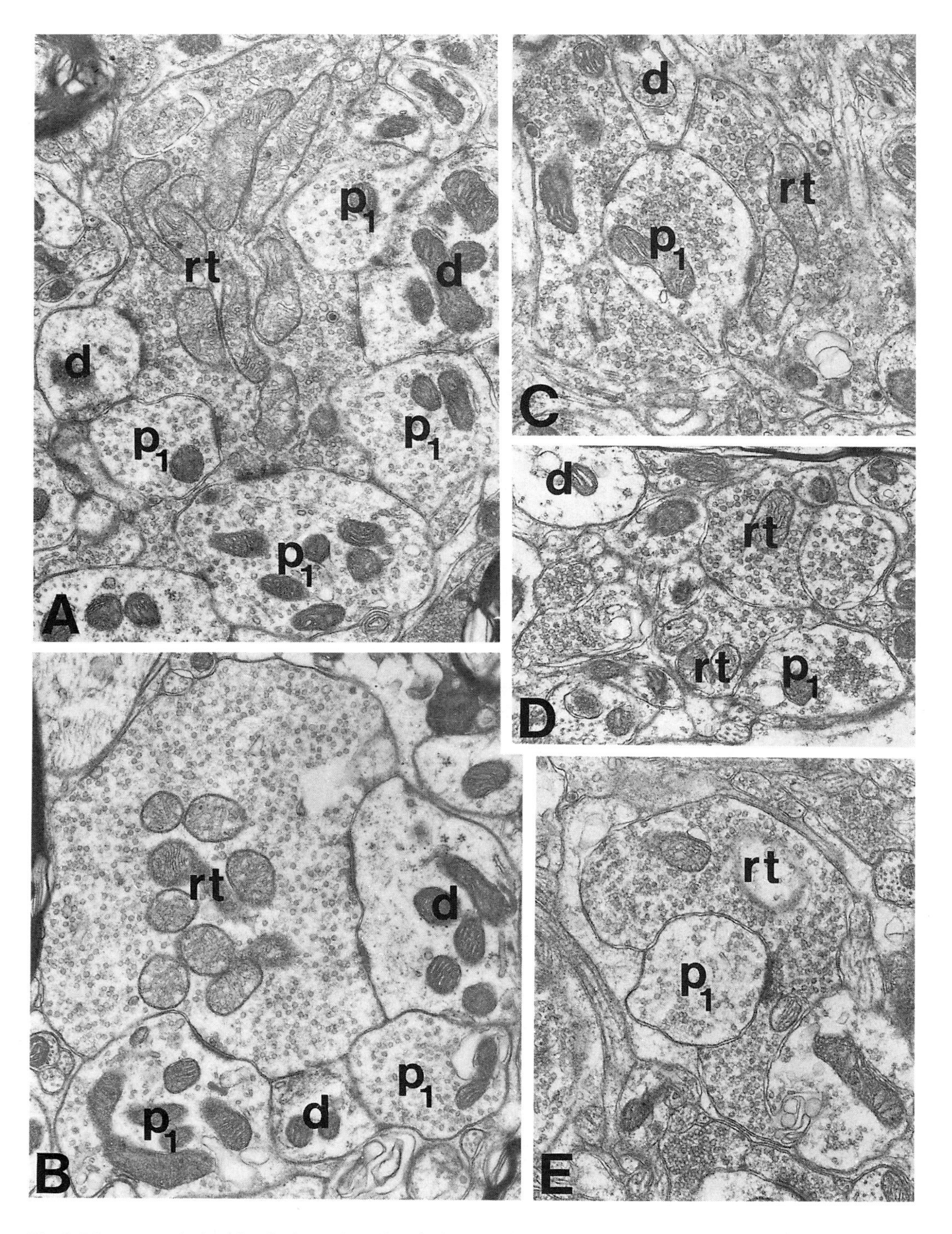

Fig. 2. P1 presynaptic dendrites in the cat lateral geniculate complex. *A*. A layer of the dorsal lateral geniculate nucleus; *B*. medial interlaminar nucleus; *C*,*D*, ventral lateral geniculate nucleus; *E*, parvocellular C laminae of the dLGN. Retinal terminals (rts) contact these P1s in each structure. Conventional dendrites (d) are also found.

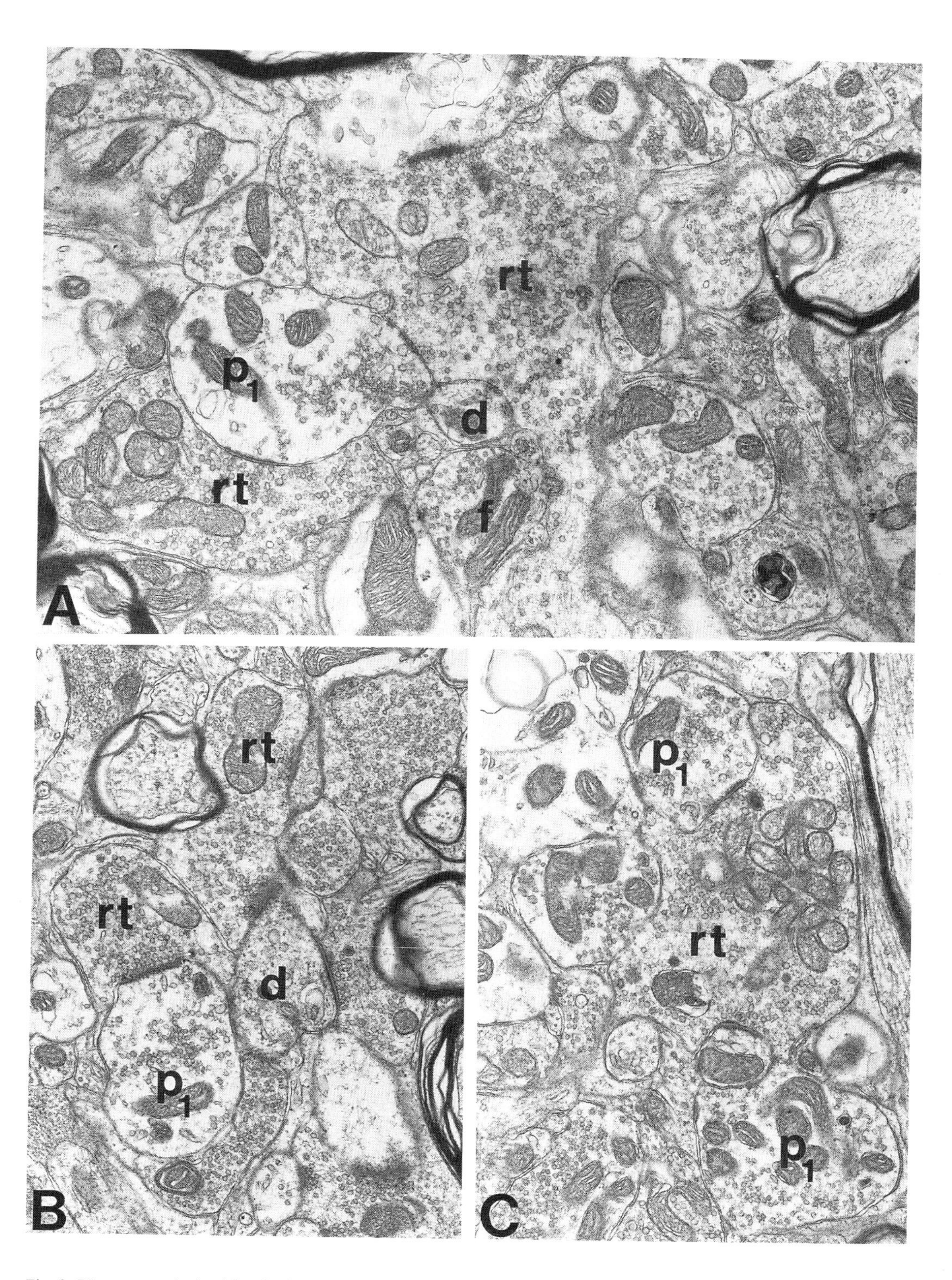

Fig. 3. P1 presynaptic dendrites in the cat pretectum. *A*, olivary pretectal nucleus; *B,C*, nucleus of the optic tract.

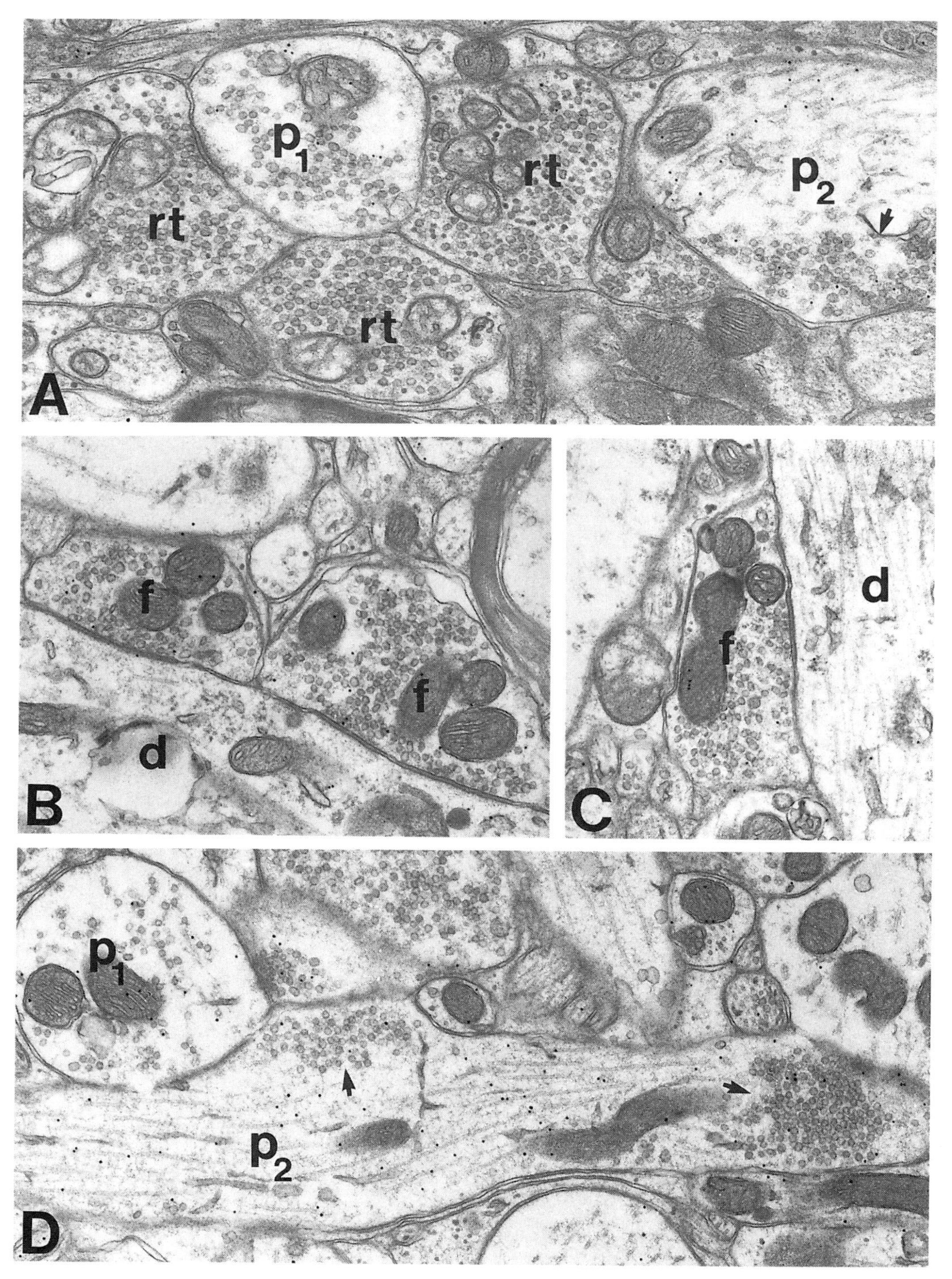

Fig. 4. GABA containing profiles in the rabbit superior colliculus. *A*, P1 presynaptic dendrite surrounded by retinal terminals (rts). *B*,*C*, F type axon terminals; *D*, P2 presynaptic dendrite. The profiles were labeled by colloidal gold-conjugated GABA antibodies using post-embedding immunocytochemistry.

In fact, the loose clustering of the synaptic vesicles is one "signature" of these profiles that distinguishes them from other GABA containing profiles described below. The P1 and its associated triad is therefore a ubiquitous and fundamental GABAergic circuit in the SVS.

The F axon terminal

A second major class of GABA synapse is formed by F profiles. F profiles are called F1s in dLGN (Guillery) and usually F profiles in other SVS structures (OPN, NOT, SC). F profiles contain pleomorphic synaptic vesicles, some of which are dramatically flattened. These vesicles are densely and uniformly distributed within a dark cytoplasmic ground (Figs. 4*B*,*C*). They form symmetric synaptic contacts that have thin post-synaptic densities, often with an apposition area greater than that of P1s (Figs. 4*B*,*C*). Fs are easily distinguished from P1 profiles both because of the numbers and density of flattened synaptic vesicles and because they are always presynaptic, not postsynaptic, to other structures. F profiles arise from a variety of extrinsic sources. In the dLGN, they are the terminal axons of neurons in the perigeniculate nucleus, the thalamic reticular nucleus, and the pretectum (Uhlrich and Cucchiaro, 1992). In SC, F profiles in the deep layers arise from the substantia nigra, zona incerta, the pretectum, and a number of brainstem nuclei (Mize, 1992). In the NOT, they arise from the medial terminal nucleus (MTN) of the accessory optic system and probably also from SC (Van der Want et al., 1992). F type terminals also are thought to arise intrinsically from interneurons because some GABAergic local circuit neurons in the LGN, SC and NOT have axons.

The terminal morphology of F profiles is quite homogeneous. F axon terminals have simpler synaptic arrangements than P1s and they are sometimes stacked along the surfaces of proximal and distal dendrites. They rarely participate in serial synapses and are rarely found within glomeruli. There is evidence that F axon terminals are selectively associated with a subunit of the GABAa/benzodiazepine receptor in both the dLGN (Soltesz and Crunelli, 1992) and SC (Mize, 1992).

P2 presynaptic dendrites

The third profile type that contains GABA in the SVS is a second class of presynaptic dendrite, called a P2 in this review. P2s have been most thoroughly characterized in the cat SC. P2s are large calibre dendrites that have small, discrete clusters of densely packed synaptic vesicles often distributed at regular intervals along the dendrite (Fig. 4*D*) (Mize, 1992). These dendrites have been termed H profiles in SC because reconstructions in cat reveal that they are the dendrites of horizontal cells (Mize, 1992). P2 profiles can be differentiated from P1 profiles by the diameter of the dendrite and the compact but densely packed clustering of their synaptic vesicles. Reconstructions of retinal synaptic islands in monkey SC have shown that P2s receive synaptic input from both P1 and F synapses (Mize, 1992), and thus participate as postsynaptic elements in GABA upon GABA synaptic circuits.

P2 profiles also have been identified in the cat dLGN (Hamos et al., 1985), the tree shrew LGN (called F3 profiles) (see Uhlrich and Cucchiaro, 1992), and in the rabbit NOT (Nunes Cardozo et al., 1993). In cat LGN, they are dendritic shafts with punctate clusters of vesicles that receive mostly Y retinal input and contact Y relay cells (see Uhlrich and Cucchiaro, 1992). Large calibre dendrites with punctate clusters of pleomorphic synaptic vesicles have also been illustrated in the NOT (Fig. 4C of Nunes-Cardozo et al., 1994). Whether this profile type is present in other SVS structures has not yet been established.

Physiological actions of GABA in the SVS

GABA interneurons have been implicated in a variety of inhibitory receptive field properties, including surround antagonism, directional selectivity, movement sensitivity, binocular inhibition and length tuning (see Sillito, 1992, for review). Because the receptive field properties of cells in the LGN, pretectum and SC vary so dramatically in type, the presence of the same three GABAergic synaptic units in all of these regions was unexpected. Recently, however, it has been argued that intrinsic GABA circuits are not involved in directional

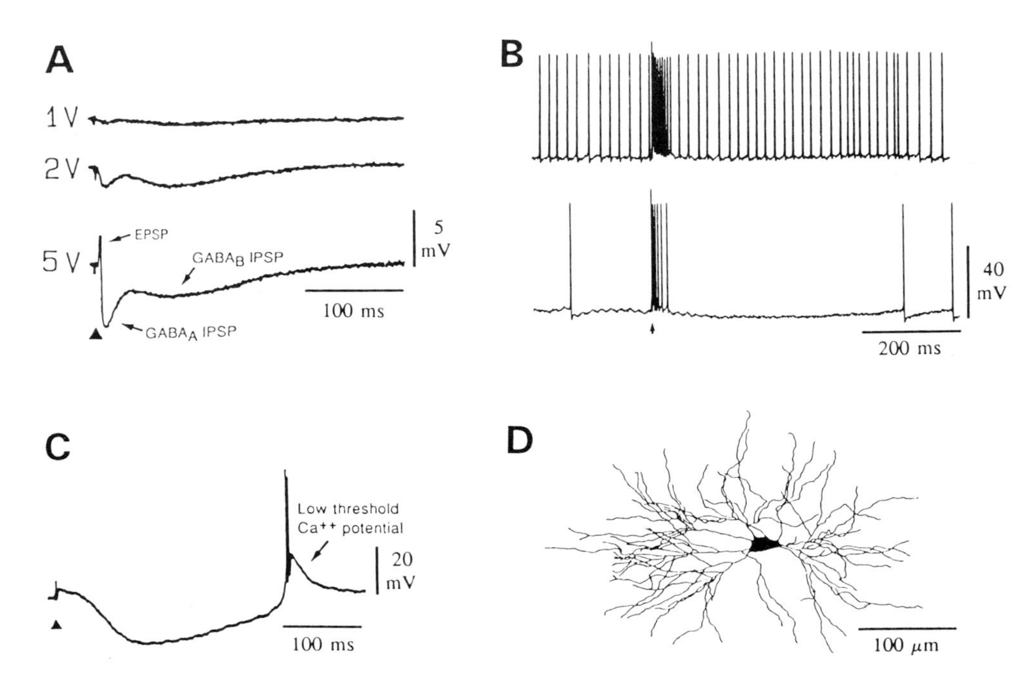

Fig. 5. IPSPs recorded in vitro from relay neurons in the dorsal lateral geniculate nucleus. *A*, Intracellular voltage traces illustrate a GABAa IPSP of short latency and duration and a GABAb IPSP of longer latency and duration. *B*, Intracellular traces showing that the GABAb IPSP has little effect on discharge when the cell is depolarized to –35 mV. *C*, A low threshold Ca^{2+} spike generates a burst of action potentials even after GABAa is blocked by bicuculline. *D*, Drawing of an HRP filled relay cell in layer *A* from which such recordings were obtained (from Soltesz and Crunelli, 1992).

selectivity in cat SC (Mize, 1992) or rabbit NOT (Van der Want et al., 1992) nor in surround antagonism nor binocular inhibition in the LGN (Norton and Godwin, 1992). The GABAergic circuits common to the SVS may therefore be involved in modulating cell membrane conductances rather than in shaping the receptive field properties of these cells. Modulation of membrane conductance could be common to many neurons in the SVS.

In vitro slice experiments illustrate at least two and probably three types of GABA-mediated inhibition in the lateral geniculate nucleus (Fig. 5) (reviewed by Soltesz and Crunelli, 1992). Optic tract stimulation induces two hyperpolarizing potentials. The first has a short latency onset, peak and duration, increases conductance in a chloride channel, produces a large decrease in input resistance, is sensitive to muscimol, is reversibly blocked by bicuculline, and is thus a GABAa receptor mediated event (Fig. 5). The second has a longer onset, peak and duration, increases conductance of a potassium channel, produces less decrement in input resistance, is baclofen sensitive, is reversibly blocked by phaclofen and saclofen, and is thus a GABAb receptor mediated event (Fig. 5). A third IPSP is a "miniature" GABAa hyperpolarization that has been identified only in other thalamic nuclei. GABA can also evoke a fast depolarization at certain resting membrane potentials (Soltesz and Crunelli, 1992).

GABAa and GABAb hyperpolarizing potentials have also been recorded in the superior colliculus (see Arakawa and Okada, 1988; Kraszewski and Grantyn, 1992) and both GABAa and GABAb receptors have been identified by receptor autoradiography and antibody immunocytochemistry throughout the dLGN and the superior colliculus (for reviews, see Soltesz and Crunelli, 1992; Mize, 1992).

The role of the P1, F and P2 synaptic units in these receptor mediated processes is uncertain. The P1 profile of the retinal triad is thought to mediate feed forward inhibition in which the P1, activated by retinal input, hyperpolarizes the same D element that has been transiently depolarized by the RT. The inhibition thus feeds forward onto the output neuron. The P1 feedforward synapse is positioned spatially to inhibit retinal excitation after a brief synaptic delay.

Sherman and Koch (1986) suggest that the retinal triad mediates a shunting inhibition which increases the conductance and decreases the input resistance of the membrane only at the site of the triad. Consistent with this hypothesis, Bloomfield and Sherman (1988) have shown that the dendrites of interneurons that form F2 (i.e. P1) synapses conduct current poorly because their processes are thin relative to their length. They are thus electrically isolated units that do not effect the membrane potential of the soma. This type of shunting inhibition is thought to be mediated by postsynaptic GABAa receptors that produce an IPSP with minimal hyperpolarization of the membrane (Sherman and Koch, 1986).

F axon terminals are thought to mediate feed backward inhibition in which the output neuron feeds back onto its input, either directly or indirectly, to modify that input. Both extrinsic and intrinsic GABAergic axons may feed back onto relay cells in dLGN (Uhlrich and Cucchiaro, 1992). F axon terminals probably also act on GABAa receptors, but they may also act at the GABAb receptor. GABAb inhibition is less localized and can cancel EPSPs at some distance from the site of synapse (Sherman and Koch, 1986). GABAb inhibition in the LGN has a longer onset and protracted duration (200–300 ms) and could provide tonic inhibition to dLGN cells. GABAb receptors have also been implicated in low threshold Ca^{2+} spikes (Fig. 5) and the associated burst firing that occurs in the LGN when the membrane is hyperpolarized (for review, see Soltesz and Crunelli, 1992).

The role of the P2 profile is entirely unknown. In SC, P2s are the thick calibre, primary dendrites of horizontal cells that extend for up to 2–4 mm along the surface of that structure (Mize, 1992). Because of the extensive distribution of their dendrites, they have been implicated in lateral and/or remote inhibition that occurs in SC (for review, see Mize, 1992). The large calibre and smooth contour of their dendrites suggests that P2s could behave electrotonically like the dendrites of dLGN relay cells. Modeling studies show that these dendrites are electrically compact, have low impedances, and allow current to spread for long distances without decrement (Bloomfield and Sherman, 1989). For the horizontal cell, input at one site along the dendrite could release transmitter at synaptic sites all along the P2 dendrite.

Conclusions

SVS structures have two and usually three basic GABAergic synaptic units: F axon terminals and two types of presynaptic dendrites (P1 and P2s). Despite differences in frequency and size, the synaptic "signatures" that distinguish these three units are remarkably homogeneous throughout the SVS. F profiles have distinctively flattened, densely packed vesicles in a compact bouton. The "signature" of the P1 profile includes its loose and uniformly dispersed synaptic vesicles and its intermediate position in retinal triads. The P2 is equally distinctive in the compact clustering of its synaptic vesicles and the smooth, large calibre of its dendrites.

As the receptive field properties and global functions of these regions are so different, it is possible to conclude that these common inhibitory synaptic units underlie fundamental membrane and channel properties that are found in many relay neurons of the SVS. Typical F axon terminals and P1 presynaptic dendrites have also been found in the optic tectum of birds and frogs (Streit et al., 1978; Antal, 1991) and many of the same GABA receptor mediated events also occur in these species. It is therefore likely that these GABAergic synaptic units appear early in vertebrate evolution and may be conserved across multiple brain structures and in many species. If so, the diverse inhibitory receptive field characteristics in the SVS may be determined by the specificity of synaptic connections rather than the structure of the synapse itself.

Acknowledgments

This work was supported by USPHS grant NIH EY-02973 from the National Eye Institute. Grace Butler prepared the figures. Dr. Robert Spencer contributed some of the cat LGN and pretectum tissue. Drs. Johannes Van der Want and Bob Nunes-Cardozo provided the tissue from the rabbit superior colliculus. We thank A. Sillito and V. Crunelli for providing us with Figs. 1 and 5.

References

Antal, M. (1991) Distribution of GABA immunoreactivity in the optic tectum of the frog: a light and electron microscopic study. *Neuroscience*, 42: 879–891.

Arakawa, T. and Okada, Y. (1988) Excitatory and inhibitory action of GABA on synaptic transmission in slices of guinea pig superior colliculus. *Eur. J. Pharmacol.*, 158: 217–224.

Bloomfield, S.A. and Sherman, S.M. (1989) Dendritic current flow in relay cells and interneurons of the cat's lateral geniculate nucleus. *Proc. Natl. Acad. Sci. USA*, 86: 3911–3914.

Campbell, G. and Lieberman, A.R. (1985) The olivary pretectal nucleus: experimental anatomical studies in the rat. *Philos. Trans. R. Soc. London Ser B*, 310: 573–609.

Guillery, R.W. (1969) The organization of synaptic interconnections in the laminae of the dorsal lateral geniculate nucleus of the cat. *Z. Zellforsch. Mikrosk. Anat.*, 96: 1–38.

Hamos, J.E., Van Horn, S.C., Raczkowski, D., Uhlrich, D.J. and Sherman, S.M. (1985) Synaptic connectivity of a local circuit neuron in the lateral geniculate nucleus of the cat. *Nature*, 317: 618–621.

Kraszewski, K. and Grantyn, R. (1992) Development of GABA-ergic connections in vitro: increasing efficacy of synaptic transmission is not accompanied in miniature currents. *J. Neurobiol.*, 23: 766–781.

Mize, R.R. (1992) The organization of GABAergic neurons in the mammalian superior colliculus. *Prog. Brain Res.*, 90: 219–248.

Mize, R.R. and Horner, L.H. (1984) Retinal synapses of the cat medial interlaminar nucleus and ventral lateral geniculate nucleus differ in size and synaptic organization. *J. Comp. Neurol.*, 224: 579–590.

Mize, R.R., Spencer, R.F. and Horner, L.H. (1986) Quantitative comparison of retinal synapses in the dorsal and ventral (parvicellular C) laminae of the cat dorsal lateral geniculate nucleus. *J. Comp. Neurol.*, 248: 57–73.

Mize, R.R., Marc, R.E. and Sillito, A.M. (1992) GABA in the retina and central visual system. *Prog. Brain Res.*, 90: 1–545.

Montero, V.M. (1989) The GABA-immunoreactive neurons in the interlaminar regions of the cat lateral geniculate nucleus: light and electron microscopic observations. *Exp. Brain Res.*, 75: 497–512.

Norton, T.T. and Godwin, D.W. (1992) Inhibitory GABAergic control of visual signals at the lateral geniculate nucleus. *Prog. Brain Res.*, 90: 193–217.

Nunes-Cardozo, B., Mize, R. and Van der Want, J. (1994) GABAergic and non-GABAergic projection neurons from the nucleus of the optic tract to the superior colliculus in the rabbit. *J. Comp. Neurol.*, in revision.

Raczkowski, D., Hamos, J.E. and Sherman, S.M. (1988) Synaptic circuitry of physiologically identified w-cells in the cat's dorsal lateral geniculate nucleus. *J. Neurosci.*, 8: 31–48.

Sherman, S.M. and Koch, C. (1986) The control of retinogeniculate transmission in the mammalian lateral geniculate nucleus. *Exp. Brain Res.*, 63: 1–20.

Sherman, S.M. and Spear, P.D. (1982) Organization of the visual pathways in normal and visually deprived cats. *Physiol. Rev.*, 62: 738–855.

Sillito, A.M. (1992) GABA mediated inhibitory processes in the function of the geniculo-striate system. *Prog. Brain Res.*, 90: 349–384.

Soltesz, I. and Crunelli, V. (1992) GABAa and pre- and post-synaptic GABAb receptor-mediated responses in the lateral geniculate nucleus. *Prog. Brain Res.*, 90: 151–169.

Streit, P., Knecht, E., Reubi, J.C., Hunt, S.P. and Cuenod, M. (1978) GABA specific presynaptic dendrites in pigeon optic tectum: a high resolution autoradiographic study. *Brain Res.*, 149: 204–210.

Uhlrich, D.J. and Cucchiaro, J.B. (1992) GABAergic circuits in the lateral geniculate nucleus of the cat. *Prog. Brain Res.*, 90: 171–192.

Van der Want, J.J.L., Nunes-Cardozo, J.J. and Van der Togt, C. (1992) GABAergic neurons and circuits in the pretectal nuclei and the accessory optic system of mammals. *Prog. Brain Res.*, 90: 283–305.

F. Bloom (Editor)
Progress in Brain Research, Vol. 100

CHAPTER 17

Hypothalamic magnocellular neurosecretory neurons: intrinsic membrane properties and synaptic connections

Leo P. Renaud

Neurosciences Unit, Loeb Research Institute, Ottawa Civic Hospital and University of Ottawa, Ottawa, Ontario, Canada K1Y 4E9

Introduction

In mammals, hormone release from the neurohypophysis is centrally regulated by hypothalamic magnocellular "neurosecretory" cells (MNCs). This chapter reviews recent progress on CNS mechanisms regulating posterior pituitary secretion, with a particular emphasis on the cellular neurophysiology and synaptic neuropharmacology of MNCs in the rat. MNCs synthesizing either vasopressin (VP) or oxytocin (OT) are located in the supraoptic (SON) and paraventricular nuclei (PVN) and in several accessory hypothalamic nuclei. Among mammalian peptidergic neurons, the accessibility of MNCs for in vivo and in vitro exploration has led to a better understanding of their molecular biology, stimulus-secretion coupling at axon terminals, activity patterns in relation to hormone release, intrinsic membrane properties, chemical neuroanatomy and neuropharmacology of afferent connections. In the intact organism, "extrinsic" factors that influence MNC activity include plasma osmolality, blood volume and pressure, visceral afferents (chemo-, gastro- and baroreceptors), nausea and emetic agents, emotion and sensory (nipple, vaginal and nociceptive) stimulation. This information is conveyed to MNCs through humoral and central pathways, changing their excitability and firing patterns in accordance with the integration of synaptic and intrinsic conductances. In this system, outcome is reflected by plasma levels of VP or OT.

Intrinsic properties

Passive properties

Data collected from intracellular recordings in hypothalamic explant and slice preparations (reviewed by Renaud and Bourque, 1991) indicate that MNCs have relatively high input resistances (range 50–350 MΩ) and long membrane time constants (~14 ms). The linearity of their *I–V* relationships near resting membrane potentials (–55 to –75 mV), high input resistances and isopotential behavior suggest that current spread between soma and dendrites is efficient. In addition, their relatively long membrane time constants are likely to prolong postsynaptic (and post-spike) potentials, thereby promoting temporal summation of intrinsic and synaptic conductances contributing to the generation of electrical impulses.

Action potentials

In the rat, MNCs have somatic action potentials generated by rapidly inactivating and tetrodotoxin-sensitive Na^{+}, and voltage-dependent Ca^{2+} conductances. The activation of high threshold Ca^{2+} channels is responsible for a conspicuous shoulder on their spike repolarization phase, and this conductance displays a unique time- and frequency-dependence. Interestingly, neurohypophysial axon terminal membrane demonstrates similar characteristics (Bourque, 1990), and one might speculate that this ionic conductance might contribute to a form of local presynaptic plasticity and/or be the target for modulation by endogenous transmitters or peptides.

Afterpotentials
The presence of a 50–100 ms hyperpolarizing afterpotential (HAP) immediately following each somatic action potential results from the activation of a transient Ca^{2+}-dependent outward K^+ conductance. This conductance, termed I_{to}, has features similar to I_A, and both I_{to} and the HAP are blocked by 4-aminopyridine. The HAP functions as a form of "intrinsic inhibition", setting an upper limit on the maximum firing frequency during trains of action potentials; it can also explain a transient "silent period" in cell excitability in MNCs following their antidromic activation.

Another prominent but pharmacologically distinct afterhyperpolarization, termed AHP, follows current-evoked bursts of action potentials. This AHP results from the activation of a slow Ca^{2+}-dependant K^+ conductance (I_{AHP}) whose amplitude and duration is proportionate to the number of impulses in the train, and is selectively blocked by apamin, a bee venom polypeptide. The AHP serves to stabilize the steady state firing frequency of continuously firing cells, and possibly their patterns of activity since this conductance can shunt or mask the occurrence of a late depolarizing afterpotential, or DAP (see below).

Activity patterns
In the rat, a majority of VP-immunoreactive MNCs display a unique phasic bursting activity, generated by an intrinsic mechanism. The basis for phasic bursting appears to be a slow DAP that may follow the HAP, and results from the activation of a Ca^{2+} dependent inward membrane current (I_{DAP}). In SON MNCs, activation of I_{DAP} induces a region of negative slope resistance which crosses spike threshold in the post-spike I–V relation; when the post-spike membrane voltage enters this region, a suprathreshold depolarization will occur, ensuring that repetitive firing will be sustained. Individual phasic bursts are sustained by a small (<10 mV) depolarizing plateau which arises from summation of DAPs. Any prevailing extrinsic depolarizing (e.g. synaptic, osmotic) drive contributes to sustaining this depolarizing plateau. With time, the conductance underlying the plateau potential becomes inactivated, possibly due to a Ca^{2+} dependent inactivation, and this along with any synaptic inhibitory influences may produce a cessation in firing and collapse of the plateau. Membrane hyperpolarization, triggered for example by a barrage of inhibitory postsynaptic potentials (IPSPs) may also trigger collapse of a plateau and cessation of firing. Such an event is proposed to underly the cessation of phasic firing observed during in vivo recordings from VP-secreting MNCs in response to a transient rise in arterial blood pressure that activates peripheral baroreceptors (Jhamandas and Renaud 1986).

Osmosensitivity
MNCs in SON demonstrate an intrinsic osmosensitivity, with a threshold near +8 mOsm. In a recent report on the transduction process, Oliet and Bourque (1993) suggest that changes in membrane tension associated with osmotically evoked volume changes directly modulate the activity of stretch-inactivated cationic channels, endowing MNCs with an intrinsic osmosensitivity. In vivo, the osmoreception appears more complex and likely involves additional inputs from intrinsically osmosensitive neurons in two circumventricular organs, the subfornical organ (SFO) and organum vasculosum lamina terminalis (OVLT; see Nissen et al., 1992).

Synaptic neuropharmacology and ligand gated conductances

Anatomical tracer, immunocytochemical and electrophysiological approaches have contributed to major advances in our understanding of the chemical neuroanatomy and function of pathways conveying humoral, osmotic and neural information to MNCs. The caudal nucleus tractus solitarii (cNTS) contains a primary synapse for viscerosensory information from gastric mechanoreceptors, hepatic portal osmoreceptors, renal afferents, baroreceptors, chemoreceptors, the area postrema and somatosensory afferents (see Cunningham and Sawchenko, 1991). NTS neurons may relay information directly to MNCs but more commonly, MNCs receive their information indirectly via relays in various brainstem structures. Current investigations are focused on the mechanisms whereby specific information about different modalities reaches VP- and/or OT-secreting MNCs. The following sections attempt to link neuropharmacology with afferents to MNCs.

Amino acids: GABAergic receptors

MNCs have a prominent GABAergic innervation (Decavel and Van Den Pol, 1992). In SON MNCs, intracellular observations in vivo and in vitro reveal a tonic barrage of spontaneous IPSPs mediated by postsynaptic $GABA_A$ receptors (MNCs appear to lack postsynaptic $GABA_B$ receptors) and activated through mono- or polysynaptic pathways, as illustrated by the following examples.

Median preoptic nucleus (MnPO)
The MnPO, located at the midpoint of the lamina terminalis, projects directly to SON and PVN and participates in hydromineral regulation. Animals bearing MnPO lesions are unable to respond appropriately with secretion of VP in response to a hyperosmotic challenge (see Gardiner et al., 1985), suggesting a facilitatory influence for MnPO on MNC excitability. However, recent investigations using electrical or chemical microstimulation in MnPO reveal a bicuculline-sensitive (? $GABA_A$-mediated) inhibition in both VP- and OT-secreting MNCs (Nissen and Renaud, 1994). The MnPO receives input from the SFO, whose neurons sense circulating angiotensin II (AII) and send angiotensin-immunoreactive fibers to MnPO (Lind et al., 1985), and the OVLT, whose neurons display an intrinsic osmosensitivity (Nissen et al., 1993a). In addition, MnPO receives prominent medullary catecholaminergic afferents. Thus, MnPO is well positioned to integrate information from humoral, osmotic and visceral sources. It remains unclear as to how the observed inhibitory input to MNCs from MnPO is involved functionally in hydromineral regulation.

Diagonal band of Broca (DBB) and baroreceptor inhibition
Whereas stimulation in DBB also induces a GABAergic postsynaptic inhibition in MNCs, there are two important differences: first, anatomical tracer studies indicate that the connection is indirect, with non-GABAergic DBB neurons projecting to the lateral hypothalamic perinuclear zone (PNZ) where there is a proposed synapse with GABAergic interneurons (Nissen et al, 1993b); second, the DBB-evoked inhibition is selective, and preferentially affects putative VP-synthesizing MNCs (Jhamandas and Renaud, 1986). Moreover, since the integrity of DBB and its catecholaminergic innervation are critical for baroreceptor-initiated inhibition of VP-secreting MNCs (Cunningham et al., 1992), the DBB is considered to represent a major component in a multisynaptic central baroreflex pathway.

Amino acids: glutamatergic receptors

A substantial number of glutamate-immunoreactive synapses have been detected on MNCs (Decavel and Van Den Pol, 1992), in keeping with observations that these cells contain glutamate receptors of both NMDA- and non-MNDA types (see Renaud and Bourque, 1991). Some of these synapses originate in OVLT where stimulation evokes monosynaptic excitatory postsynaptic potentials (EPSPs) mediated by both non-NMDA and NMDA receptors in VP-synthesizing MNCs (Yang et al., 1994). OVLT participates in two important challenges that involve VP, i.e. fever and plasma hyperosmolality, and the detection of an excitatory pathway from OVLT to VP-secreting MNCs provides functional evidence for this linkage; whether the OVLT $\rightarrow$ SON pathway actually transmits information related to these modalities awaits confirmation.

Catecholamines: focus on noradrenergic receptors

MNCs receive a dense catecholaminergic input from the A1 noradrenergic cells in the caudal ventrolateral medulla, with a smaller contribution from the A2 noradrenergic cells of the caudal NTS (Swanson and Sawchenko, 1983). Electrophysiological data derived from both in vivo and in vitro studies indicate that this is mainly an excitatory innervation, mediated through postsynaptic $\alpha 1$ adrenergic receptors whose activation causes reduction in a K^+ conductance, possibly I_A, and initiates bursting activity (Day, 1989; Renaud and Bourque, 1991). However this is not a simple innervation. Whereas exogenous noradrenaline enhances the excitability of both VP- and OXY-secreting MNCs, the A1 → MNC pathway appears to be selective for VP-secreting MNCs, whereas the A2 → MNC pathway appears to be selective for OT-secreting MNCs (Raby and Renaud, 1989). Moreover, exogenous noradrenaline has an excitatory action when applied in micromolar concentrations, mediated via postsynaptic $\alpha 1$ adrenoreceptors, but a depressant action when applied at higher (millimolar) concentrations, mediated via presynaptic $\alpha 2$ and postsynaptic β adrenergic receptors (Renaud and Bourque, 1991; Khanna et al., 1993). In addition, the neuropeptide NPY, which co-exists within a majority of A1 neurons, appears to have postsynaptic NPY_1 receptors that potentiate the actions of noradrenaline and presynaptic NPY_2 receptors to reduce the input from A1 neurons (Day, 1989). Finally, ATP may contribute to the excitation of SON MNCs evoked by A1 stimulation (Day et al., 1993).

Peptides: focus on angiotensin AT_1 and cholecystokinin B type receptors

Angiotensin II-like immunoreactivity is observed both in MNCs themselves and in their afferents. Exogenously applied angiotensin II induces a slow onset, prolonged membrane depolarization in SON MNCs, involving an AT_1 type receptor and mediated through a non-selective cationic conductance (Yang et al., 1992). A major source for angiotensinergic afferents is the SFO, and electrophysiological data suggest that angiotensin may be the neurotransmitter underlying a long duration increase in excitability in SON MNCs following SFO stimulation (Jhamandas et al., 1989). In the rat, CCK influences MNC function through independent peripheral and central mechanisms. Systemically administered CCK octapeptide (CCK-8) acts at peripheral A type receptors to trigger a vagally mediated and selective increase in the excitability of OT-secreting MNCs (Renaud et al., 1987). The central pathways and transmitters for this response remain undefined. CCK has separate central actions on MNCs. CCK immunoreactivity is present in fibers located near MNCs, CCK is co-synthesized in a population of MNCs, and exogenous applications of CCK produce membrane depolarizations in both VP- and OT-secreting MNCs, mediated via B-type receptors (Jarvis et al., 1992). Future studies are needed to assess the role(s) of CCK in specific afferent pathways, CCK's role as a co-existing and co-released neurotransmitter, the possibility of local release from somata-dendrites of MNCs and paracrine actions, mechanisms that regulate CCK synthesis and storage in MNCs, its release from neurohypophysial axon terminals and influence on neurosecretion of other peptides.

Comment

As the intrinsic properties and synaptic connectivity of MNCs become more clearly defined and characterized, a profile of the integrative properties of MNCs will emerge. There remains the elucidation of signal transduction mechanisms that govern immediate changes in cell excitability and long-term changes in gene expression and protein synthesis.

Acknowlegement

The author acknowledges the sustained support of the Canadian MRC and Heart and Stroke Foundation.

References

Bourque, C.W. (1990) Intraterminal recordings from the rat neurohypophysis in vitro. *J. Physiol.*, 421: 247–262.

Cunningham, E.T. and Sawchenko, P.E. (1991) Reflex control of magnocellular asopressin and oxytocin secretion. *Trends Neurosci.*, 14: 406–411.

Cunningham, J.T., Nissen, R. and Renaud, L.P. (1992) Catecholamine depletion of the diagonal band reduces baroreflex inhibition of supraoptic neurons. *Am. J. Physiol.*, 263: R363–R367.

Day, T.A. (1989) Control of neurosecretory vasopressin cells by noradrenergic projections of the caudal ventrolateral medulla. *Prog. Brain Res.*, 81: 303–317.

Day, T.A., Sibbald, J.R. and Khanna, S. (1993) ATP mediates an excitatory noradrenergic neuron input to supraoptic vasopressin cells. *Brain Res.*, 607: 341–344.

Decavel, C. and Van Den Pol, A.N. (1992) Converging GABA- and glutamate-immunoreactive axons make synaptic contact with identified hypothalamic neurosecretory neurons. *J. Comp. Neurol.*, 316: 104–116.

Gardiner, G.W., Verbalis, J.G. and Stricker, E.M. (1985). Impaired secretion of vasopressin and oxytocin in rats after lesions of nucleus medianus. *Am. J. Physiol.*, 269: R681–R688.

Jarvis, C.R., Bourque, C.R. and Renaud, L.P. (1992) Depolarizing action of cholecystokinin on rat supraoptic neurones in vitro. *J. Physiol.*, 458: 621–632.

Jhamandas, J.H. and Renaud, L.P. (1986) A gamma aminobutyric acid-mediated baroreceptor input to supraoptic vasopressin neurones in the rat. *J. Physiol.*, 381: 595–606.

Jhamandas, J.H., Lind, R.W. and Renaud, L.P. (1989) Angiotensin II may mediate excitatory neurotransmission from the subfornical organ to the hypothalamic supraoptic nucleus: an anatomical and electrophysiological study in the rat. *Brain Res.*, 487: 52–61.

Khanna, S., Sibbald, J.R. and Day, T.A. (1993) α2-adrenoceptor modulation of A1 noradrenergic neuron input to supraoptic vasopressin cells. *Brain Res.*, 613: 164–167.

Lind, R.W., Swanson, L.W. and Ganten, D. (1985) Organization of angiotensin II immunoreactive cells and fibers in the rat central nervous system. *Neuroendocrinology*, 40: 2–24.

Nissen, R. and Renaud, L.P. (1994) GABA receptors mediate median preoptic nucleus-evoked inhibition of vasopressin and oxytocin neurones in rat supraoptic nucleus. *J. Physiol.*, in press.

Nissen, R., Bourque, C.R. and Renaud, L.P. (1993a) Membrane properties of organum vasculosum lamina terminalis neurons recorded in vitro. *Am. J. Physiol.*, 264: R811–R815.

Nissen, R., Cunningham, J.T. and Renaud, L.P. (1993b) Lateral hypothalamic lesions alter baroreceptor-evoked inhibition of rat supraoptic vasopressin neurones. *J. Physiol.*, 470: 751–766.

Oliet, S.H.R. and Bourque, C.W. (1993) Mechanosensitive channels transduce osmosensitivity in supraoptic neurons. *Nature*, 364: 341–343.

Raby, W.N. and Renaud, L.P. (1989) Nucleus tractus solitarius innervation of supraoptic nucleus: anatomical and electrophysiological studies in the rat suggest differential innervation of oxytocin and vasopressin neurons. *Prog. Brain Res.*, 81: 319–327.

Renaud. L.P. and Bourque, C.W. (1991) Neurophysiology and neuropharmacology of hypothalamic magnocellular neurons secreting vasopressin and oxytocin. *Prog. Neurobiol.*, 36: 131–169.

Renaud, L.P., Tang, M., McCann, M.J., Stricker, E.M. and Verbalis, J.G. (1987) Cholecystokinin and gastric distention activate oxytocinergic cells in rat hypothalamus. *Am. J. Physiol.*, 253: R661–R665.

Swanson, L. E. and Sawchenko, P.E. (1983) Hypothalamic integration: organization of the paraventricular and supraoptic nuclei. *Annu. Rev. Neurosci.*, 6: 269–324.

Yang, C.R., Phillips, M.I. and Renaud. L.P. (1992) Angiotensin II receptor activation depolarizes rat supraoptic neurons in vitro. *Am. J. Physiol.*, 263: R1333–R1338.

Yang, C.R., Senatorov, V.V. and Renaud, L.P. (1994) Organum vasculosum lamina terminalis-evoked post-synaptic responses in rat supraoptic neurones in vitro. *J. Physiol.*, in press.

F. Bloom (Editor)
Progress in Brain Research, Vol. 100

CHAPTER 18

Molecular principles from neuroendocrine models: steroid control of central neurotransmission

George Fink

MRC Brain Metabolism Unit, University Department of Pharmacology, 1 George Square, Edinburgh EH8 9JZ, UK

Introduction

Neuroendocrinology, the study of the interactions between the nervous and endocrine systems, has generated several principles crucial for our understanding of chemical neurotransmission. The peripheral neuroendocrine system is comprised mainly of the gastrointestinal system and its innervation, and the adrenal medulla. The pioneering studies of Gaddum, von Euler, Mutt and Gregory on the gastrointestinal system provided the basis for our knowledge of the amino acid residue sequence of many peptides which also have an important role in the central nervous system (Dockray, 1988). The adrenal medulla continues to provide an important model for exocytosis and the intracellular signals involved in the neural control of endocrine secretion (Johnson, 1988).

The central neuroendocrine system is subdivided into the hypothalamo-neurohypophysial system and the hypothalamo-adenohypophysial system. The main function of the hypothalamo-neurohypophysial system is to synthesize and secrete into the systemic circulation, two important nonapeptides, vasopressin and oxytocin. The main physiological role of vasopressin is to reduce the volume of urine produced by the kidneys (hence the synonym, antidiuretic hormone); oxytocin stimulates milk ejection in response to suckling and coordinates and potentiates uterine contractions during parturition. Several properties make the hypothalamo-neurohypophysial system an excellent model system which has generated important principles for neurophysiology, microphysiology, neuropeptide synthesis, axonal transport and exocytosis.

First, vasopressin and oxytocin are synthesized in large (magnocellular) neurons of the hypothalamic paraventricular and supraoptic nuclei (PVN and SON). Their large size and the convenient anatomical location of the PVN and SON magnocellular neurons makes them excellent targets for neurophysiological and microphysiological studies, especially since the readily accessible neurohypophysis facilitates reliable neuronal identification by antidromic stimulation (Cross et al., 1975). Secondly, the rate of synthesis and turnover of AVP and oxytocin is several orders of magnitude greater than that of other transmitters and this makes it relatively easy to use them as models for neuropeptide synthesis and axonal transport (Cross et al., 1975; Sachs et al., 1969). Thirdly, the fact that both peptides contain a cysteine residue allows the use of ^{35}S cysteine as a convenient marker for studies of synthesis, transport and turnover. Fourthly, the neurohypophysis is virtually a "bag" of nerve terminals derived from the PVN and SON and is, therefore, an excellent model for studying the detailed mechanism of neuropeptide release. Finally, the homozygous Brattleboro rat, a mutant which renders AVP mRNA defective so that it cannot be translated, has provided an excellent model for studies of gene transcription and translation and for studies on the effects of AVP gene "knockout" (Sokol and Valtin, 1982; North et al., 1993).

The hypothalamo-adenohypophysial system also provides an excellent model for studying neuropeptide synthesis release and action (Fink, 1976; Fink, 1988; Fink and Sheward, 1989). Both of the two hypothalamo-hypophysial systems have been reviewed extensively (especially in the *Progress in Brain Research* series) and so here attention is focused on the way central neuroendocrine systems can be used to study the profound effects that steroid hormones exert on central neurotransmission by actions on individual brain nuclei and complex neural circuits in which one or more synapses allow for switching of signal sign and thus the mechanism of disinhibition.

Steroids exert either inhibitory or stimulatory effects on neuronal systems. Broadly, the stimulatory actions of steroids have a long latency (hours to days) while the latency of inhibitory actions is short (seconds to minutes) (Fink, 1979a). Membrane (extragenomic) actions of "neurosteroids", mainly pregnenolone and its derivatives, have recently excited considerable interest (Baulieu, 1993). The most compelling evidence for this type of action comes from studies of the steroid anaesthetic alphaxalone which acts through $GABA_A$ receptors (Fink et al., 1982; Majewska, 1987). The present review, however, focuses on the well-established classical action of steroids involving activation of specific steroid receptors which, by way of an action on DNA steroid response elements, perhaps in association with other transcription factors, either stimulate or inhibit gene transcription.

The inhibitory actions of steroids in negative feedback control of the hypothalamic-pituitary-gonadal and adrenal systems have been by far the most well-studied (Fink, 1979a) although the intracellular mechanism of action has yet to be established. Steroids also exert powerful stimulatory actions, examples of which include the facilitation by glucocorticoids of the activity of tyrosine hydroxylase and tryptophan hydroxylase, the rate-limiting enzymes for catecholamine and indolamine synthesis, respectively, and phenylethanolamine *N*-methyl transferase which converts noradrenaline to adrenaline (McEwen et al., 1986). Here attention is focused on two important examples of the stimulatory actions of gonadal steroids illustrating important principles for our understanding of how steroids evoke clear neuroendocrine signals and behaviours through apparently complex actions on brain cells and circuits.

Positive feedback of oestrogen on the hypothalamic-pituitary-gonadotropin system

The spontaneous ovulatory surge of luteinizing hormone (LH) is generated by a positive feedback cascade in which a surge of oestradiol-17β (E_2) acts on the brain to trigger a surge of luteinizing hormone releasing hormone (LHRH), and on the anterior pituitary gland to increase pituitary responsiveness to LHRH (Fink, 1979b, 1988) (Fig. 1). The increase in pituitary responsiveness to LHRH is further potentiated by progesterone. Luteinizing hormone releasing hormone has the unique capacity to increase responsiveness (by at least sevenfold in vivo) to itself. This property of LHRH, which we termed the priming effect of LHRH,

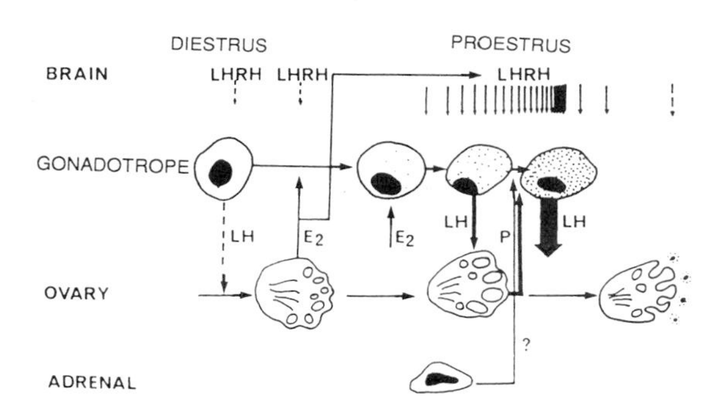

Fig. 1. Schematic diagram which shows the cascade of events which produce the spontaneous ovulatory LH surge in the rat. The increase in plasma concentrations of oestradiol-17β (E_2, the ovarian signal) increases the responsiveness of the pituitary gonadotrophs (increased stippling) to LHRH and also triggers the surge of LHRH. Pituitary responsiveness to LHRH is further increased by progesterone (P) secreted from the ovary in response to the LH released during the early part of the LH surge and by the priming effect of LHRH, the unique capacity of the decapeptide to increase pituitary responsiveness to itself. The priming effect of LHRH coordinates the surges of LHRH with increasing pituitary responsiveness so that the two events reach a peak at the same time. The conditions are thereby made optimal for a massive surge of LH. This cascade, which represents a form of positive feedback, is terminated by destruction of a major component of the system in the form of the rupture of the ovarian follicles (ovulation). From Fink (1979b) with the permission of the British Council.

is probably important to further potentiate pituitary responsiveness to LHRH and to coordinate the increased release of LHRH and pituitary responsiveness to LHRH so that both reach a peak at the same time thereby ensuring a massive spontaneous, ovulatory surge of LH.

Does oestradiol-17β stimulate LHRH biosynthesis?

To determine whether E_2 stimulates the biosynthesis of LHRH, we used the same paradigm as that used to prove that oestradiol stimulated the LHRH surge (Sarkar and Fink, 1979a; Rosie et al., 1990). That is, rats were ovariectomized on the morning of di-oestrus in order to eliminate the spontaneous surge of E_2 and injected s.c. with either oestradiol benzoate (EB) or vehicle (0.2 ml sesame oil). The animals were killed between 1600 and 1700 h of the next day (presumptive pro-oestrus). LHRH mRNA was determined by quantitative in situ hybridization in serial coronal sections of the hypothalamus. The concentrations of LHRH mRNA in perikarya in the medial preoptic area, diagonal band of Broca and medial septum were significantly greater in EB compared with oil-treated control animals (Rosie et al., 1990). Thus, oestradiol in its positive feedback mode stimulates the synthesis of LHRH mRNA. This increased synthesis of LHRH mRNA correlates with increased hypothalamic content of LHRH which can be detected in the adult (Chiappa and Fink, 1977), but is seen much more clearly during the first pro-oestrus (Sarkar and Fink, 1979b).

How does oestradiol stimulate LHRH biosynthesis and release?

Since LHRH neurons do not contain oestradiol receptors (Shivers et al., 1983), the action of E_2 must be mediated by interneurons. Pharmacological and in situ hybridization data show that at least two potent stimulatory and two disinhibitory mechanisms could mediate the positive feedback action of E_2 on LHRH mRNA synthesis and LHRH release. With respect to the stimulatory mechanisms, pharmacological studies (Dow et al., 1994) carried out using the specific α_1 adrenoreceptor antagonist, prazosin, confirmed our earlier findings with phenoxybenzamine (Sarkar and Fink, 1981) that an α_1 adrenoreceptor mechanism is involved in E_2 stimulation of LHRH and LH release. Prazosin also reduced the total number of LHRH mRNA containing cells in the preoptic area suggesting that an α_1 adrenoreceptor mechanism also mediates E_2 stimulation of LHRH mRNA synthesis (Rosie et al., 1994). Prazosin also blocked the E_2-induced decrease of POMC mRNA expression in the arcuate nucleus (Rosie et al., 1994). Thus, stimulation by E_2 of an α_1 adrenoreceptor mechanism could exert a direct effect on LHRH neurons and, for reasons outlined below, could also stimulate LHRH synthesis and release by the inhibition of arcuate POMC neurons which are known to project to LHRH cell bodies and send projections to the median eminence where they are juxtaposed to and may inhibit LHRH neurons.

Oestradiol-17β stimulation of LHRH biosynthesis and release also involves 5-HT which has been shown to play an important role in stimulating LH release under conditions of high levels of E_2 (Weiner et al., 1988). Guided by our in situ hybridization studies which showed that in our E_2 positive feedback model (Rosie et al., 1990), there was a massive and apparently selective stimulation of 5-HT_2 receptor mRNA in the dorsal raphe nucleus (Sumner and Fink, 1993), we focused attention on the role of 5-HT_2 receptors, and found that the specific 5-HT_2 receptor antagonist, ritanserin, blocked the spontaneous LH surge in pro-oestrous rats (Dow et al., 1994). The mixed 5-HT_2/α_1 adrenoreceptor antagonist, ketanserin, not only blocked the LH surge, but also blocked basal LH release. Taken together, these data suggest the E_2 stimulation of LHRH biosynthesis and release is mediated by an α_1 adrenoreceptor and a 5-HT_2 receptor mechanism (Fig. 2).

With respect to disinhibitory mechanisms, the most parsimonious hypothesis consists of an opioid and dopaminergic "clamp" whereby arcuate proopiomelanocortin (POMC) containing neurons and dopaminergic (DA) neurons inhibit LHRH neurons (Fig. 3). Both DA and POMC neurons are known to project to the external layer of the median eminence where these terminals are close to those of LHRH neurons. Proopiomelanocortin (β-endorphin) terminals also

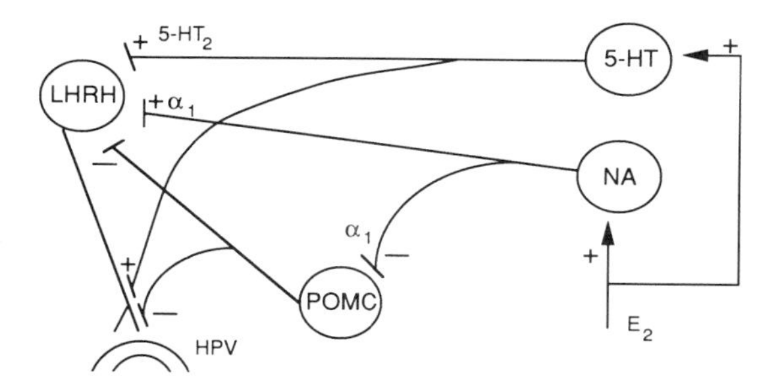

Fig. 2. Schematic diagram illustrating how E_2 could stimulate LHRH biosynthesis and release into the hypophysial portal vessels (HPV) by stimulating noradrenergic (NA) and 5-hydroxytryptamine (5-HT) neurons. The NA and 5-HT neurons (shown schematically in midbrain nuclei) could have a direct action on LHRH neurons by way of α_1 and 5-HT$_2$ receptors, respectively (Dow et al., 1994; Rosie et al., 1994). Our data also suggest that NA neurons could mediate E_2 inhibition of pro-opiomelanocortin (POMC) neurons. Since POMC neurons, at low E_2 levels, normally inhibit LHRH biosynthesis and release, inhibition of POMC neurons would disinhibit LHRH neurons. Other neurotransmitters (e.g. GABA and neuropeptide Y) and more complex neural circuits may also be involved. With respect to circuits, Morello and Taleisnik (1988) and Morello et al. (1989), for example, suggest that dorsal raphe 5-HT neurons stimulate LH release by an adrenergic mechanism involving the locus coeruleus and that medial raphe 5-HT neurons inhibit LH release by way of a GABAergic mechanism. (+, stimulation; —, inhibition).

represent about 9% of all synaptic input to LHRH neurons (Chen et al., 1989). The hypothesis (Fink, 1988) is that in the presence of low levels of oestradiol-17β, arcuate POMC and DA neurons inhibit LHRH biosynthesis and release; when plasma oestradiol-17β concentrations rise to surge levels, POMC and DA neurons are switched off and as a consequence LHRH neurons are disinhibited leading to LHRH biosynthesis and release. Since DA is known to inhibit prolactin release, the inhibition of DA neurons by oestrogen could also result in the spontaneous pre-ovulatory prolactin surge which occurs concurrently with the ovulatory gonadotrophin surge in rat and man (Djahanbakhch et al., 1984; Fink, 1988).

Numerous previous data show that opioids inhibit LH release (Kalra, 1986) and that the selective destruction of tuberoinfundibular DA neurons stimulates LHRH and LH release (Sarkar et al., 1981). With respect to the role of arcuate POMC neurons, E_2 significantly reduced POMC mRNA levels in the anterior region of the arcuate nucleus (Thomson et al., 1990) and E_2 can significantly reduce POMC transcription within 60 min (Roberts et al., 1986). These results together with those of Wise et al. (1990) are compatible with the possibility that the stimulation of LHRH biosynthesis and release by oestrogen positive feedback may be mediated, in part, by disinhibition of LHRH neurons as a consequence of inhibiting POMC biosynthesis in arcuate neurons that project to and, at low E_2 levels, inhibit LHRH neurons. As pointed out above, the action of E_2 on POMC neurons may be mediated by an α_1 adrenoreceptor mechanism.

This hypothesis is probably grossly oversimplified in that other neurotransmitters, such as GABA, excitatory amino acids and other neuropeptides (e.g. neuropeptide Y) may also modulate LHRH biosynthesis and release. Nonetheless, this hypothesis and the data are important for our understanding of how E_2 regulates LHRH biosynthesis and release, and in addition provide important principles that are relevant to central neurotransmission in general. In particular, the hypothesis focuses attention on inhibitory mechanisms that are common (probably the most common mechanism) within the central nervous system. Corroborative evidence for the role of the inhibitory and disinhibitory mechanisms in LHRH release comes from metabolic studies which showed that 2-[^{14}C]deoxyglucose utilization was markedly reduced in the arcuate nucleus, median eminence and preoptic area around the time of the spontaneous ovulatory LH surge (McQueen and Fink, 1988).

The mechanism of E_2 action: principles from the LHRH priming effect

The priming effect of LHRH is briefly discussed here because recent studies have pointed to a possible common mechanism of action of LHRH and oestrogen. The priming effect of LHRH (for definition see above) differs from the simple releasing action of LHRH in that it (i) cannot be mimicked by K^+ depolarization or Ca^{2+} ionophores, (ii) is independent of normal extracellular Ca^{2+} concentrations, (iii) involves an elongation and change in orientation of the microfilaments, (iv) involves the movement of secretory granules towards the plasma membrane of im-

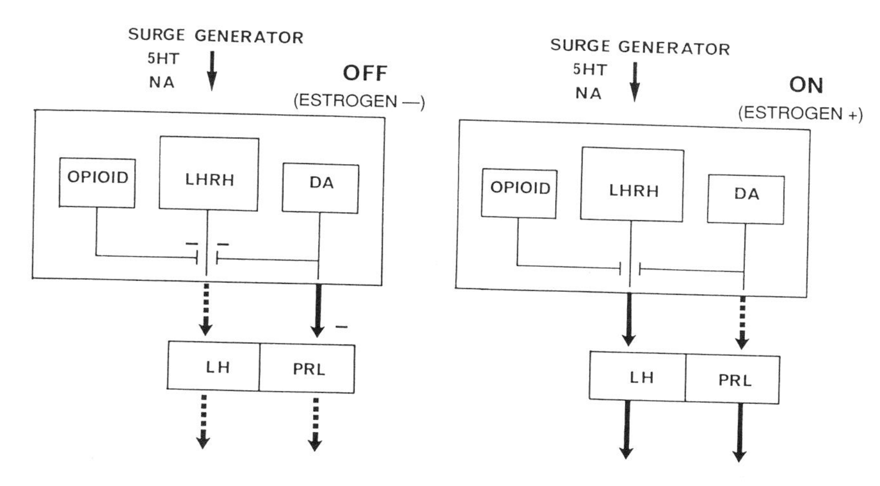

Fig. 3. A highly schematic diagram illustrating the hypothesis of the "opioid and dopamine (DA) clamp". The hypothesis is based on (i) the close juxtaposition of DA, POMC (β-endorphin) and LHRH neurons in the median eminence, (ii) the fact that both DA and opioids inhibit LHRH release, and (iii) that E_2 inhibits POMC gene transcription and reduces POMC mRNA levels in arcuate neurons. Thus, at low E_2 concentrations, the opioid and DA neurons are active and inhibit LHRH release, while at high E_2 concentrations, presumed to reduce opioid (β-endorphin) and DA neuronal activity, LHRH neurons are disinhibited and, therefore, release LHRH. A similar mechanism may operate for LHRH biosynthesis (see text). Neuronal inhibition and disinhibition are powerful and precise control mechanisms which operate widely throughout the nervous system. The diagram also shows that at low E_2 concentrations, the presumed higher release of DA into hypophysial portal blood may inhibit prolactin (PRL) release, while at high E_2 concentrations, DA release is reduced thereby allowing PRL release to occur. This is one of several mechanisms which can explain the concurrence of the LHRH/LH and PRL surges in the rat and the human. From Fink (1988) with permission.

muno-identified gonadotrophs ("margination") and (v) involves the synthesis of a new protein (Fink, 1988).

Our studies on LHRH priming show two important principles for the action of E_2. First, the pI and M_r (70 kDa) of the LHRH-induced protein were identical to those of an E_2-induced protein associated with mating behaviour in the female rat (Mobbs et al., 1990a). The amino acid residue sequences of the N-termini of the LHRH and E_2 induced proteins are identical to one another and with what was originally thought to be phospholipase $C\alpha$ (PLC-α) (Mobbs et al., 1990a,b). However, the existence of PLC-α is controversial and so with the aid of DNA complementary to the LHRH/E_2-induced protein, further studies are now in progress to determine its precise role.

Secondly, it is important to stress that massive changes in pituitary responsiveness cannot simply be attributed to changes in LHRH binding to LHRH receptors (Clayton et al., 1986). That is, the efficacy of LHRH in evoking the LH surge is dependent upon potentiation of post receptor intracellular signalling events. Thus, for example, we have shown that LHRH priming involves a massive potentiation of (i) the IP_3 released in pituitary tissue in response to LHRH, and (ii) the efflux of Ca^{2+} from pituitary slices (indicative of increased release of Ca^{2+} from intracellular stores) (Mitchell et al., 1988).

LHRH is also known to stimulate lordosis (mating) behaviour about 1 h after injection into the mid brain central grey. Thus, the 70 kDa protein may be an important component of a mechanism common to both LHRH and oestrogen which is involved in massive potentiation of pituitary responsiveness to LHRH and also of lordosis behaviour. In terms of intracellular signalling, the priming effect of LHRH may also provide a useful model for studies of long-term potentiation in brain, in that both phenomena may be dependent on activation of protein kinase C by a "conditioning stimulus" (Johnson et al., 1992).

Long-term reversible effects of steroids

Steroid effects on AVP neurons of the bed nucleus of the stria terminalis (BNST)

De Vries et al. (1986) first demonstrated that there is a marked sex difference in the vasopressinergic innervation of the rat brain outside the classical hypothalamic-neurohypophysial system. These sex steroid-dependent differences are most pronounced in the lateral habenula and lateral septum in which there is a dense plexus of AVP containing fibres which project from BNST perikarya thought to play a crucial role in "social memory" (Dantzer et al., 1988). We confirmed these findings in the mouse and showed that no immunoreactive AVP fibres are present in the lateral habenula of the LHRH deficient mutant hypogonadal (*hpg*) mouse (Mayes et al., 1988). AVP-containing fibres in the lateral habenula can be induced in *hpg* mice by the insertion of a hypothalamic graft from normal mice into the third ventricle. The graft innervates the hypophysial portal vessels with LHRH-containing terminals which stimulate gonadotrophin and thereby testosterone secretion which in turn stimulates the male pattern of AVP-containing fibres in the habenula (Mayes et al., 1988). The administration of testosterone or oestrogen, but not the non-aromatizable androgen, 5α-dihydrotestosterone, to male *hpg* mice also results in the normal development of AVP fibres in the lateral habenula. The substantially higher concentrations of AVP fibres in male compared with female brain suggests that the amount of oestrogen in the male brain is higher than in the female presumably because the amount of testosterone available for conversion to oestrogen by aromatase exceeds significantly the combined effect of small amounts of androgen and relatively small amounts (by comparison with testosterone in the male) of oestrogen in the female.

Our recent studies were designed to determine whether the action of the gonadal sex steroids was due to stimulation of AVP biosynthesis as assessed by determination of AVP mRNA by in situ hybridization using a ^{35}S-labelled 49 mer oligonucleotide probe complementary to the 5′ end of the glycopeptide coding domain of AVP mRNA (Rosie et al., 1993). Exposure to supraphysiological concentrations of testosterone for 6–12 days induced an exponential, 50-fold increase in the number of cells that expressed AVP mRNA in the BNST in *hpg* mice (Fig. 4). Similar results were obtained in the adult rat (Miller et al., 1989). The action of testosterone is "all-or-none" in

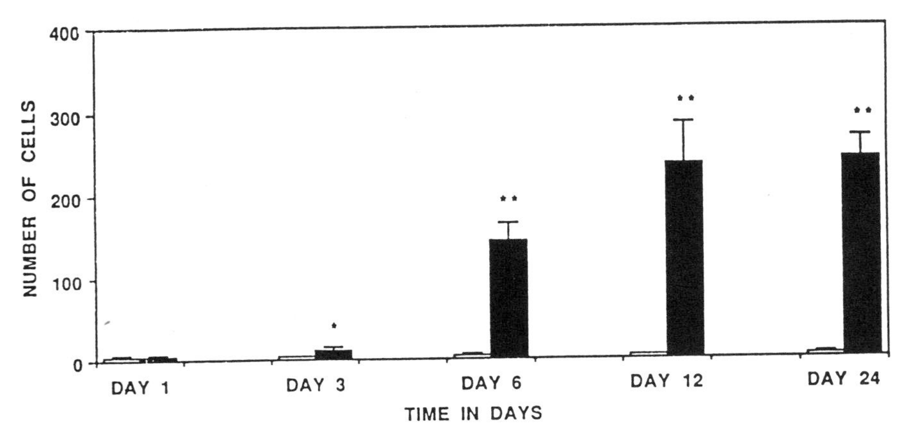

Fig. 4. The mean (±SEM) number of cells expressing AVP mRNA in the BNST of hypogonadal (*hpg*) mice at different times after implanting either empty (open bars) or testosterone-propionate filled (closed bars) silicone elastomer capsules. Note the exponential increase in the number of cells expressing AVP mRNA between days 3 and 6 of implanting the testosterone-propionate filled capsules. Significance of differences (Mann-Whitney *U*-test): $*P < 0.05$; $**P < 0.01$. From Rosie et al. (1993) with permission.

that the level of AVP mRNA in the AVP-positive cells was similar at all times after testosterone implantation and also similar or less than that in the few AVP-positive cells in animals not treated with testosterone (Rosie et al., 1993). The relatively long time taken for testosterone to exert its effect on AVP mRNA in BNST neurons is consistent with its effect on AVP immunoreactivity in the lateral habenula and lateral septum and on behaviour and suggests that the action of testosterone may be mediated by indirect or slow intracellular transduction mechanisms. Relevant to this are the preliminary findings (J Quinn, unpublished) in our laboratory that the 5′ promoter region of the AVP gene contains a steroid response element, an AP1 site and three E-boxes (members of the helix-loop-helix family of proteins often involved in tissue-specific expression) flanked by CT elements. The latter suggest that the action of E_2 on AVP gene expression may involve a complex interaction of activated E_2 receptor and other transcription factors.

Conclusions

Because of their potent actions on brain-pituitary functions, and because so much is already known about steroid receptors and their interactions with steroid response elements on genes in general, the study of steroid action on brain and pituitary offers a powerful method of establishing key principles about the molecular mechanisms that underlie central neurotransmission. Attention here is focused on acute and longer-term stimulatory effects of gonadal steroids: (i) the stimulation by E_2 of LHRH biosynthesis and the LHRH surge, and (ii) the biosynthesis of AVP in the bed nucleus of the stria terminalis. The action of E_2 is not due to a direct action on LHRH neurons, but is mediated by interneurons involving at least four possibly separate mechanisms; two stimulatory (α_1 adrenoreceptor and 5-HT_2 receptor) and at least two disinhibitory (POMC and dopaminergic). Teleologically, the need for four systems to control E_2 induced LHRH biosynthesis and release is probably because the secretion of LHRH is essential for fertility and reproduction of the species. The mechanism of action of oestradiol in increasing pituitary responsiveness to LHRH is not yet established, but the synthesis of new proteins and the potentiation of post-receptor mechanisms are involved. Oestrogen is also essential for mating behaviour and for increased motor activity that both reach a peak at the time of ovulation early on oestrus (Fink, 1988). There is a remarkable economy in this system in which the same steroid secreted by the ovary activates in a precisely timed manner the mechanisms (amplifier cascades) required for the ovulatory surge of LH and for mating. The stimulation by E_2 of LHRH biosynthesis, the surge-release and action of LHRH, and the stimulation by gonadal steroids of AVP biosynthesis in the BNST provide excellent models for studies of the role of the regulation of gene expression in central neurotransmission which underlies phenomena that involve relatively lengthy (hours to days) biochemical cascades. The two models also provide the basis for precise correlative studies between steroid control of CNS gene expression, central neurotransmission and behaviour.

References

Baulieu, E.E. (1993) Neurosteroids: a function of the brain. In: R. Mornex, C. Jaffiol and J. Leclère (Eds.), *Progress in Endocrinology, 9th Int. Congress of Endocrinology*, Nice, Parthenon Publishing, pp. 147–151.

Chen, W.-P., Witkin, J.W. and Silverman, A.-J. (1989) Beta-endorphin and gonadotropin-releasing hormone synaptic input to gonadotropin-releasing hormone neurosecretory cells in the male rat. *J. Comp. Neurol.*, 286: 85–95.

Chiappa, S.A. and Fink, G. (1977) Hypothalamic luteinizing hormone releasing factor and corticotrophin releasing activity in relation to pituitary and plasma hormone levels in male and female rats. *J. Endocrinol.*, 72: 195–210.

Clayton, R.N., Young, L.S., Naik, S.I., Detta, A. and Abbot, S.D. (1986) Pituitary GnRH receptors – recent studies and their functional significance. In: G. Fink, A.J. Harmar and K.W. McKerns (Eds.), *Neuroendocrine Molecular Biology*, Plenum Press, New York, pp. 429–440.

Cross, B.A., Dyball, R.E.J., Dyer, R.G., Jones, C.W., Lincoln, D.W., Morris, J.F. and Pickering, B.T. (1975) Endocrine neurons. *Recent Prog. Horm. Res.*, 31: 243–286.

Dantzer, R., Koob, G.F., Bluthé, R.-M. and Le Moal, M. (1988) Septal vasopressin modulates social memory in male rats. *Brain Res.*, 457: 143–147.

De Vries, G.J., Duetz, W., Buijs, R.M., van Heerikhuize, J. and Vreeburg, J.T.M. (1986) Effects of androgens and estrogens on the vasopressin and oxytocin innervation of the adult rat brain. *Brain Res.*, 399: 296–302.

Djahanbakhch, O., McNeilly, A.S., Warner, P.M., Swanston, I.A.

and Baird, D.T. (1984) Changes in plasma levels of prolactin, in relation to those of FSH, oestradiol, androstenedione and progesterone around the preovulatory surge of LH in women. *Clin. Endocrinol.*, 20: 463.

Dockray, G.J. (1988) Regulatory peptides and the neuroendocrinology of gut-brain relations. *Q. J. Exp. Physiol.*, 73: 703–727.

Dow, R.C., Williams, B.C., Bennie, J., Carroll, S. and Fink, G. (1994) A central 5-HT_2 receptor mechanism plays a key role in the proestrus surge of luteinizing hormone but not prolactin in the rat. *Psychoneuroendocrinology*, in press.

Fink, G. (1976) The development of the releasing factor concept. *Clin. Endocrinol.*, 5: 245s–260s.

Fink, G. (1979a) Feedback actions of target hormones on hypothalamus and pituitary with special reference to gonadal steroids. *Annu. Rev. Physiol.*, 41: 571–585.

Fink, G. (1979b) Neuroendocrine control of gonadotrophin secretion. *Br. Med. Bull.*, 35: 155–160.

Fink, G. (1988) The G.W. Harris Lecture. Steroid control of brain and pituitary function. *Q. J. Exp. Physiol.*, 73: 257–293.

Fink, G. and Sheward, W.J. (1989) Neuropeptide release in vivo: measurement in hypophysial portal blood. In: G. Fink and A.J. Harmar (Eds.), *Neuropeptides: A Methodology*, Wiley, Chichester, UK, pp. 157–188.

Fink, G., Sarkar, D.K., Dow, R.C., Dick, H., Borthwick, N., Malnick, S. and Twine, M. (1982) Sex difference in response to alphaxalone anaesthesia may be oestrogen dependent. *Nature*, 298: 270–272.

Johnson Jr, R.G. (1988) Accummulation of biological amines into chromaffin granules: a model for hormone and neurotransmitter transport. *Physiol. Rev.*, 68: 232–307.

Johnson, M.S., Mitchell, R. and Thomson, F.J. (1992) The priming effect of luteinizing hormone-releasing hormone (LHRH) but not LHRH-induced gonadotropin release, can be prevented by certain protein kinase C inhibitors. *Mol. Cell. Endocrinol.*, 85: 183–193.

Kalra, S.P. (1986) Neural circuitry involved in the control of LHRH secretion: a model for preovulatory LH release. In: W.F. Ganong and L. Martini (Eds.), *Frontiers in Neuroendocrinology*, Raven Press, New York, pp. 31–75.

Majewska, M.D. (1987) Steroids and brain activity: essential dialogue between body and mind. *Biochem. Pharmacol.*, 36: 3781–3788.

Mayes, C.R., Watts, A.G., McQueen, J.K., Fink, G. and Charlton, H.M. (1988) Gonadal steroids influence neurophysin II distribution in the forebrain of normal and mutant mice. *Neuroscience*, 25: 1013–1022.

McEwen, B.S., de Kloet, E.R. and Rostene, W. (1986) Adrenal steroid receptors and actions in the nervous system. *Physiol. Rev.*, 66: 1121–1188.

McQueen, J.K. and Fink, G. (1988) Changes in local cerebral glucose utilization associated with the spontaneous ovulatory surge of luteinizing hormone in the rat. *Neuroendocrinology*, 47: 551–555.

Miller, M.A., Urban, J.H. and Dorsa, D.M. (1989) Steroid dependency of vasopressin neurons in the bed nucleus of the stria terminalis by *in situ* hybridization. *Endocrinology*, 125: 2335–2340.

Mitchell, R., Johnson, M., Ogier, S.-A. and Fink, G. (1988) Facilitated calcium mobilization and inositol phosphate production in the priming effect of LH-releasing hormone in the rat. *J. Endocrinol.*, 119: 293–301.

Mobbs, C.V., Fink, G. and Pfaff, D.W. (1990a) HIP-70: a protein induced by estrogen in the brain and LH-RH in the pituitary. *Science*, 247: 1477–1479.

Mobbs, C.V., Fink, G. and Pfaff, D.W. (1990b) HIP-70: An isoform of phosphoinositol-specific phospholipase C-α. *Science*, 249: 566–567.

Morello, H. and Taleisnik, S. (1988) The inhibition of proestrous LH surge and ovulation in rats bearing lesions of the dorsal raphe nucleus is mediated by the locus coeruleus. *Brain Res.*, 440: 227–231.

Morello, H., Caligaris, L., Haymal, B. and Taleisnik, S (1989) Inhibition of proestrous LH surge and ovulation in rats evoked by stimulation of the medial raphe nucleus involves a GABA-mediated mechanism. *Neuroendocrinology*, 50: 81–87.

North, W.G., Moses, A.M. and Share, L. (Eds.) (1993) The neurohypophysis: a window on brain function. *Ann. N. Y. Acad. Sci.*, 689: pp. 701.

Roberts, J.L., Wilcox, J.N. and Blum, M. (1986) The regulation of proopiomelanocortin gene expression by estrogen in the rat hypothalamus. In: G. Fink, A.J. Harmar and K.W. McKerns (Eds.), *Neuroendocrine Molecular Biology*, Plenum Press, New York, pp. 261–270.

Rosie, R., Thomson, E. and Fink, G. (1990) Oestrogen positive feedback stimulates the synthesis of LHRH mRNA in neurons of the rostral diencephalon of the rat. *J. Endocrinol.*, 124: 285–289.

Rosie, R., Wilson, H. and Fink, G. (1993) Testosterone induces an all-or-none, exponential increase in arginine vasopressin mRNA in the bed nucleus of stria terminalis of the *hypogonadal* mouse. *Mol. Cell. Neurosci.*, 4: 121–126.

Rosie, R., Sumner, B.E.H. and Fink, G. (1994) An α_1 adrenergic mechanism mediates estradiol stimulation of LHRH mRNA synthesis and estradiol inhibition of POMC mRNA synthesis in the hypothalamus of the prepubertal female rat. *J. Steroid Biochem. Mol. Biol.*, 49: in press.

Sachs, H., Fawcett, P., Takabatake, Y. and Portanova, R. (1969) Biosynthesis and release of vasopressin and neurophysin. *Recent Prog. Horm. Res.*, 25: 447–484.

Sarkar, D.K. and Fink, G. (1979a) Effects of gonadal steroids on output of luteinizing hormone releasing factor into pituitary stalk blood in the female rat. *J. Endocrinol.*, 80: 303–313.

Sarkar, D.K. and Fink, G. (1979b) Mechanism of the first spontaneous gonadotrophin surge and that induced by pregnant mare serum and effects of neonatal androgen in rats. *J. Endocrinol.*, 83: 339–354.

Sarkar, D.K. and Fink, G. (1981) Gonadotropin-releasing hormone surge: possible modulation through postsynaptic α-adrenoreceptors and two pharmacologically distinct dopamine receptors. *Endocrinology*, 108: 862–867.

Sarkar, D.K., Smith, G.C. and Fink, G. (1981) Effect of manipulating central catecholamines on puberty and the surge of luteinizing hormone and gonadotropin releasing hormone induced by pregnant mare serum gonadotropin in female rats. *Brain Res.*, 213: 335–349.

Shivers, B.D., Harlan, R.E., Morrell, J.E. and Pfaff, D.W. (1983) Absence of oestradiol concentration in cell nuclei of LHRH-immunoreactive neurones. *Nature*, 304: 345–347.

Sokol, H.W. and Valtin, H. (Eds.) (1982) The Brattleboro rat. *Ann. N. Y. Acad. Sci.*, 394: pp. 828.

Sumner, B.E.H. and Fink, G. (1993) Effects of acute estradiol on 5-hydroxytryptamine and dopamine receptor subtype mRNA expression in female rat brain. *Mol. Cell. Neurosci.*, 4: 83–92.

Thomson, E., Rosie, R., Blum, M., Roberts, J.L. and Fink, G. (1990) Estrogen positive feedback reduces arcuate pro-opiomelanocortin mRNA. *Neuroendocrinology*, 52: P3.17.

Weiner, R.I., Findell, P.R. and Kordon, C. (1988) Role of classic and peptide neuromediators in the neuroendocrine regulation of LH and prolactin. In: E. Knobil and J. Neill (Eds.), *The Physiology of Reproduction,* Raven Press, New York, pp. 1235–1281.

Wise, P.M., Scarbrough, K., Weiland, N.G. and Larson, G.H. (1990) Diurnal pattern of proopiomelanocortin gene expression in the arcuate nucleus of proestrous, ovariectomized, and steroid-treated rats; a possible role in cyclic luteinizing hormone secretion. *Mol. Endocrinol.* 4: 886–892.

F. Bloom (Editor)
Progress in Brain Research, Vol. 100

CHAPTER 19

Resolving a mystery: progress in understanding the function of adrenal steroid receptors in hippocampus

Bruce S. McEwen, Heather Cameron, Helen M. Chao, Elizabeth Gould, Victoria Luine[1], Ana Maria Magarinos, Constantine Pavlides, Robert L. Spencer, Yoshifumi Watanabe and Catherine Woolley

Laboratory of Neuroendocrinology, Rockefeller University, 1230 York Avenue, New York, NY 10021, USA; and [1]Department of Psychology, Hunter College, New York 10021, NY, USA

Introduction

In 1968, we reported that the adrenal glucocorticoid, corticosterone, was taken up and retained in high levels by the hippocampal formation of adrenalectomized rats (McEwen et al., 1968). The hippocampal localization of corticosterone was a complete surprise, since sex hormones, such as estradiol, were known to be taken up and retained by the hypothalamus and preoptic area of gonadectomized rats. Moreover, the hippocampus was not known to be directly associated with neuroendocrine function, whereas the hypothalamus and preoptic area are designated as the "hypophysiotrophic area" and are involved in control of gonadotrophin secretion and sexual behavior.

The hippocampus is recognized for its involvement in learning and memory, and for its plasticity in relation to spatial orientation and working memory (Eichenbaum and Otto, 1992). It has been only very recently that we have begun to understand some of the significance of adrenal steroid action on the hippocampus. The evolution of this knowledge is the subject of this short review. What we and others have found is that adrenal steroids play multiple roles in the structural and functional plasticity of the hippocampal formation.

Two adrenal steroid receptor types in the brain

Uptake and retention of ^{3}H steroid hormones by cells is the result of binding of the hormone by intracellular receptors that end up in the cell nuclear compartment, where these receptors bind to specific nucleotide sequences on DNA known as "hormone responsive elements" (Miner and Yamamoto, 1991). The uptake and retention of adrenal steroids by the hippocampus, first found in the rat, has been demonstrated in species as divergent as the duck and the rhesus monkey and this seems to be a widespread and evolutionarily stable trait of the hippocampal region (reviewed in McEwen et al., 1986).

Although it was first thought to reflect a single intracellular receptor type, we now know that two distinct adrenal steroid receptor types are present not only in hippocampus but also in other brain regions and in other tissues of the body. There were four steps in recognizing this fact (for review, see McEwen et al., 1986):

(1) Studies on the kidney in the early 1970s revealed two adrenal steroid receptor types, referred to as Type I and Type II;
(2) Radioimmunoassay of cell nuclear corticosterone in rats exposed in vivo to low, intermediate and

high levels of corticosterone revealed high affinity and lower affinity nuclear retention of the steroid;

(3) Type I and Type II receptors were definitively demonstrated in hippocampus using specific agonist and antagonist steroids in specific binding assays;

(4) The cloning of mineralocorticoid (Type I) and glucocorticoid (Type II) receptors demonstrated each form to be the product of a distinct gene; the mRNA for both receptors can be demonstrated in the hippocampus (Fig. 1). It is the Type I (mineralocorticoid) receptor that is concentrated in the hippocampal formation; this receptor, because of its high affinity for ^{3}H corticosterone, picks up and retains tracer levels of steroid injected into ADX rats, as we found in 1968 (see above).

Receptors in search of a function

Clues to the role of adrenal steroid receptors in hippocampus emerged slowly at first, but progress has accelerated during the past 10 years. The first effects of adrenal steroids concerned destruction of neural tissue, but later findings pointed to additional effects that protect neurons and enhance plasticity.

Neuronal destruction

The first clue regarding an effect of adrenal steroids in the hippocampus was the finding that ACTH or cortisone treatment of guinea pigs caused necrosis of pyramidal neurons of the hippocampus (for review, see

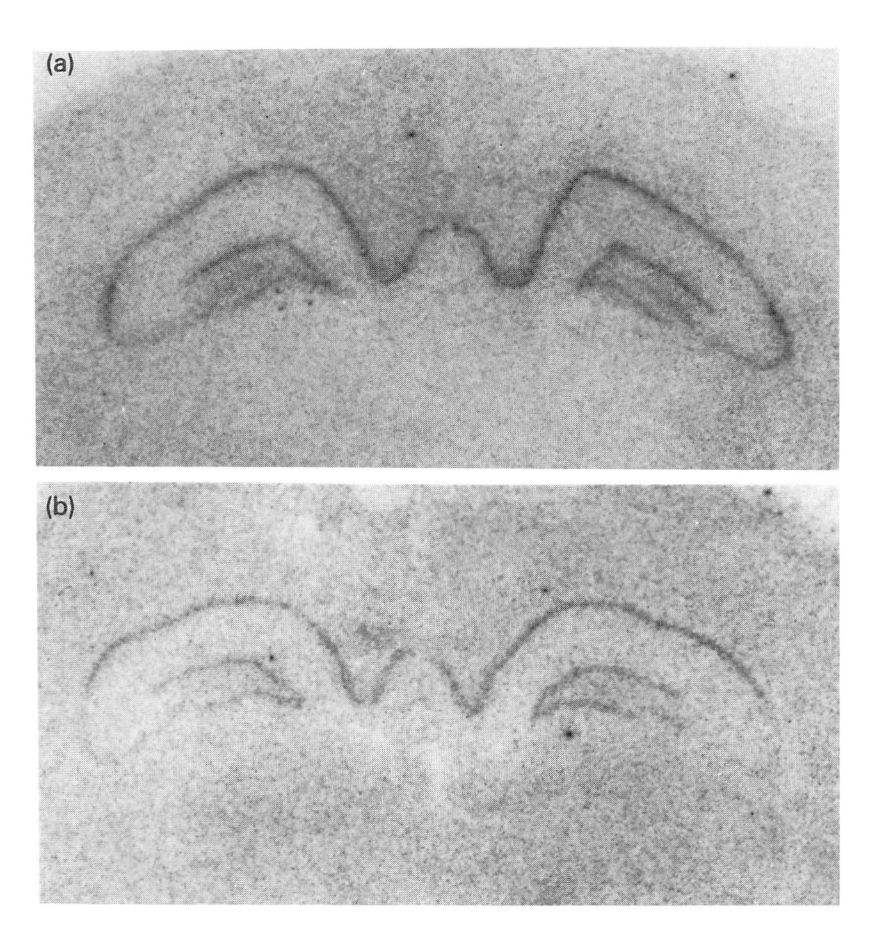

Fig. 1. In situ hybridization to rat hippocampus with riboprobes corresponding to (*A*) mineralocorticoid (Type I) receptor, MR; and (*B*) glucocorticoid (Type II) receptor, GR. Details of the hybridization procedure may be found in Chao et al. (1994).

Sapolsky, 1992). Subsequently, Landfield and coworkers found that aging in the rat results in some pyramidal neuron loss in hippocampus that can be retarded by adrenalectomy in mid-life (Landfield, 1987). Sapolsky then demonstrated that daily corticosterone injections into young adult rats over 12 weeks mimicked the pyramidal neuron loss seen in aging; he went on to demonstrate that excitatory amino acids play an important role in the cell loss by showing, first, that corticosterone exacerbates kainic acid-induced damage to hippocampus, as well as ischemic damage; and, second, that glucocorticoids potentiate excitatory amino acid-killing of hippocampal neurons in culture (Sapolsky, 1992).

In order to examine what actually happens to hippocampal neurons as a result of high levels of glucocorticoids, we have recently used the single section Golgi technique to demonstrate that, after 21 days of daily corticosterone exposure, the apical dendritic tree of CA3 pyramidal neurons has atrophied (Gould et al., 1991) (see Fig. 2). Moreover, this atrophy is prevented by the anti-epileptic drug, phenytoin, given prior to corticosterone each day, thus implicating the release and actions of excitatory amino acids in the process (Watanabe et al., 1992). The effect of corticosterone on dendritic length and branching is not found on basal dendrites of CA3 pyramidal neurons, nor on CA1 pyramidal neurons or dentate gyrus granule neurons; this specificity suggests that the mossy fiber input from the dentate gyrus may be involved (Gould et al., 1991). In this connection, the sensitivity of the CA3 pyramidal neurons is reminiscent of the specific damage to CA3 neurons as a result of kainic acid infusion or of epileptigenic stimulation of the perforant pathway, both of which are attributable to the mossy fiber input to CA3 from the dentate gyrus (for review, see McEwen et al., 1993).

In order to find out if the dendritic atrophy occurs under physiological conditions, we investigated the effects of repeated stress, which is known to induce glucocorticoid secretion. Restraint stress for 21 days produced atrophy that could be blocked by two agents: (1) cyanoketone, which inhibits adrenal steroid formation (Magarinos and McEwen, 1993); and (2) phenytoin, which prevents excitatory amino acid release and actions via T-type calcium channels (for data and review, see Watanabe et al., 1992). These findings support the notion that the stress-induced surge of glucocorticoids facilitates the release of excitatory amino acids, which, in turn, produces the morphological changes. Evidence in support of a causal link between adrenocortical secretion and glutamate release, measured indirectly by lactic acid formation, has been provided recently (Krugers et al., 1992).

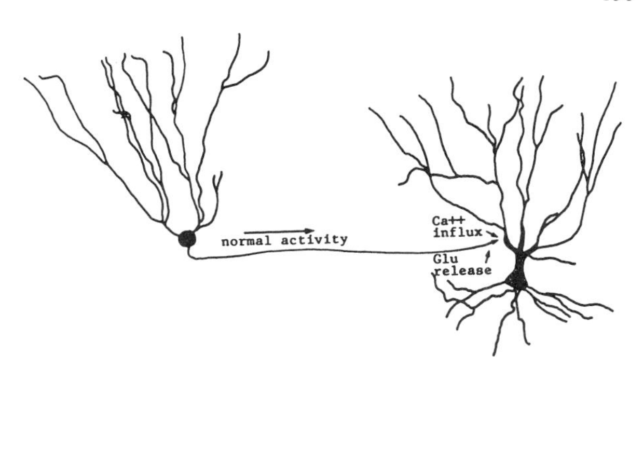

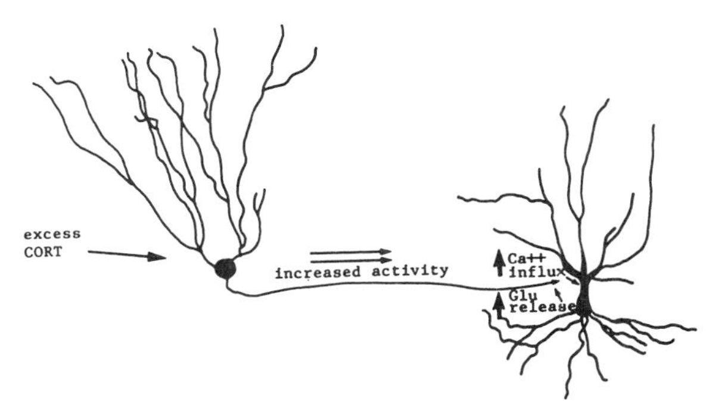

Fig. 2. Schematic diagram showing atrophy of apical dendrites of CA3 pyramidal neurons of hippocampus as a result of the synergistic interaction between excitatory amino acids released by mossy fibers and circulating glucocorticoids released by stress or administered by injection. Reprinted from Gould et al., 1991, by permission.

More severe and prolonged stress (e.g. cold swim; and social, i.e. dominance/subordinance hierarchies in rats) produce actual loss of hippocampal neurons (Mizoguchi et al., 1992; Sapolsky, 1992). We do not yet know the relationship of the dendritic atrophy to cell loss: i.e. whereas it is attractive to suppose that atrophy may represent the first stage of cell damage, the fact that this atrophy is reversible (Magarinos, unpublished data) and that it occurs on apical, but not on basal, dendrites argues that it may be an adaptive process by viable cells.

Death and replacement of dentate gyrus granule neurons

In contrast to pyramidal neurons of Ammons horn, granule neurons of the adult dentate gyrus depend on adrenal steroids for their survival (see Fig. 3). Moreover, continuous replacement of granule neurons occurs in the adult dentate gyrus; and both neurogenesis and cell death increase following adrenalectomy suggesting that it may be the death of neurons that signals the formation of replacement neurons (Gould and McEwen, 1993).

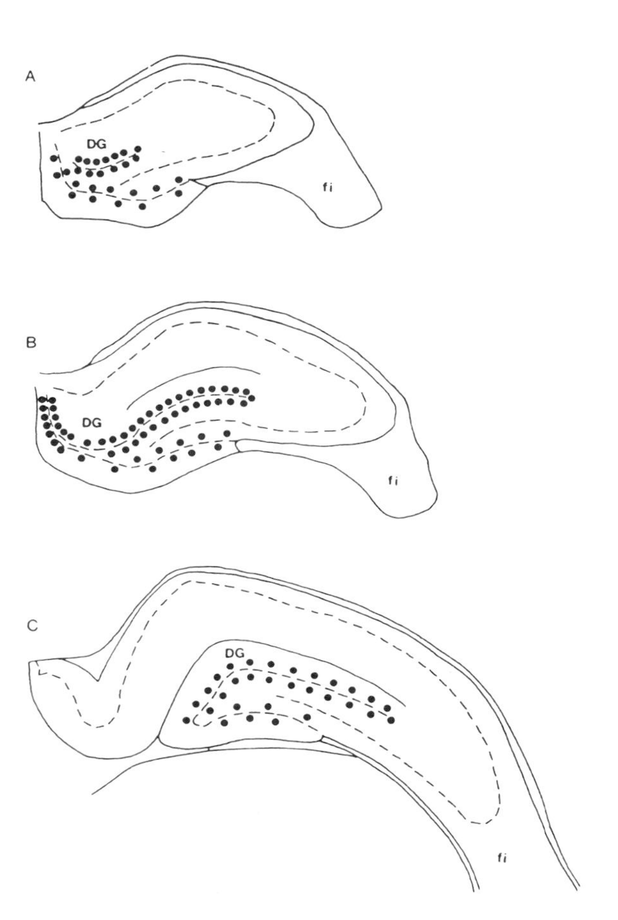

Fig. 3. Pyknotic granule neurons in dentate gyrus are distributed throughout entire hippocampal formation 7 days after bilateral adrenalectomy in adult life. Reprinted from Gould et al. (1990) by permission.

The first clue as to the magnitude of the adrenalectomy effect on neuronal death in dentate gyrus was the finding that, 3 months after ADX of adult rats, some rats showed almost total loss of dentate gyrus granule neurons (Sloviter et al., 1989). This finding attracted much attention because it contrasted with the prevalent view of stress- and glucocorticoid-induced neuronal death in Ammons horn. It was also very puzzling because only some ADX rats showed the effect.

In retrospect, we now understand both of these unusual aspects. First, the selectivity of the ADX effect on killing of dentate gyrus granule neurons has been verified in that it does not occur in other neuronal populations. Moreover, it may be related to the unusual nature of the dentate gyrus, which arises postnatally and continues to generate new neurons in adult life (Gould and McEwen, 1993). Second, the occurrence of cell death only in some rats is related to accessory adrenal tissue; when not removed at the time of ADX, this tissue can supply enough adrenal steroids to prevent neuronal loss; indeed, very low levels of adrenal steroids, sufficient to occupy Type I adrenal steroid receptors (see above) completely blocks dentate gyrus neuronal loss.

Adrenalectomy is not a natural event, and thus one must consider the physiological role of this type of plasticity. One explanation for why the dentate gyrus can make new cells in adult life, as well as get rid of them, is that the hippocampus may change during adult life in relation to its function in spatial orientation and memory (Sherry et al., 1992). Birds that use space around them to hide and locate food, and voles that use their environment to find mates, have larger hippocampal volumes than species that do not; moreover, there are indications that hippocampal volume may change under some circumstances, e.g. with the season of the year (Sherry et al., 1992). It remains to be seen whether it is the dentate gyrus that changes under these changing conditions.

Long-term potentiation

A much more rapid form of plasticity is long-term potentiation. A single burst of high frequency stimulation can immediately alter the responsiveness of neurons to further stimuli, an effect lasting over a time course of many hours to days. A number of in vitro studies have demonstrated, in the hippocampal CA1 field, that acute stress produces an impairment in LTP or in primed-burst potentiation (PBP) (for references, see Diamond et al., 1992; Pavlides et al., 1993a). In these studies, a negative correlation was also seen between the degree of LTP induced and plasma corticosterone levels. More recently, an inverted U-shaped curve was found with regard to induction of PBP in CA1 in relation to circulating glucocorticoid levels (Diamond et al., 1992). That is, both insufficiency as well as high levels of corticosterone had a detrimental effect on PBP; optimal potentiation was obtained at an intermediate level (10–20 μg/dl) of circulating corticosterone.

Within the dentate gyrus, an acute administration of corticosterone has also been shown to produce a decrement in LTP (for data and references, see Pavlides et al., 1993a). The most recent data, however, indicates that LTP can be modulated biphasically by adrenal steroids acting, respectively, via Type I and Type II receptors. In the dentate gyrus, LTP was found to be modulated by adrenal steroids acting via Type I and Type II receptors. Type I receptor stimulation in ADX rats was found to enhance LTP within an hour, whereas Type II receptor stimulation in ADX rats was found to suppress LTP (Pavlides et al., 1993b) (see Fig. 4). These biphasic effects help to explain a biphasic dose response curve with respect to induction of primed-burst potentiation (PBP) in the CA1 region of Ammon's horn, in which low levels of corticosterone within the daily variation of basal secretion facilitate PBP and stress levels of corticosterone inhibit PBP (Diamond et al., 1992).

That the CA1 responds in a similar manner as the dentate gyrus to corticosteroids is not surprising, for two reasons. First, Type I and Type II receptors coexist in pyramidal neurons and dentate gyrus granule neurons (for review, see McEwen et al., 1986). Second, in CA1 pyramidal neurons, corticosteroids biphasically modulate excitability: i.e. Type I receptor stimulation facilitates excitability by disinhibiting a 5HT1A receptor input; while Type II receptor stimulation inhibits excitability by suppressing a beta receptor-mediated noradrenergic input (for review, see Joels and DeKloet, 1992).

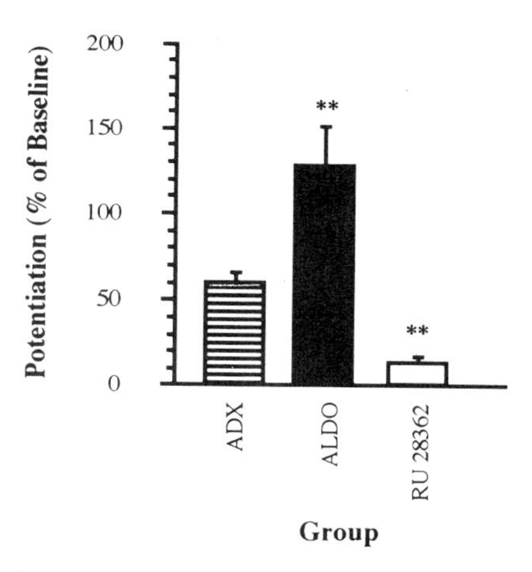

Fig. 4. Effects of Type I and Type II adrenal steroid receptors on dentate gyrus long-term potentiation. Extracellular field potentials were recorded in the dentate gyrus granule cell layer with stimulation of the medial perforant pathway. Animals were ADX (2 days prior to the experiment) and received an acute peripheral injection of either vehicle, the Type I agonist, aldosterone (100 μg/kg) or the Type II agonist, Ru 28362 (100 μg/kg). In comparison to the ADX controls, high frequency stimulation in the aldosterone injected animals produced a significant ($p < 0.001$) enhancement of long-term potentiation. In contrast, Ru 28362 elicited a pronounced suppression of LTP. Specific Type I and Type II antagonists reversed these effects (data not shown; see Pavlides et al., 1993a).

Conclusions

It appears that adrenal steroid receptors in the hippocampus are not an enigma, after all, and that they mediate at least three types of morphological and functional changes in the hippocampus. Thus, the role of the hippocampus in learning and memory processes may be subject to regulation by circulating adrenal steroids over both short and long time periods. What is the possible physiological or pathophysiological significance of each of these types of plasticity?

Long-term potentiation and primed-burst potentiation, as noted above, are biphasically regulated by adrenal steroids, with basal levels of corticosterone facilitating, and stress levels of corticosterone inhibiting, the potentiation of the response. Type I and Type II receptors appear to mediate the rising and falling limbs, respectively. Based on the work of Diamond et al. (1992), it is tempting to relate the rising limb of the biphasic curve to the increase in basal corticosterone secretion that precedes waking activity and the falling limb to the response to stressful stimuli. Since LTP or PBP may be related to learning and memory, it is important to note that there are indications that both adrenocortical insufficiency and exposure to excess adrenal steroids adversely affect cognitive function in experimental animals and humans (Martigioni et al., 1992; Conrad and Roy, 1993) and that the waking period is associated with greater sensitivity as far as hippocampal evoked responses than the sleeping phase of the diurnal cycle (for review, see McEwen et al., 1992).

However, the effects of adrenocortical excess or insufficiency may not be due to acute actions of the hormone, but rather to the longer-term changes in hippocampal morphology that have also been described. Repeated restraint stress, which causes atrophy of apical dendrites of CA3 pyramidal neurons, impairs the initial learning of an 8-arm radial maze (Villegas et al., 1993); and Cushing's disease has been linked to a reduction in hippocampal volume, which, in turn, is correlated with decreased verbal memory function (Starkman et al., 1992). These two preliminary studies need to be followed up by extensive investigations of glucocorticoid effects on both rodent spatial maze learning and human cognitive function.

Adrenalectomy, which causes increased turnover of dentate granule neurons, has been shown to result in impaired spatial learning (Armstrong et al., 1993; Conrad and Roy, 1993; Vaher et al., 1993). However, it is not yet clear whether neuronal loss per se is responsible for the behavioral deficits. At the same time, acute effects of adrenal steroid replacement on spatial learning have been described, involving differential effects mediated by Type I and Type II receptors, indicating that rapid actions of adrenal steroids must always be reckoned with (Oitzl and DeKloet, 1992).

Thus the story of adrenal steroid receptors in the hippocampus continues to unfold in ever-widening circles that encompass other aspects of neuroscience and other aspects of neuroendocrinology. The adult hippocampus is also the site of two other types of hormonally directed plasticity. Estradiol and progesterone regulate the cyclic formation and breakdown of excitatory synapses on CA1 pyramidal neurons (Woolley and McEwen, 1992, 1993); and excess thyroid hormone promotes a reduction in dendritic spines on CA1 pyramidal neurons, whereas transient elevation of thyroid hormone neonatally produces a long-term increase of dendritic branching and spine density in the CA3 region of the hippocampal formation (for review, see Gould et al., 1991). It remains to be seen what interactive roles thyroid hormone balance, sex hormones and gender play, together with the actions of adrenal steroids, in determining the ability of the hippocampus to change its structure and function in response to the external environment.

Acknowledgments

Research support for work described in this review was obtained from NIH Grants NS 07080 and MH41256 to BMc and MH49184 to EG, as well as Air Force Grant F 49620 to BMc and CP. AMM was supported by Fogarty Fellowhship F05TW04499.

References

Armstrong, J.D., McIntyre, D.C., Neubort, S. and Sloviter, R.S. (1993) Learning and memory after adrenalectomy-induced hippocampal dentate granule cell degeneration in the rat. *Hippocampus*, 3: 359–371.

Chao, H., Blanchard, D.C., Blanchard, R.J., McEwen, B.S. and Sakai, R.R. (1994) The effect of social stress on hippocampal gene expression. *Mol. Cell. Neurosci.*, 4: 543–548.

Conrad, C.D. and Roy, E.J. (1993) Selective loss of hippocampal granule cells following adrenalectomy: implications for spatial memory. *J. Neurosci.*, 13: 2582-2590.

Diamond, D.M., Bennet, M.C., Fleshner, M. and Rose, G.M. (1992) Inverted-U relationship between the level of peripheral corticosterone and the magnitude of hippocampal primed burst potentiation. *Hippocampus*, 2: 421–430.

Eichenbaum, H. and Otto, T. (1992) The hippocampus – what does it do? *Behav. Neural Biol.*, 57: 2–36.
Gould, E. and McEwen, B.S. (1993) Neuronal birth and death. *Curr. Opinion Neurobiol.*, 3: 676–682.
Gould, E., Woolley, C. and McEwen, B.S. (1991) The hippocampal formation: morphological changes induced by thyroid, gonadal and adrenal hormones. *Psychoneuroendocrinology*, 16: 67–84.
Gould, E., Woolley, C.S. and McEwen, B.S. (1990) Short-term glucocorticoid manipulations affect neuronal morphology and survival in the dentate gyrus. *Neuroscience* 37: 367–375.
Joels, M. and DeKloet, E.R. (1992) Control of neuronal excitability by corticosteroid hormones. *Trends Neurosci.*, 15: 25–30.
Krugers, H., Jaarsma, D. and Korf, J. (1992) Rat hippocampal lactate efflux during electroconvulsive shock or stress is differently dependent on entorhinal cortex and adrenal integrity. *J. Neurochem.*, 58: 826–830.
Landfield, P. (1987) Modulation of brain aging correlates by long-term alterations of adrenal steroids and neurally-active peptides. *Prog Brain Res.*, 72: 279–300.
Magarinos, A. and McEwen, B.S. (1993) Blockade of corticosterone synthesis by cyanoketone prevents chronic stress-induced dendritic atrophy of hippocampal neurons in the rat. *Abstr. Soc. Neursci.*, 17: 168, no. 71.1.
Martignoni, E., Costa, A., Sinforiani, E., Liuzzi, A., Chiodini, P., Mauri, M., Bono, G. and Nappi, G. (1992) The brain as a target for adrenocortical steroids: cognitive implications. *Psychoneuroendocrinology*, 17: 343–354.
McEwen, B.S., Weiss, J. and Schwartz, L. (1968) Selective retention of corticosterone by limbic structures in rat brain. *Nature*, 220: 911–912.
McEwen, B.S., DeKloet, R. and Rostene, W. (1986) Adrenal steroid receptors and actions in the nervous system. *Physiol. Rev.*, 66: 1121–1188.
McEwen, B., Angulo, J., Cameron, H., Chao, H., Daniels, D., Gannon, M., Gould, E., Mendelson, S., Sakai, R., Spencer, R. and Woolley, C. (1992) Paradoxical effects of adrenal steroids on the brain: protection versus degeneration. *Biol. Psychol.*, 31: 177–179.
McEwen, B.S., Spencer, R.L. and Sakai, R.R. (1993) Adrenal steroid actions upon the brain: versatile hormones with good and bad effects. In: J. Schulkin (Ed.), *Hormonally Induced Changes in Mind and Brain*, Academic Press, New York, pp. 157–189.
Miner, J. and Yamamoto K. (1991) Regulatory crosstalk at composite response elements. *Trends Biochem. Sci.*, 16: 423–426.
Mizoguchi, K., Kunishita, T., Chui, D.-H. and Tabira, T. (1992) Stress induces neuronal death in the hippocampus of castrated rats. *Neurosci. Lett.*, 138: 157–160.
Oitzl, M. and DeKloet, R. (1992) Selective corticosteroid antagonists modulate specific aspects of spatial orientation learning. *Behav. Neurosci.*, 106: 62–71.
Pavlides, C., Watanabe, Y. and McEwen, B.S. (1993a) Effects of glucocorticoids on hippocampal long-term potentiation. *Hippocampus*, 3: 183–192.
Pavlides, C., Watanabe, Y., Magarinos, A.M. and McEwen, B.S. (1993b) Opposing role of Type I and Type II adrenal steroid receptors in hippocampal long term potentiation. Unpublished.
Sapolsky R. (1992) *Stress, the Aging Brain and the Mechanisms of Neuron Death*, MIT Press, Cambridge, MA, 423 pp.
Sherry, D.F., Jacobs, L.F. and Gaulin, S.J. (1992) Spatial memory and adaptive specialization of the hippocampus. *Trends Neurosci.*, 15: 298–303.
Sloviter, R.S., Valiquette, G., Abrams, G. M., Ronk, C., Sollas, A.I., Paul, L.A. and Neubort, S.L. (1989) Selective loss of hippocampal granule cells in the mature rat brain after adrenalectomy. *Science*, 243: 535–538.
Starkman, M., Gebarski, S., Berent, S. and Schteingart, D. (1992) Hippocampal formation volume, memory dysfunction, and cortisol levels in patients with Cushing's syndrome. *Biol. Psychol.*, 32: 756–765.
Vaher, P., Luine, V.N., Gould, E. and McEwen, B.S. (1993) Spatial memory deficit after adrenalectomy in rats. *Abstr. Soc. Neurosci.*, 17: 366, no. 152.9.
Villegas, M., Martinez, C., Luine, V.N. and McEwen, B.S. (1993) Chronic stress impairs spatial memory in rats. *Abstr. Soc. Neurosci.*, 17: 1621, no. 664.18.
Watanabe, Y., Gould, E., Cameron, H., Daniels, D. and McEwen, B.S. (1992) Phenytoin prevents stress- and corticosterone-induced atrophy of CA3 pyramidal neurons. *Hippocampus*, 2: 431–436.
Woolley, C. and McEwen, B.S. (1992) Estradiol mediates fluctuation in hippocampal synapse density during the estrous cycle in the adult rat. *J. Neurosci.*, 12: 2549–2554.
Woolley, C. and McEwen, B.S. (1993) Roles of estradiol and progesterone in regulation of hippocampal dendritic spine density during the estrous cycle in the rat. *J. Comp. Neurol.*, 336: 293–306.

F. Bloom (Editor)
Progress in Brain Research, Vol. 100

CHAPTER 20

Neural-immune interactions

Suzanne Y. Felten and David L. Felten

Department of Neurobiology and Anatomy, Box 603, University of Rochester School of Medicine and Dentistry, 601 Elmwood Avenue, Rochester, NY. 14642, USA

Introduction

At the heart of the emerging discipline of psychoneuroimmunology is the contention that the nervous system can modulate the actions of the immune system. Further implications of this assumption are that stimuli from the environment of the organism that impact on and are processed by the nervous system, such as sensory stimuli, stress, and various psychosocial factors, can be translated by the nervous system into signals that may modulate the function of the immune system, thereby affecting health and potential illness from pathogens or tumors. Because the sum of experience, emotion, and environment come together in the CNS (limbic and cortical processing), the most provocative implication is that the "mind" (brain) may be able to influence health related to immunologic function.

Behavioral studies

Conditioning

In 1975, Ader and Cohen (reviewed in Ader and Cohen, 1991) reported that immune suppression caused by the drug cyclophosphamide could be conditioned using classical behavioral conditioning. This was not simply a conditioned stress response caused by glucocorticoid release, because it also could be elicited in adrenalectomized animals. Ader and colleagues further demonstrated that conditioned immunosuppression could be used to maintain autoimmune mice into old age with far less immunosuppressive drug than was required to keep unconditioned autoimmune mice alive. This work provides evidence that the CNS can detect immunologic reactivity of some types and can, in turn, generate immunoregulatory signals to the periphery. For these reasons, their work became one of the cornerstones of the emerging discipline of psychoneuroimmunology.

Stress

A variety of physical and psychological stressors have been found to alter cellular and humoral immunity (reviewed in Ader et al., 1990); sometimes positively, and often negatively, depending upon the nature and timing of the stressor. Some models of disease have shown a poorer response to challenge following stress (e.g. influenza challenge in C57BL/6 mice). In some models, stress-induced alterations in specific measures of immune response appear to act through glucocorticoid interactions, some appear to act through sympathetic noradrenergic signaling, and others appear to act through neurotransmitter or hormonal systems not yet fully identified (e.g. Cunnick et al., 1990). Subtle variations in the timing and nature of the stressor can alter the pattern of outflow of neurotransmitters and hormones from the CNS, which in turn, can differentially influence a variety of measures of immunologic responsiveness. The most important questions yet to be answered are to what extent these altered measures of immunologic responsiveness can effect an animal in a given state when challenged by a virus, other pathogens, tumors, or autoimmune challenge, and to what extent we can intervene during a stressor with pharmacologic agents modulating the relevant mediators to block adverse immunologic consequences of the stressor.

Endocrine system

Since the early demonstration that hypophysectomy can alter the development and function of the immune system, neurohormones have been investigated for immunomodulatory activity. Physicians have been using glucocorticoids as anti-inflammatory agents for decades. The observations that many autoimmune diseases are more prevalent in women than in men suggests an interaction of the sex hormones in this process. Virtually all anterior pituitary hormones, and most peripheral target organ hormones can alter specific immunologic activity, sometimes profoundly. These complexities are well beyond this review, and are summarized in many chapters in *Psychoneuroimmunology*, 2nd edition (Ader et al., 1991). Some cells of the immune system, including T lymphocytes, can produce classical anterior pituitary hormones in response to viral stimulation or releasing factor challenge, further blurring the boundaries between cytokines and neurotransmitters/neurohormones. An important unanswered question regarding this novel production of classical hormones by cells of the immune system is the extent of their involvement in ongoing immune response to a disease challenge.

The most commonly acknowledged hormonal system that influences immunologic reactivy is the hypothalamo-pituitary-adrenal (HPA) axis via glucocorticoids. Although glucocorticoids inhibit a variety of macrophage, T cell, and B cell functions, they are not universally inhibitory (Munck and Guyre, 1991). They can activate latent viruses, and shift the CD4 cell response from cellular (T_{H1}) to humoral immunity (T_{H2} predominant). This may be an important factor in the progression of HIV infection. Recent evidence indicates that immune responses to virus and other challenges (mainly via IL-1 and perhaps other cytokines such as IL-6 and TNF-α) can themselves induce CRF-ACTH-glucocorticoid secretion (Besedovsky and del Rey, 1991). This may act as a counter-regulatory signal to dampen proliferation and other immunologic activation that could be damaging to the organism (e.g. autoimmune provoking). The role of the HPA axis in autoimmunity, viral infections, and the balance between cellular and humoral immunity is an important area of clinical interest in neural-immune signaling. The balance between immunologically induced and CNS stress-induced HPA axis activation may be critical in the progression and severity of relevant diseases.

A further system that has provoked interest is growth hormone (GH). GH administration to old mice can restore the weight and cellularity to the involuted thymus, and can restore T cell proliferation, cytokine production and functioning (Kelley, 1991).

Autonomic nervous system

Sympathetic innervation

Both primary (bone marrow and thymus) and secondary (spleen, lymph nodes, mucosal associated) lymphoid organs are innervated by noradrenergic (NA) and peptidergic nerve fibers. Fluorescence histochemical studies of spleen and thymus innervation done in our laboratories (e.g. Williams and Felten, 1981) found NA innervation present in smooth muscle compartments such as the vascular, and capsular/trabecular areas and also in the parenchyma with no clear vascular association. Immunocytochemical studies in the rat revealed that, NA innervation was associated primarily with areas occupied by T cells (e.g. the periarteriolar lymphatic sheath in the spleen, the paracortical zone of the lymph node, and the thymic cortex), and areas with many macrophages (e.g. the marginal zone of the spleen, the medullary cords of the lymph node), but tended to avoid B cell zones (lymphoid follicles) (Felten et al., 1987a; reviewed in Felten and Felten, 1991).

Examination of the innervation of the spleen at the electron microscopic level revealed tyrosine hydroxylase positive nerve terminals in direct contact with T lymphocytes (Felten and Olschowka, 1987; Felten and Felten, 1991; Felten et al., 1992), in both adventitial zones and deeper regions populated by lymphoid cells. NE released from more distant terminals can diffuse and interact with receptors on lymphocytes, resulting in modulation of their function. The role of these close neuro-effector contacts on a small subset of lymphoid cells is unclear. It has not yet been determined whether these lymphocytes have special characteristics that cause them to have contact with nerve terminals. It is possible that they serve as intermediates in translating neuronal signals into lymphocyte signals (cytokines), but it is equally likely that they are in contact with neurons to provide cytokine signals to regulate release of neurotransmitters (e.g. IL-1 or IL-2 influencing NE release) from local nerve terminals.

Parasympathetic and peptidergic innervation

Although parasympathetic (cholinergic) innervation has been reported in the thymus, its extent is not clear, nor has it been demonstrated in other lymphoid organs (reviewed in Felten and Felten, 1991). However, extensive substance P, calcitonin gene-related peptide, vasoactive intestinal peptide, neuropeptide Y (often co-localized with NE) and other peptides have been found in nerves in the parenchyma of bone marrow, thymus, spleen, lymph nodes and mucosal-associated lymphoid tissues (for reviews, see *Brain, Behavior, and Immunity* Special Issue, and Bellinger et al., 1990).

Receptors

Lymphocytes and macrophages have receptors for a number of neurotransmitters (reviewed in Ader et al., 1990; Carr and Blalock, 1991), including α and β adrenergic receptors, muscarinic cholinergic receptors and receptors for vasoactive intestinal polypeptide (VIP), somatostatin, and substance P. Other receptors have been suggested, but are not yet well characterized.

Perhaps the best receptor characterization is of the β_2-adrenergic receptor, which has been demonstrated on both T and B lymphocytes and on macrophages. β-receptor agonists have been shown to increase cAMP production (reviewed in Roszman and Carlson, 1991) as would be expected. However, such agonists can also act synergistically with activators of lymphocytes such as phytohemagglutinin (PHA) which alone does not increase cAMP, to greatly augment cAMP production in T lymphocytes. β-Agonists also can synergize cAMP generation in T lymphocytes activated by T cell receptor stimulation with anti-CD3 antibody.

Functional significance

Two major strategies have been used to investigate the functional role of neurotransmitters in immunomodulation: (1) pharmacologic interventions; and (2) denervation of lymphoid organs. The NA sympathetic system has been studied in greatest detail. Following the transmitter or nerve manipulation, immunologic responses have been measured (reviewed in Felten et al., 1993), including: (1) individual cellular functions of specific subsets (mainly T cells and B cells), including proliferation, differentiation, production and release of specific products (immunoglobulin isotypes, cytokines), or trafficking; (2) collective cellular interactions such as primary and secondary antibody responses, cytotoxic T cell responses, delayed type hypersensitivity responses, natural killer responses; and (3) response of the organism to challenge by viruses, bacteria, tumors, or autoimmunity. This hierarchical approach has permitted investigation of the role of neurotransmitters in modulating disease-relevant immune functions, while pursuing more mechanistic and molecular aspects of such interactions.

In vitro studies have shown that NE or other β-agonists, when present at the beginning of cell culture, can enhance primary antibody responses and cellular immune responses (e.g. cytotoxic T cell responses), probably through enhancement of the initiative phase of immune responses (reviewed in Madden and Livnat, 1991). In contrast, when NE or β-agonists are added at the effector cell phase, these effector cells are inhibited (e.g. antibody production, cytotoxic T cell activity are decreased). NE can stimulate the pre-B cells that develop into IgM-producing cells. Thus, timing of the signal is very important, and both the cell subset and its stage of activation can determine the response to the signal.

Denervation of non-immune (i.e. non-stimulated murine lymph nodes of NE nerve fibers with 6-hydroxydopamine results in increased cell proliferation, altered cellularity (diminished T cells, enhanced B cells) and mitogen responses, and an immunoglobulin isotype switch (in the absence of antigen challenge) from IgM to IgG, accompanied by increased secretion of IFN-γ by T_{H1} T helper cells (Madden et al., 1994a,b). In challenged mice, NA denervation results in diminished primary antibody responses (80% decrease in spleen challenged systemically, 97% decrease in LNs challenged in the draining site), enhanced secondary antibody responses, diminished cytotoxic T lymphocyte responses (50%) and IL-2 secretion, diminished delayed-type hypersensitivity responses, and enhanced natural killer cell responses (for review, see Bellinger et al., 1992). Thus, it appears that NE stimulates T_{H1} responses (cellular immunity) while NA denervation diminishes them. However, humoral immunity is also affected, perhaps with additional signaling to B cells, with NE favoring IgM responses and NA denervation shifting away from IgM towards IgG. These studies suggest that NE signaling is neither simplistic nor unidirectional, but may differentially influence multiple cell types at multiple stages during a reaction. Working out the nature and timing of these interactions is one of the major challenges of the field. Further, during an infection, neurotransmitters may influence three separate sites: (1) the site of initial presentation and continuing presentation as a replicating antigen; (2) the site of generation of an immune response (usually a secondary lymphoid organ); and (3) the site of inflammation or response to the infection.

Other neurotransmitters have also been studied for immunomodulatory activity. For example, substance P can stimulate T cell proliferation, enhance antibody production of IgA and IgM, enhance cellular immune responses, and stimulate macrophage chemotaxis (Payan, 1992). During an ongoing behavior, many sets of nerve fibers, releasing many neurotransmitters differentially at a variety of sites of immunologic reactivy, may simultaneously impinge on many subsets of lymphoid cells to achieve a collective effect. Sorting out this signaling will be as complex as sorting out cytokine influences on immune responses.

Autoimmune disease

Although denervation studies have demonstrated that the absence of NE is often related to a decreased immune response, especially of T cells, studies using several models of autoimmune disease demonstrate that sympathetic denervation also may exacerbate the development of experimental autoimmune disease (such as experimental allergic encephalomyelitis or experimentally induced arthritis) in susceptible animals (reviewed in Levine et al., 1991; Bellinger et al., 1993). Although this could be interpreted superficially as indicating that NE has an inhibitory effect on immune responses that is removed with sympathectomy, thereby causing augmented responses leading to autoimmunity, it is much more accurate to view sympathectomy as removing an important regulatory influence that normally promotes an optimal response. Autoimmune disease should not be viewed as an increased immune response, but rather as an inappropriate response. The ability of NE to suppress this abnormal response should be seen as an enhancement of normal function.

Aging

Our investigations of sympathetic innervation of lymphoid organs in aging Fischer 344 rats (Felten et al., 1987b; reviewed in Ackerman et al., 1991; Bellinger et al., 1993), have shown that the spleen and lymph nodes of 27-month-old rats has greatly diminished sympathetic innervation compared with their younger counterparts. These aging rats showed diminished T cell proliferation, IL-2 secretion, and cellular immune responses. The similarity of immune response in normal aging (with age-related loss of NE nerves) and younger rats experimentally denervated of NA nerve fibers, suggests that some aspects of immunosenescence (mainly diminished T cell responses) may be causally related to loss of NA innervation of secondary organs.

Conclusions

Although we have learned much about neurotransmitter and neurohormone influences on immune responses, we have just scratched the surface. We view the following areas of investigation to be important steps in further expanding our understanding of neural modulation of immune responses. (1) Further identification of chemically specific nerve fibers that distribute to lymphoid organs and tissues, the specific cell types with which they associate, the receptor expression for those neurotransmitters on specific subsets of lymphoid cells, and variability by gender, age, strain, and other characteristics. (2) Dermination of the specific immunologic responses, in vitro and in vivo, that are induced by each neurotransmitter or neurohormone in question (dose-response studies necessary, for physiological correlations). This includes looking at synergistic interactions of these molecules with each other and with cytokines. (3) Determination of the pattern of neurotransmitters and neurohormones resulting from specific behavioral paradigms (and how those patterns change with time), thereby cataloging "stress" responses according to the pattern of efferent signaling rather than the sensory characteristics of the stimulus. (4) Determination of the role of these signal molecules, both singly and in combination, during ongoing challenges from viruses, bacteria, tumors, autoimmune reactivity, and other immunologic challenges. (5) Use of neurotransmitter agonists or antagonists to deliberately enhance or inhibit specific reactions of specific cells of the immune system during a specific immunologic challenge, with the prospect of intervening for the benefit of patients. (6) Investigation of immunologic development and neurotransmitter or neurohormone signaling, continuing throughout adulthood into immunosenescence. (7) Distribution of cytokines (constituitive and inducible) and cytokine receptors in the nervous system. (8) Effects of cytokines on specific CNS structures and influences on subsequent outflow back to the immune system, studied both as individual cytokines and collective groups of cytokines that occur during infections or other immunologic reactions. (9) Pathways in the CNS utilized during stressors, conditioning, or other immunologically relevant states, to achieve signaling to the immune system. (10) A complete mapping of any specific neural-immune interactions from the initial behavior, to the CNS cells and reactions induced by that behavior, to the pathways activated by that behavior, to the collective timing and output of the signal molecules secreted or released, to the interactions with specific cells of the immune system at specific sites, to the altered collective interactions of those cells, to the influence exerted by these cells on health and illness during a specific challenge to the immune system. With this type of comprehensive sequential understanding of specific neural-immune interactions, we will be at the threshold of explaining psychoneuroimmunology at the level of biological signaling. However, the important extension of these observations to intervention with neural agents in infections, tumors, autoimmunity, and other immune-relevant diseases will depend on a better understanding of how individual cell responses contribute to the overall reactivity of the immune system. Perhaps one of the most important contributions of neural-immune signaling is the conspicuous necessity of requiring a hierarchical understanding of the immune system itself from the molecular and cellular level through collective cellular interactions, to the overall immunophysiology of the entire system in a challenged host.

Acknowledgements

This work was supported in part by NS 25223 from the National Institutes of Health, and MH 00899, P50 MH 40381, and MH 42076 from the National Institutes of Mental Health.

References

Ackerman, K.D., Bellinger, D.L., Felten, S.Y. and Felten, D.L. (1991) Ontogeny and senescence of noradrenergic innervation of the rodent thymus and spleen. In: R. Ader, D.L. Felten and N. Cohen (Eds.), *Psychoneuroimmunology*, 2nd edition, Academic Press, San Diego, pp. 72–125.

Ader, R. and Cohen, N. (1991) The influence of conditioning on immune responses. In: R. Ader, D.L. Felten and N. Cohen (Eds.), *Psychoneuroimmunology*, 2nd edition, Academic Press, San Diego, pp. 611–646.

Ader, R., Felten, D.L. and Cohen, N. (1990) Interactions between the brain and the immune system. *Annu. Rev. Pharmacol. Toxicol.*, 30: 561–602.

Ader, R., Felten, D.L. and Cohen, N. (Eds.) (1991) *Psychoneuroimmunology*, 2nd edition, Academic Press, San Diego.

Bellinger, D.L., Lorton, D., Romano, T., Olschowka, J.A., Felten, S.Y. and Felten, D.L. (1990) Neuropeptide innervation of lymphoid organs. *Ann. N. Y. Acad. Sci.*, 594: 17–33.

Bellinger, D.L., Felten, S.Y. and Felten, D.L. (1992) Neural-immune interactions: Neurotransmitter signaling of cells of the immune system. In: A. Tasman and M.B. Riba (Eds.), *Review of Psychiatry*, American Psychiatric Press, Washington, DC, pp. 127–144.

Bellinger, D.L., Felten, S.Y., Ackerman, K.D., Lorton, D., Madden, K.S. and Felten, D.L. (1993) Noradrenergic sympathetic innervation of lymphoid organs during development, aging, and in autoimmune disease. In: F. Amenta (Ed.), *Aging of the Autonomic Nervous System*, CRC Press, Boca Raton, FL, pp. 243–284.

Besedovsky, H.O. and del Rey, A. (1991) Physiological implications of immune-neuro-endocrine network. In: R. Ader, D.L. Felten and N. Cohen (Eds.), *Psychoneuroimmunology*, 2nd edition, Academic Press, San Diego, pp. 589–608.

Carr, D.J.J. and Blalock, J.E. (1991) Neuropeptide hormones and receptors common to the immune and neuroendocrine systems: Bidirectional pathway of intersystem communication. In: R. Ader, D.L. Felten and N. Cohen (Eds.), *Psychoneuroimmunology*, 2nd edition, Academic Press, New York, pp. 573–588.

Cunnick, J.E., Lysle, D.T., Kucinski, B.J. and Rabin, B.S. (1990) Evidence that shock-induced immune suppression is mediated by adrenal hormones and peripheral b-adrenergic receptors. *Pharmacol. Biochem. Behav.*, 36: 645–651.

Felten, S.Y. and Felten, D.L. (1991) Innervation of lymphoid tissue. In: R. Ader, D.L. Felten and N. Cohen (Eds.), *Psychoneuroimmunology*, 2nd edition, Academic Press, New York, pp. 27–68.

Felten, S.Y. and Olschowka, J.A. (1987) Noradrenergic sympathetic innervation of the spleen: II. Tyrosine hydroxylase (TH)-positive nerve terminals form synaptic-like contacts on lymphocytes in the splenic white pulp. *J. Neurosci. Res.*, 18: 37–48.

Felten, D.L., Ackerman, K.D., Wiegand, S.J. and Felten, S.Y. (1987a) Noradrenergic sympathetic innervation of the spleen: I. Nerve fibers associate with lymphocytes and macrophages in specific compartments of the splenic white pulp. *J. Neurosci. Res.*, 18: 28–36.

Felten, S.Y., Bellinger, D.L., Collier, T.J., Coleman, P.D. and Felten, D.L. (1987b) Decreased sympathetic innervation of spleen in aged Fischer 344 rats. *Neurobiol. Aging*, 8: 159–165.

Felten, S.Y., Felten, D.L., Bellinger, D.L. and Olschowka, J.A. (1992) Noradrenergic and peptidergic innervation of lymphoid organs. In: J.E. Blalock (Ed.), *Chemical Immunology: Neuroimmunoendocrinology*, 2nd edition, S. Karger, Basel, pp. 25–48.

Felten, D.L., Felten, S.Y., Bellinger, D.L. and Madden, K.S. (1993) Fundamental aspects of neural-immune signaling. *Psychother. Psychosom.*, 60: 46–56.

Kelley, K.W. (1991) Growth hormone in immunobiology. In: R. Ader, D.L. Felten and N. Cohen (Eds.), *Psychoneuroimmunology*, 2nd edition, Academic Press, San Diego, CA, pp. 337–402.

Levine, J.D., Goetzl, E.J. and Basbaum, A.I. (1991) Contribution of the nervous system to the pathophysiology of rheumatoid arthritis and other polyarthritides. *Rheum. Dis. Clin. N. Am.*, 13: 369–383.

Madden, K.S. and Livnat, S. (1991) Catecholaminergic influences on immune reactivity. In: R. Ader, D.L. Felten and N. Cohen (Eds.), *Psychoneuroimmunology*, 2nd edition, Academic Press, San Diego, CA, pp. 283–310.

Madden, K.S., Felten, S.Y., Felten, D.L., Hardy, C.A. and Livnat, S. (1994a) Sympathetic nervous system modulation of the immune system. II. Induction of lymphocyte proliferation and migration in vivo by chemical sympathectomy. *J. Neuroimmunol.*, 49: 67–75.

Madden, K.S., Moynihan, J.a., Brenner, G.J., Felten, S.Y., Felten, D.L. and Livnat, S. (1994b) Sympathetic nervous system modulation of the immune system. III. Alterations in T and B cell proliferation and differentiation in vitro following chemical sympathectomy. *J. Neuroimmunol.*, 49: 77–87.

Munck, A. and Guyre, P.M. (1991) Glucocorticoids and immune function. In: R. Ader, D.L. Felten and N. Cohen (Eds.), *Psychoneuroimmunology*, 2nd edition, Academic Press, San Diego, CA, pp. 447–474.

Payan, D.G. (1992) The role of neuropeptides in inflammation. In: J.I. Gallin, I.M. Goldstein and R. Snyderman (Eds.), *Inflammation: Basic Principles and Clinical Correlates*, 2nd edition, Raven Press, New York, pp. 177–192.

Roszman, T.L. and Carlson, S.L. (1991) Neurotransmitters and molecular signaling in the immune response. In: R. Ader, D.L. Felten and N. Cohen (Eds.), *Psychoneuroimmunology*, 2nd edition, Academic Press, San Diego, CA, pp. 311–335.

Special Issue (1991) Peptidergic localization and innervation of lymphoid tissue. *Brain Behav. Immun.*, 5: 1–147.

Williams, J.M. and Felten, D.L. (1981) Sympathetic innervation of murine thymus and spleen: a comparative histofluorescence study. *Anat. Rec.*, 199: 531–542.

F. Bloom (Editor)
Progress in Brain Research, Vol. 100

CHAPTER 21

Human cortical functions revealed by magnetoencephalography

Riitta Hari

Low Temperature Laboratory, Helsinki University of Technology, 02150 Espoo, Finland

Introduction

Magnetoencephalography, MEG, complements the traditional microlevel and macrolevel approaches to unravelling the functions of the human brain. MEG recordings are based on non-invasive detection of weak magnetic fields produced by cerebral electric currents. Signals from various cortical regions can be differentiated with good temporal and spatial resolution. MEG studies have so far focused on functional mapping of the healthy human cerebral cortex but some clinical applications are also emerging. For details of the MEG method, see previous reviews (Hari and Lounasmaa, 1989; Sato, 1990; Hämäläinen et al., 1993).

MEG is closely related to EEG, the recording of electric potential differences on the scalp. MEG patterns caused by multiple simultaneous sources are often more straightforward to interpret than the corresponding EEG distributions. One reason for this is that concentric electric inhomogeneities do not affect the magnetic field. A combination of MEG and EEG is necessary for identifying all active brain areas and all orientations of the source currents (cf. Wikswo et al., 1993).

Measurements

Electric currents flowing in the brain generate weak magnetic signals, typically 100–1000 fT (fT = femto-Tesla = 10^{-15} T), i.e. only one part in 10^9 or 10^8 of the Earth's geomagnetic field. The signals are first picked up by superconducting flux transformers (Fig. 1*A*) and then detected by SQUIDs (Superconducting QUantum Interference Devices), which are extremely sensitive to magnetic fields. The configuration of the flux transformer (a magnetometer, axial gradiometer, or a planar gradiometer; Fig. 1*B*) affects the instrument's sensitivity to various brain versus noise sources.

To localize the underlying neural activity, the field pattern outside the head must be determined in detail. Instead of earlier time-consuming sequential mapping with (1–37)-channel instruments, it is now possible to record signals simultaneously over the entire cortex. The first whole-head helmet-type magnetometers were constructed recently by Neuromag Ltd (Espoo, Finland; 122 channels, Fig. 1*C*) and by CTF Systems Inc. (Vancouver, Canada; 64 channels).

Origin and interpretation of MEG signals

An inherent limitation of the MEG (and EEG) method is the non-uniqueness of the inverse problem. This means that, in principle, several current distributions may produce identical signal patterns outside the brain. Therefore, source models are used for data interpretation. The most common model of a local source is a tangential current dipole within a sphere. The orientation, strength and three-dimensional location of the equivalent current dipole (ECD), which best explains the measured field pattern, are determined by means of a least-squares fit to the data. The

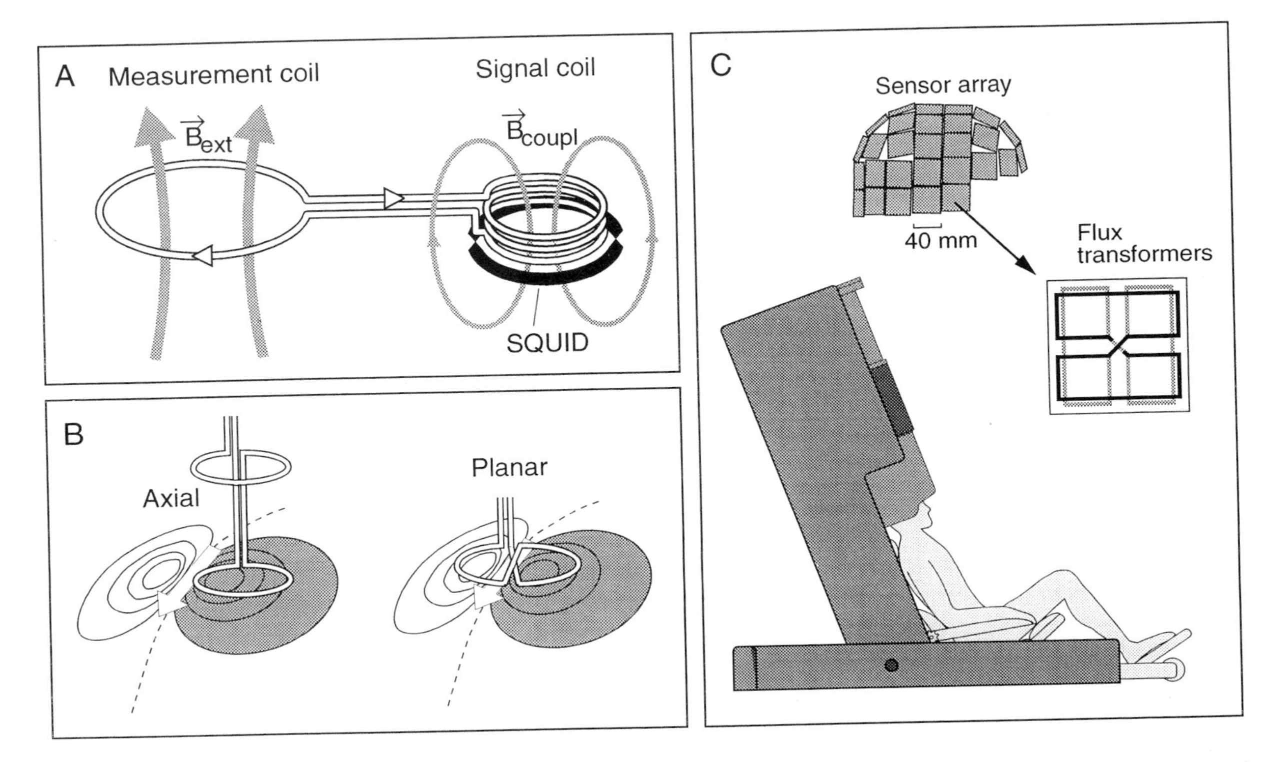

Fig. 1. *A*, Principle of MEG signal detection. The external magnetic field induces into the superconducting flux transformer a current that couples the magnetic field from the signal coil to the SQUID. *B*, Two flux transformer configurations above a schematic magnetic field pattern produced by a current dipole. The axial gradiometer picks the maximum signal at both field extrema whereas the planar figure-of-eight transformer senses the strongest signal (gradient) just above the dipole. *C*, Schematic illustration of a whole-head neuromagnetometer. The subject's head is inside the dewar, which contains the SQUID sensors immersed in liquid helium. With this 122-channel device (Neuromag-122^{TM}), the gradient of the magnetic field is picked up outside the head simultaneously at 61 locations of the helmet-shaped sensor array (upper insert); at each location, two orthogonal field gradients are measured with planar flux transformers of figure-of-eight configuration (lower insert). The experiments are carried out inside a magnetically shielded room.

MEG signals are caused mainly by intracellular synaptic currents in the fissural cortex (cf. Hämäläinen et al., 1993). Physiologically, the ECD reflects the centre of gravity of an active cortical layer smaller than 2 cm in diameter. Multiple simultaneous sources can be modelled with time-varying multidipole models.

The solutions can be improved considerably by incorporating the anatomical constraints obtained, for example, from magnetic resonance images (MRIs). Such a combination of structural (MRI) and functional (MEG) information is especially useful for clinical purposes. Future developments will also include integration of MEG and positron emission tomography (PET) data.

Examples

Auditory responses and sensory memory

Figure 2*A* shows auditory evoked magnetic fields elicited by short tones. The most prominent deflection, N100m, peaks about 100 ms after sound onset and has dipolar field patterns over both temporal lobes (Fig. 2*B*). The signals are slightly larger and earlier in the left (contralateral) than the right hemisphere. A two-dipole model, one ECD in each supratemporal cortex, explained the signal distribution satisfactorily. Superposition of the ECDs on the 3D-MRI reconstruction (Fig. 2*C*) implies that N100m may receive a contribution from Heschl's gyrus (cf. Tissari et al., 1993), although its main source is probably in the cortex of planum temporale.

Sensory memory in its simple form can be studied by determining recovery cycles of different MEG responses. The N100m amplitude increases with the interstimulus interval (ISI) and reaches a plateau at ISIs of about 8 s. As shown in Fig. 2*D*, the ISI dependence is similar in the left and right hemispheres, implying that tones leave traces of similar duration into auditory cortices of both hemispheres. The storage of auditory information for a few seconds can also be demonstrated by presenting infrequent deviant sounds among a series of monotonously repeated standard sounds.

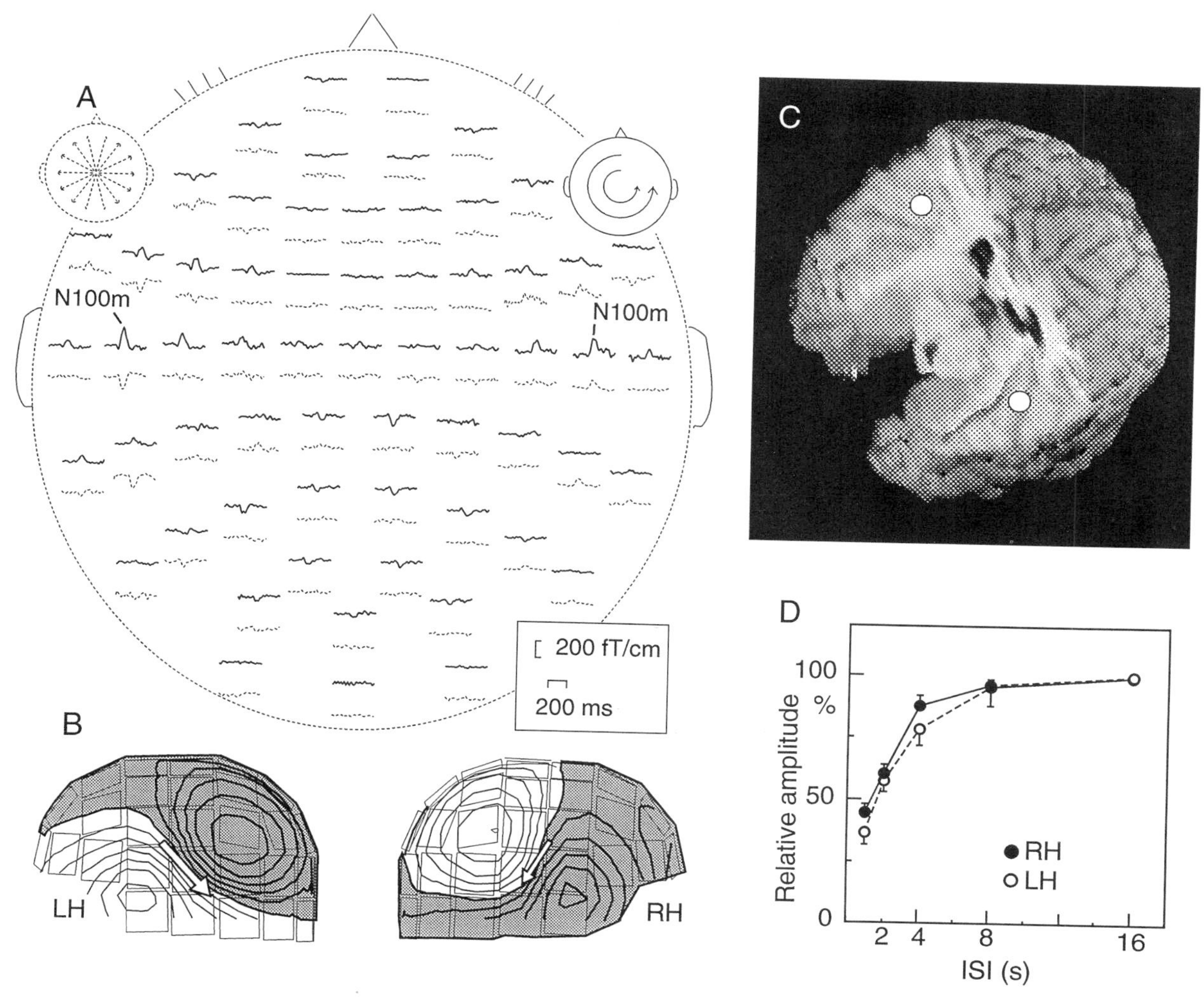

Fig. 2. *A*, AEFs to 50-ms tones presented to the right ear once every 4 s. In each response pair, the upper (solid) curves indicate the azimuthal and the lower (dotted) ones the polar derivatives (illustrated in the right- and left-sided insert heads, respectively); the relative amplitudes of the two traces indicate the direction of the source current at each location. The head is viewed from above, with the nose pointing to the top of the figure. The traces start 50 ms before the tone onset. Passband is 0.03–40 Hz. *B*, Field patterns during N100m over the left (LH) and right (RH) hemispheres. The maps have been drawn over the helmet-shaped sensor array. The shadowed areas indicate magnetic flux out of the skull. The arrows show the sites and orientations of the equivalent current dipoles best explaining the signal patterns. The isocontours are separated by 100 fT. *C*, 3-D MRI reconstruction of the subject's brain. The frontal lobes have been removed to expose the surfaces of the temporal lobes. The white circles indicate the sources of N100m. Adapted from Tissari et al. (1993). *D*, Mean (± SEM; 9 subjects) relative amplitudes of N100m over the left and right hemispheres to contralateral stimulation. Adapted from Mäkelä et al. (1993).

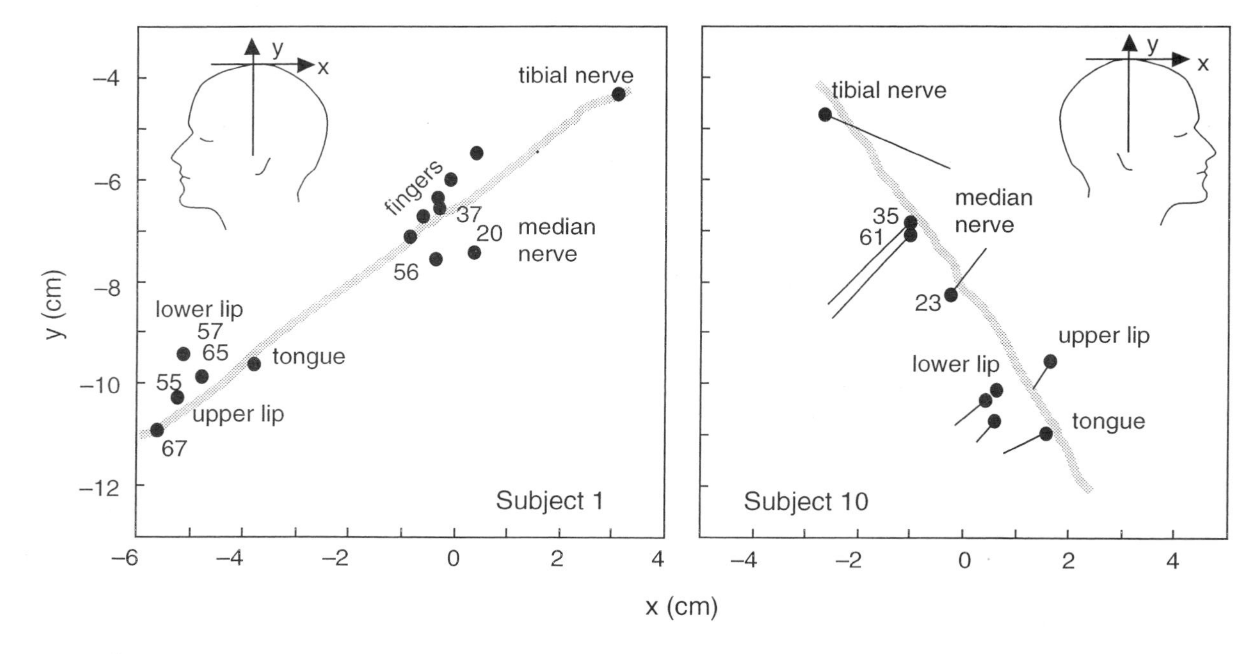

Fig. 3. ECD locations for stimulation of different parts of the body in two subjects. The numbers indicate response latencies in ms. For Subject 10, dipole orientations are also shown. The thick grey line illustrates the approximate course of the Rolandic fissure, determined on the basis of ECD locations and orientations. Adapted from Hari et al. (1993b).

Such recordings suggest that the auditory sensory memory lasts about 10 s (Sams et al., 1993).

Somatosensory responses and somatotopy

Recordings of somatosensory evoked fields (SEFs) can be employed to study functional organization of the somatosensory cortices. In the contralateral primary somatosensory cortex SI, a clear somatotopical order is evident, with different body parts represented at different locations (Fig. 3). The non-invasively identified Rolandic fissure is an important functional landmark in studies of epileptic foci and in preoperative planning of epilepsy surgery.

SEF recordings also indicate activation of other cortical areas. At 100–140 ms, signals peak in the secondary somatosensory cortices SII of both hemispheres, with slightly longer latencies for the ipsi- than the contralateral stimuli (Hari et al., 1993b). A novel source area was recently found in the posterior parietal cortex, medial and posterior to the SI hand area (Forss et al., 1994); this source was most active 70–110 ms after the stimulus.

Signals following voluntary blinking

The eye acts as a strong electric dipole and eye movements and blinks are serious sources of artefacts in EEG and MEG recordings. EEG responses have been observed in the posterior parts of the brain about 200 ms after blink artefacts; the signals were interpreted as visual evoked responses caused by luminance changes (Heuser-Link et al., 1992).

Figure 4 (left) shows magnetic field patterns associated with voluntary blinking (Hari et al., 1994); the magnetic signals were averaged using the vertical electro-oculogram as the trigger. Strong signals occur close to both orbits during a blink. About 200 ms later, another signal with a dipolar field pattern emerges in the posterior head areas. Superposition of its ECD on the brain surface (Fig. 4, right; white circles) indicates activation of the parieto-occipital midline. The signals disappeared when blinking occurred in complete darkness. However, since the source location was clearly distinct from that of early visually evoked fields (Fig. 4, right; black circle), the blink-related posterior response does not seem to be an ordinary visual evoked response but rather a sign of activation of the posterior parietal cortex (PPC).

The PPC is known to be strongly associated with the control of eye movements and fixation. In monkey, eye-movement-related neurons in this area react preferentially to large, salient stimuli which "signal the presence of an important stimulus in the visual environment" (Goldberg and Robinson, 1977). Interestingly, the PPC is strongly connected to prefrontal cortical areas underlying spatial working memory (Wilson et al., 1993). The observed response in the human PPC might thus be related to updating of the spatial working memory, necessary for maintaining a continuous image of the environment, despite interruption of the visual input for a tenth of a second during each blink.

Clinical perspectives and conclusions

During preoperative evaluation of epileptic patients, irritative foci have been located in relation to functional landmarks at cortical projection areas, determined on the basis of magnetic evoked responses. A good agreement has been reported in several patients between MEG foci and intraoperative corticographic findings. Determining the temporal relationships between several foci may also be clinically important. For example, in some patients, it is possible to find the callosal conduction time between foci in homologous

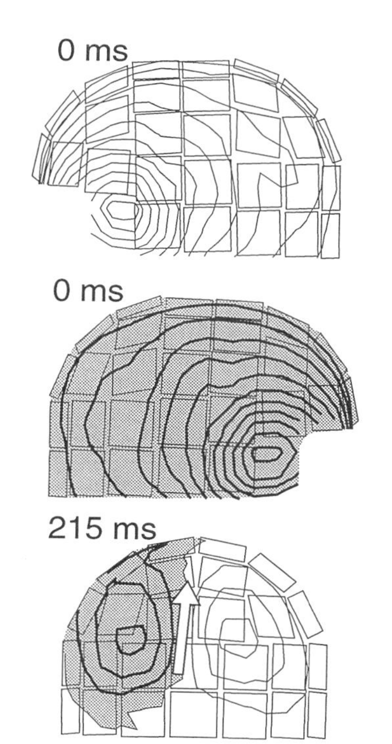

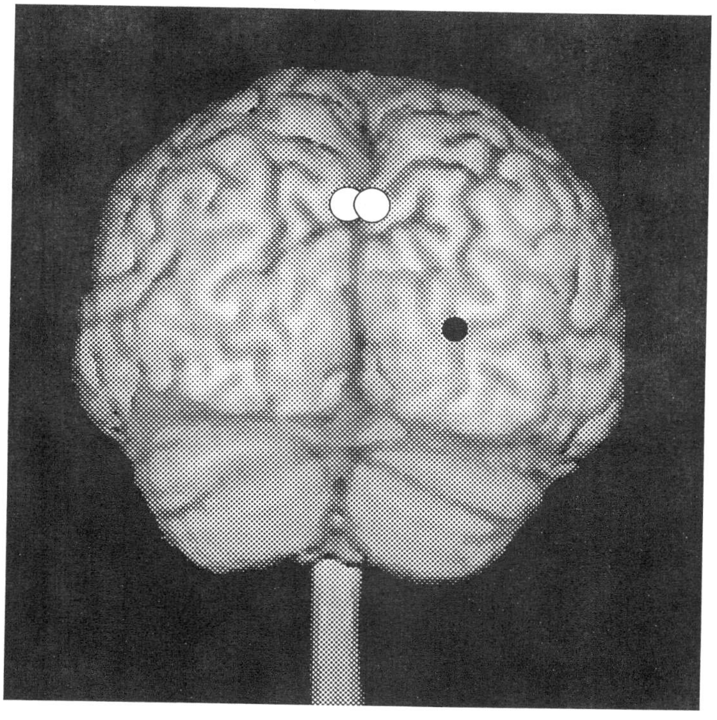

Fig. 4. *Left:* Field patterns over the left and right hemispheres during the maximum blink signal (0 ms) and over the back of the head during the peak of the posterior response (215 ms). The isocontours are separated by 200 fT (upper two patterns) and by 100 fT (the lowest pattern). *Right:* ECDs superimposed on a MRI surface reconstruction of the brain, viewed from the back. The white circles indicate ECDs for the blink-associated signals, determined from two measurements; the dipole orientations are perpendicular to the course of the parieto-occipital sulcus. The black circle shows the ECD for visual evoked field. Modified from Hari et al. (1994).

areas of the two hemispheres (Hari et al., 1993a). Other clinical applications aim at a better understanding of the pathophysiology of different neurological diseases (Hari, 1993). More extensive clinical applications necessitate installation of high-quality multichannel instruments into hospital environments.

MEG complements information obtained from other brain imaging methods in studying human cortical functions. The MEG approach lacks the accuracy of single-cell recordings but is non-invasive and gives information on an analysis level that is relevant for an understanding of the macroscopic functional organization of the cerebral cortex.

Whole-head MEG recordings allow comparison of hemispheric differences in the processing of complex stimuli, such as speech sounds. This is of paramount importance for research into the neural basis of cognitive activity in humans, such as studies of brain dynamics and related disorders during complex problem solving. One recent whole-head MEG study revealed the activation sequence of several cortical areas associated with overt and covert picture naming (Salmelin et al., 1994). In addition to evoked response recordings, the whole-head neuromagnetometers provide a global picture of ongoing spontaneous activity, which is of great interest for both basic neuroscience and for clinical applications.

Acknowledgments

This study has been financially supported by the Academy of Finland and by the Sigrid Jusélius Foundation. The MRIs were obtained at the Department of Radiology, Helsinki University Central Hospital. I thank Olli V. Lounasmaa, Linda McEvoy and Riitta Salmelin for comments on the manuscript.

References

Forss, N., Hari, R., Salmelin, R., Ahonen, A., Hämäläinen, M., Kajola, M., Knuutila, J. and Simola, J. (1994) Activation of the human parietal cortex by median nerve stimulation. *Exp. Brain Res.*, in press.

Goldberg, M. and Robinson, D. (1977) Visual mechanisms underlying gaze: function of the cerebral cortex. In: R. Baker and A. Berthoz (Eds.), *Control of Gaze by Brain Stem Neurons. Developments in Neuroscience*, Vol. 1, Elsevier, pp. 469–476.

Hämäläinen, M., Hari, R., Ilmoniemi, R., Knuutila, J. and Lounasmaa, O.V. (1993) Magnetoencephalography – theory, instrumentation, and applications to noninvasive studies of the working human brain. *Rev. Modern Phys.*, 41: 413–497.

Hari, R. (1993) Magnetoencephalography as a tool of clinical neurophysiology. In: E. Niedermeyer and F. Lopes da Silva (Eds.), *Electroencephalography. Basic Principles, Clinical Applications and Related Fields*, 3rd edition, Williams & Wilkins, Baltimore, pp. 1035–1061.

Hari, R. and Lounasmaa, O.V. (1989) Recording and interpretation of cerebral magnetic fields. *Science,* 244: 432–436.

Hari, R., Ahonen, A., Forss, N., Granström, M.-L., Hämäläinen, M., Kajola, M., Knuutila, J., Mäkelä, J.P., Paetau, R., Salmelin, R. and Simola, J. (1993a) Parietal epileptic mirror focus detected with a whole-head neuromagnetometer. *NeuroReport*, 5: 45–48.

Hari, R., Karhu, J., Hämäläinen, M., Knuutila, J., Salonen, O., Sams, M. and Vilkman, V. (1993b) Functional organization of the human first and second somatosensory cortices: a neuromagnetic study. *Eur. J. Neurosci.*, 5: 724–734.

Hari, R., Salmelin, R., Tissari, S., Kajola, M. and Virsu, V. (1994) Visual stability during eyeblinks. *Nature*, 367: 121–122.

Heuser-Link, M., Dirlich, G., Berg, P., Vogl, L. and Scherg, M. (1992) Eyeblinks evoke potentials in the occipital brain region. *Neurosci. Lett.*, 143: 31–34.

Mäkelä, J., Ahonen, A., Hämäläinen, M., Hari, R., Ilmoniemi, R., Kajola, M., Knuutila, J., Lounasmaa, O.V., McEvoy, L., Salmelin, R., Sams, M., Simola, J., Tesche, C. and Vasama, J.-P. (1993) Functional differences between auditory cortices of the two hemispheres revealed by whole-head neuromagnetic recordings. *Human Brain Mapping,* 1: 48–56.

Salmelin, R., Hari, R., Lounasmaa, O.V. and Sams, M. (1994) Dynamics of brain activation during picture naming. *Nature*, in press.

Sams, M., Hari, R., Rif, J. and Knuutila, J. (1993) The human auditory sensory memory trace persists about 10 s: neuromagnetic evidence. *J. Cogn. Neurosci.*, 5: 363–370.

Sato, S. (Ed.) (1990) *Magnetoencephalography. Advances in Neurology*, Vol. 54, Raven Press, New York.

Tissari, S., Hämäläinen, M., Hari, R. and Mäkelä, J. (1993) Sources of auditory evoked fields superimposed on 3D-reconstruction of temporal lobes. In: *9th Int. Conf. Biomagnetism*, Vienna, 1993, Volume of Abstracts, pp. 138–139.

Wikswo, J.J., Gevins, A. and Williamson, S. (1993) The future of EEG and MEG. *Electroencephalogr. Clin. Neurophysiol.*, 87: 1–9.

Wilson, F.A.W., Ó Schalaidhe, S.P. and Goldman-Rakic, P.S. (1993) Dissociation of object and spatial processing domains in primate prefrontal cortex. *Science*, 260: 1955–1958.

F. Bloom (Editor)
Progress in Brain Research, Vol. 100

CHAPTER 22

The prefrontal system: a smorgasbord

Ivan Divac

Department of Medical Physiology, Panum Institute, University of Copenhagen, Blegdamsvej 3C, DK-2200 Copenhagen, Denmark

Introduction

Three decades ago, Rosvold[1] and Szwarcbart (1964) published a paper in which they argued that in addition to sensory and motor systems, there is the prefrontal system (PFS) in the brain consisting of the prefrontal cortex (PFC) and a set of anatomically and functionally related formations. The concept of PFS was based mainly on the results of neurobehavioral and anatomical studies on rhesus monkeys. Joint functioning of the components of PFS was inferred from impaired performance in delayed response-type tasks following lesions or electrical stimulation anywhere in the system. The neostriatal region innervated by PFC was one such component.

This notion was just hatched when I arrived to Rosvold's laboratory in early 1963, a short time after my graduation. In retrospect, it is obvious that I became "imprinted" in Rosvold's laboratory since a large portion of my work has been devoted to the PFS, especially to relations between PFC and the neostriatum. During my stay at NIH, we obtained evidence that PFS can be divided in subsystems (Divac et al., 1967).

In 1965, I came to the Nencki Institute in Warsaw where Konorski and his group had been studying functions of the PFC in cats and dogs. There, I acquired my second major line of interest: the comparative approach.

Comparisons of PFS demand a definition of PFC that is useful across species. Attempts at such a definition are as old as the term PFC itself (Divac, 1988). According to the definition I accepted, PFC is the cortical projection area of the thalamic mediodorsal nucleus (MD) (to appreciate the complexity of this issue, see Divac and Öberg (1990) and Divac et al. (1993)). This definition of PFC obviously requires in turn a useful definition of MD. Neither topography, currently the main criterion, nor presently available chemical markers are reliable (Regidor and Divac, 1987) but hodology seems promising. A discussion of what is an entity in the thalamus has been discussed elsewhere (Divac, 1979).

Our work on the rat demonstrated the usefulness of defining PFC by MD projections to the cortex: The cortical area in rats outlined by Leonard[2] on the basis of thalamo-cortical connections was shown to mediate delayed response-type behavior, just as cortical areas, defined in the same way, do in any tested species, e.g. monkeys, dogs and cats (Divac, 1971).

My studies of the PFS provided data relevant for the neostriatum as well as the PFC. The studies which focused on the neostriatum have been reviewed elsewhere (Divac, 1984). Presently, I would like to review my work that focused on PFC. The decision to get involved in any of the studies reviewed here was made on the basis of existing data and opportunities. Opportunities mostly arose when I encountered people who shared my interest, usually with the contribution of a technical expertise that I did not possess. Availability of funds often determined certain undertakings. Thanks to generous donors I have been able to study

[1] This paper is dedicated to H.E. Rosvold.

[2] The influential studies from other laboratories will not be cited here; they are accessible in the relevant papers discussed in this review.

feral cats in Norway, wild rats on Hawaii, pigeons in Italy and echidnas in Australia. Predictably, sometimes the available data and opportunities caused whimsical turns in this line of my research.

My early neuropsychological studies in cats (Divac, 1968, 1972) and rats (Divac, 1971; Wikmark et al., 1973) established the generality of the concept of PFS across mammalian species. Much later, 2-deoxy-glucose was used to visualize the PFS in rats (Divac and Diemer, 1980). The results of that study showed coactivation of the PFC and many subcortical structures listed by Rosvold and Szwarcbart (1964). In a related experiment, delayed alternation impairment was observed after neostriatal lesions made by kainic acid. In these animals, PFC connections remained demonstrably preserved and functionally viable (Divac et al., 1978c). These two experiments dispersed (at least my) doubt that behavioral impairments seen after neostriatal lesions could be the consequence of accidental damage to cortical connections. In conclusion, the PFC and a part of the neostriatum do share the same functions and therefore do belong to the same system.

Anatomy of the prefrontal cortex

Good agreement between morphological and functional data obtained in the rat and the exciting discovery of dopaminergic innervation of the cortical area described by Leonard as prefrontal, suggested a study in which American opossums, tree shrews and rats were compared. The results showed a good overlap of the projection from the thalamic mediodorsal nucleus and dopamine-containing fibers originating in the ventral tegmental area in each species. Interestingly, the cortical area which receives this overlapping projection has different topography in each of these species (Divac et al., 1978a). These observations suggested that a dense dopaminergic innervation might be used to outline the prefrontal cortex instead of the projection from MD. On the basis of this generalization, an attempt was made to identify the bird equivalent of the PFC. Histochemical (Divac and Mogensen, 1985), biochemical (Divac et al., 1985) and behavioral data (Mogensen and Divac, 1982, 1993a) suggested that the postero-dorso-lateral "neostriatum" in pigeons could be considered equivalent to the PFC in mammals.

A later discovery, by immunohistochemical techniques, of a dense network of tyrosine hydroxylase- and dopamine-containing fibers also in non-prefrontal cortical areas in primates has restricted the use of cortical dopamine as an indicator of the presence and position of the PFC to the species with a steep gradient of distribution of dopamine in the cortex. Thus far, high density of dopaminergic innervation of the prefrontal area remains unchallenged as an indicator of PFC in non-primate brains. (The perirhinal and entorhinal areas, also densely innervated with dopamine-containing fibers, are not neocortical areas.) Although *density* of dopamine-containing fibers in the motor area of primates appears higher than in the PFC, biochemical measurements in samples of cortical tissue indicate that both the *amount* of dopamine (e.g. Björklund et al., 1978), and binding of spiroperidol to dopamine receptors (Divac et al., 1981) are at the highest level in the PFC. The differences in dopaminergic innervation of the cerebral cortex might offer a test of evolutionary position of a particular species, e.g. among prosimians.

One line of my research aimed to establish the localization and size of PFC in different species. We used mainly horseradish peroxidase as the tracer. This approach confirmed results from other laboratories that, in rats (Divac et al., 1978b, 1993), and probably in hedgehogs (I. Divac and J. Regidor, unpublished), the PFC is split into mesial and suprarhinal parts by a ventral cortical strip innervated by thalamic nucleus submedius and dorsally by cortex innervated by the nucleus ventralis lateralis. This "primitive" arrangement differs from the fronto-polar position of PFC found in tree shrews (Divac et al., 1978a; Divac and Passingham, 1980) and cats (Markowitsch et al., 1978). The same fronto-polar localization, but with a larger relative size of the PFC, exists in dogs and simians. Some evidence suggests that the same arrangement exists in echidnas (*Tachyglossus aculeatus*), the species with proportionally the largest PFC among mammals, including humans (Divac et al., 1987a,b). In bush-babies, PFC has a unique position and shape:

MD projects to the ventral surface of the frontal lobe and to the mesial, fronto-polar and lateral surfaces. A large part of the lateral frontal surface, reaching far rostrally, is not innervated by MD (Markowitsch et al., 1980). These observations suggest that PFC has undergone considerable changes during the evolution. Unfortunately, the sample of different species available so far is too small for a reliable and revealing typology.

Further anatomical studies on PFS revealed: (a) intricate patterns of numerous subcortico-cortical connections to the PFC in rats (Divac et al., 1978b, Divac, 1979) and cats (Markowitsch et al., 1978); (b) sources of cortical projections to the PFC in cats (Markowitsch et al., 1979); and (c) sources of afferents to the MD in tree shrews (Sapawi and Divac, 1978). Since MD projections define PFC, information about inputs to MD is expected to contribute to understanding the functional role of PFC.

In a few experiments, neurochemical techniques revealed interesting features of PFC: First, the capacity of glia to take up glutamate varied with the brain region from which the glia was cultured; the glia cultured from PFC showed a stronger uptake rate than that cultured from the visual cortex or cerebellum (Schousboe and Divac, 1979). Second, the performance of behaviors mediated by PFS did influence the turnover of synapses in rat PFC (Mogensen et al., 1982) but not the turnover of dopamine (Divac et al., 1984a; Mogensen et al., 1992).

Functional analysis of the prefrontal cortex

My studies of functional properties of PFC had the following general objectives: (i) to map PFS neurobehaviorally in different species; (ii) to analyze the relations between the PFC and its related neostriatal region; and (iii) to improve understanding of behavioral roles played by PFS.

A part of the neurobehavioral work on mapping PFS in cats, rats and pigeons has been reviewed above. Other experiments showed that in monkeys, the same small part of the PFC is essential for delayed responding and delayed alternation (Warren and Divac, 1972). Later, we showed that also in rats only a part of the PFC mediates delayed alternation (Larsen and Divac, 1978).

Neurobehavioral comparisons of different species showed essentially the same functions of PFS in such widely different animals as nocturnal omnivores (rats), solitary hunters (cats), pack hunters (dogs) and aboreal, daylight fruit gatherers (monkeys). Yet, one difference is striking: rats, dogs and cats can relearn delayed response type behavior, whereas monkeys do not. We showed that this difference cannot be attributed to differences either in test situations (Divac and Warren, 1971) or in completeness of the prefrontal ablations (Divac, 1973).

In some experiments, I attempted to see whether cats with PFC ablations succeed in relearning delayed responding by positioning and/or "sustained attention" in the presumed absence of short-term memory. In this attempt, two approaches were taken. In the first, lesioned cats which relearned delayed responding were trained under the following modifications of the task: during the delay period, the wire cage was covered with an opaque cylinder which had a small hole turned towards the experimenter. Through this hole, a syringe with meat paste was inserted into the cage and the cat was allowed to lick the food for about 10 s in the middle of the 30 s delay. The visual isolation and distraction had a negligible effect; no drastic and long-lasting relapse of impairment was seen (Divac, 1968). In the second approach, cats were blinded by bilateral transection of the optic nerve and after 10 weeks taught delayed responding. Retest after ablations of PFC produced the same degree of impairment as seen in cats with preserved vision. This result did not support either the hypothesis of distractibility as an important element in the impairment (blind cats would be less distracted and thus perform better than sighted cats) nor the hypothesis of compensation by visual imagery (blind cats should be more impaired in the absence of the possibility to make use of visual spatial orientation) (Divac, 1969).

In rats, we looked for a brain formation which takes over the functions after PFC ablations and established that the brain formation which mediates the performance in rats with PFC ablations is neither the parietal cortical area (Wörtwein et al., 1993) nor the entire dorsolateral isocortex (Wörtwein et al., 1994). The latter study also shows that PFC can mediate delayed alternation in absence of the cortico-cortical input.

In several experiments on cats (reviewed in more detail in Divac, 1984), it was shown that PFC and the associated part of the neostriatum are serially related. Once a cat relearns delayed responding after a lesion of either of these two structures, additional lesion of the remaining structure has no effect on the performance (Wikmark and Divac, 1973; Divac, 1974). Furthermore, combined lesion of the two formations did not induce a larger impairment than a lesion of either of them alone (Divac, 1968).

The roles of PFC in situations that mimic real life of some species were studied neurobehaviorally. Ablations of mesial PFC did not affect fear behavior in wild rats (Divac et al., 1984b) nor flight and defence behaviors in feral cats (Ursin and Divac, 1975). These results should be reconciled with the effects of frontal lesions on "emotional behaviors" in humans. Other experiments showed that no lesion in the rat frontal lobes had an effect on sexual performance of male rats (Larsson et al., 1980); that ablation only of the ventral part of the PFC in rats induced transient aphagia without adipsia (Mogensen and Divac, 1993b); and that ablation of the pigeon equivalent of the mammalian PFC impaired homing from unfamiliar sites, but not from familiar sites (Gagliardo and Divac, 1993).

In rats, the mesial, but not suprarhinal, PFC lesions impaired spontaneous alternation behavior (Divac et al., 1975b). A later experiment showed an interaction between the lesion localization and behavioral situation; the impairment was found only in animals with dorsal mesial lesions and only if the arms of the T-maze were made different by black or white walls (Mogensen and Divac, 1993b). The impairment in spontaneous alternation, when compared to that of delayed spatial alternation, demonstrates that neither the type of reinforcement nor the kind of learning interact essentially with PFC functioning.

Nauta has suggested that PFC mediates "interoceptive gnosis". This hypothesis was tested by taste aversion conditioning. No ablation within PFC affected taste aversion (Divac et al., 1975a; Mogensen and Divac 1993b).

Ablation of the orbital part of the PFC in monkeys and dogs induces "response disinhibition". We replicated the experiment using rats in which orbital cortex was removed. We saw no sign of any response disinhibition (Mogensen and Divac, 1993b).

Efforts to understand the function of the PFC have often emphasized the role of the spatial factor (as in egocentric orientation) in the tasks that reveal impairments after PFC lesions. We tested such hypotheses first by applying differential reinforcement for low response rates to rats with PFC ablations. The task does not require differential spatial responding but only withholding responses for a predetermined interval. Ablation of the PFC did impair performance in this task thus showing that PFC also plays a role in situations where the spatial factor is considerably reduced if not entirely eliminated (Rosenkilde and Divac, 1975).

Another study along the same lines led to the design of a task in which the informing cue had no spatial localization; cats were supposed to guide their responses to spatially separated feeders on the basis of the duration of confinement under a cage. Again, ablations of the PFC induced a clear impairment (Rosenkilde and Divac, 1976). This result confirms the conclusion that PFC also plays a role in behaviors with reduced spatial requirements.

Some authors postulated involvement of PFC in sequential behaviors. Instead of observing a complex behavior such as maternal, we trained animals to manipulate two objects sequentially in an operant test chamber. Against the prediction, orbital ablation impaired sequencing much more strongly than did a dorsomedial lesion (Mogensen and Divac, 1984). It is likely that the essential part of the PFC for sequential behavior is the medial part of the orbital area (ventral part of the mesial PFC in rats).

Conclusions

Demonstration of the existence of the PFC in a number of species from different orders offers a solid basis for extrapolation to other species, including humans. An extrapolation based on results obtained in a single species is more risky even when this species is considered to be closely related to humans. Indeed, participation of a part of the neostriatum in prefrontal functions inferred from animal studies, has also been found in humans studied with PET and functional MR scanning.

In my attempts to study functions of PFC, negative results were obtained at least as frequently as positive ones. We know now that the entire dorsolateral neocortex in rats plays *no* role in the recovery of delayed response-type behavior (Wörtwein et al., 1994); that cats after PFC lesions do *not* relearn delayed responding through body positioning or "sustained attention" (Divac, 1968); that the PFC does *not* mediate "visceral gnosis" (Divac et al., 1975a); that the PFC is *un*necessary for sexual performance of rats (Larsson et al., 1980); that large parts of the PFC either in cats (Ursin and Divac, 1975) or rats (Divac et al., 1984b) are *not* essentially involved in "emotional behaviors".

On the other hand, no species with lesions in the PFC has escaped impairment of delayed response-type behaviors. Across species, this lesion consequence is more reliable than limb paralysis after motor cortical lesions. Unfortunately, no stringent a priori description of such tasks is available. Sometimes a task which apparently belongs to this group is performed successfully by animals with lesions in the PFS (Mogensen et al., 1987). In the long run, careful analysis of task differences sensitive or insensitive to PFS lesions may provide hints about critical features of situations in which PFS lesions induce impairments.

The presence of the impairments in delayed response-type tasks by all species tested so far with PFC lesions contrasts with the different effects of the same lesions in other behavioral situations. Some conspicuous examples are first, as already mentioned, the inability of monkeys to relearn this performance after PFS lesions. Secondly, following prefrontal ablations, an exceptionally high behavioral activity is seen in Old World monkeys in comparison with other species, including squirrel monkeys. Thirdly, an established "response disinhibition" found after orbital PFC lesions in Old World monkeys could not be replicated in rats in spite of efforts to create similar behavioral requirements (Mogensen and Divac, 1993b). Of course, one can always wonder whether the identical situation in the judgement of humans is also the same for two other different species.

These few examples illustrate the problems of neurobehavioral analysis of PFC. A partial explanation of apparent inconsistencies and contradictions probably lies in the different functions of different subdivisions of PFC. This requires a parcellation of the PFC (and MD) comparable across species.

References

Björklund. A., Divac, I. and Lindvall, O. (1978) Regional distribution of catecholamines in monkey cerebral cortex. Evidence for a dopaminergic innervation of the primate prefrontal cortex. *Neurosci. Lett.*, 7: 115–119.

Divac, I. (1968) Effects of prefrontal and caudate lesions on delayed response in cats. *Acta Biol. Exp. (Warsaw)*, 28: 149–167.

Divac, I. (1969) Delayed response in blind cats before and after prefrontal ablation. *Physiol. Behav.*, 4: 795–800.

Divac, I. (1971) Frontal lobe system and spatial reversal in the rat. *Neuropsychologia*, 9: 175–183.

Divac, I. (1972) Delayed alternation in cats with lesions of the prefrontal cortex and the caudate nucleus. *Physiol. Behav.*, 8: 519–522.

Divac, I. (1973) Delayed response in cats after frontal lesions extending beyond the gyrus proreus. *Physiol. Behav.*, 10: 717–720.

Divac, I. (1974) Caudate nucleus and relearning of delayed alternation in cats. *Physiol. Psychol.*, 104–106.

Divac, I. (1979) Patterns of subcortico-cortical projections as revealed by somatopetal horseradish peroxidase tracing. *Neuroscience*, 4: 455–461.

Divac, I. (1984) The neostriatum viewed orthogonally. In: *Functions of the Basal Ganglia, CIBA Foundation Symposium 107*, Pitman, London, pp. 201–215.

Divac, I. (1988) A note on the history of the term 'prefrontal'. *IBRO News*, 16: No 2.

Divac, I. and Diemer, N.H. (1980) The prefrontal system in the rat visualized by means of labelled deoxyglucose. Further evidence for functional heterogeneity of the neostriatum. *J. Comp. Neurol.*, 190: 1–13.

Divac, I. and Mogensen, J. (1985) The prefrontal 'cortex' in the pigeon. Catecholamine histofluorescence. *Neuroscience*, 15: 677–682.

Divac, I. and Öberg, R.G.E. (1990) Prefrontal cortex: The name and the thing. In: W. Schwerdtfeger and P. Germroth (Eds.), *Structure and Development of the Forebrain in Lower Vertebrates, Experimental Brain Research Series*, Springer-Verlag, Berlin, pp. 213–220.

Divac, I. and Passingham, R.E. (1980) Connections of the mediodorsal nucleus of the thalamus in the tree shrew. II. Efferent connections. *Neurosci. Lett.*, 19: 21–26.

Divac, I. and Warren, J.M. (1971) Delayed response by frontal monkeys in the Nencki Testing Situation. *Neuropsychologia*, 9: 209–217.

Divac, I., Rosvold H.E., Szwarcbart, M. (1967) Behavioral effects of selective ablation of the caudate nucleus. *J. Comp. Physiol.Psychol.*, 63: 184–190.

Divac, I., Gade, A. and Wikmark, R.G.E. (1975a) Taste aversion in rats with lesions in the frontal lobes: no evidence for interoceptive agnosia. *Physiol. Psychol.*, 3: 43–46.

Divac, I., Wikmark, R.G.E. and Gade, A. (1975b) Spontaneous alternation in rats with lesions in the frontal lobes. An extension of the frontal lobe syndrome. *Physiol. Psychol.*, 3: 39–42.

Divac, I., Björklund, A., Lindvall, O. and Passingham, R.E. (1978a) Converging projections from the mediodorsal thalamic nucleus and mesencephalic dopaminergic neurons to the neocortex in three species. *J. Comp. Neurol.*, 180: 59–72.

Divac, I., Kosmal, A., Björklund, A. and Lindvall, O. (1978b) Subcortical projections to the prefrontal cortex in the rat as revealed by the horseradish peroxidase technique. *Neuroscience*, 3: 785–796.

Divac, I., Markowitsch, H.J. and Pritzel, M. (1978c) Behavioral and anatomical consequences of small intrastriatal injections of kainic acid in the rat. *Brain Res.*, 151: 523–532.

Divac, I., Bræstrup, C. and Nielsen, M. (1981) Spiroperidol, naloxone, diazepam and QNB binding in the monkey cerebral cortex. *Brain Res. Bull.*, 7: 469–477.

Divac, I., Lichtensteiger, W. and Gade, A. (1984a) Catecholamine microfluorometry of nigral perikarya and tyrosine hydroxylase assay in some telencephalic structures of rats exposed to different behavioral situations. *Acta Neurobiol. Exp.*, 44: 263–272.

Divac, I., Mogensen, J., Blanchard, R.J. and Blanchard, D.C. (1984b) Mesial cortical lesions and fear behavior in the wild rat. *Physiol. Psychol.*, 12: 271–274.

Divac, I., Mogensen, J. and Björklund, A. (1985) The prefrontal 'cortex' in the pigeon. Biochemical evidence. *Brain Res.*, 332: 365–368.

Divac, I., Holst, M.-C., Nelson, J., McKenzie, J.S (1987a) Afferents of the frontal cortex in the echidna (*Tachyglossus aculeatus*). Indication of an outstandingly large prefrontal area. *Brain Behav. Evol.*, 30: 303–320.

Divac, I., Pettigrew, J.D., Holst, M.-C. and McKenzie, J.S. (1987b) Efferent connections of the prefrontal cortex of echidna (*Tachyglossus aculeatus*). *Brain Behav. Evol.*, 30: 321–327.

Divac, I., Mogensen, J., Petrovic-Minic, B., Zilles, K., Regidor, J. (1993) Cortical projections of the thalamic mediodorsal nucleus in the rat. Definition of the prefrontal cortex. *Acta Neurobiol. Exp.*, 53: 425–429.

Gagliardo, A. and Divac, I. (1993) Effects of ablation of the presumed equivalent of the mammalian prefrontal cortex in pigeon homing. *Behav. Neurosci.*, 107: 280–288.

Larsen, J.K. and Divac, I. (1978) Selective ablations within the prefrontal cortex of the rat and performance of delayed alternation. *Physiol. Psychol.*, 6: 15–17.

Larsson, K., Öberg, R.G.E. and Divac, I. (1980) Frontal cortical ablations and sexual performance in male albino rats. *Neurosci. Lett.*, Suppl. 5: 319.

Markowitsch, H.J., Pritzel, M. and Divac, I. (1978) The prefrontal cortex of the cat: anatomical subdivisions based on retrograde labelling of cells in the mediodorsal thalamic nucleus. *Exp. Brain Res.*, 32: 335–344.

Markowitsch, H.J., Pritzel, M. and Divac, I. (1979) Cortical afferents to the prefrontal cortex of the cat: a study with the horseradish peroxidase technique. *Neurosci. Lett.*, 11: 115–120.

Markowitsch, H.J., Pritzel, M., Wilson, M. and Divac, I. (1980) The prefrontal cortex of a prosimian (*Galago senegalensis*) defined as cortical projection area of the thalamic mediodorsal nucleus. *Neuroscience*, 5: 1771–1779.

Mogensen, J. and Divac, I. (1982) The prefrontal 'cortex' in the pigeon. Behavioral evidence. *Brain Behav. Evol.*, 21: 60–66.

Mogensen, J. and Divac, I. (1984) Sequential behavior after modified prefrontal lesions in the rat. *Physiol. Psychol.*, 12: 41–44.

Mogensen, J. and Divac, I. (1993a) Behavioural effects of ablation of the pigeon-equivalent of the mammalian prefrontal cortex. *Behav. Brain Res.*, 55: 101–107.

Mogensen, J. and Divac, I. (1993b) Behavioural changes after ablation of subdivisions of the rat prefrontal cortex. *Acta Neurobiol. Exp.*, 53: 439–449.

Mogensen, J., Jørgensen, O.S. and Divac, I. (1982) Synaptic proteins in frontal and control brain regions of rats after exposure to spatial problems. *Behav. Brain Res.*, 5: 375–386.

Mogensen, J., Iversen, I.H. and Divac, I. (1987) Neostriatal lesions impaired rats' delayed alternation performance in a T-maze but not in a two-key operant chamber. *Acta Neurobiol. Exp.*, 47: 45–54.

Mogensen, J., Björklund, A. and Divac, I. (1992) Catecholamines and DOPAC in cortical and neostriatal regions during rats' learning of delayed alternation. *Acta Neurobiol. Exp.*, 52: 49–56.

Regidor, J. and Divac, I. (1987) Architectonics of the thalamus in echidna (*Tachyglossus aculeatus*): search for the mediodorsal nucleus. *Brain Behav. Evol.*, 30: 328–341.

Rosenkilde, C.E. and Divac, I. (1975) DRL performance following anteromedial cortical ablations in rats. *Brain Res.*, 95: 142–146.

Rosenkilde, C.E. and Divac, I. (1976) Time discrimination performance in cats with lesions in the prefrontal cortex and the caudate nucleus. *J. Comp. Physiol. Psychol.*, 90: 343–352.

Rosvold, H.E. and Szwarcbart, M.K. (1964) Neural structures involved in delayed-response performance. In: J.M. Warren and K. Akert (Eds.), *The Frontal Granular Cortex and Behavior*, McGraw-Hill, New York, pp. 1–15.

Sapawi, R.R. and Divac, I. (1978) Connections of the mediodorsal nucleus of the thalamus in the tree shrew. I. Afferent connections. *Neurosci. Lett.*, 7: 183–189.

Schousboe, A. and Divac, I. (1979) Differences in glutamate uptake in astrocytes cultured from different brain regions. *Brain Res.*, 177: 407–409.

Ursin, H. and Divac, I. (1975) Emotional behavior in feral cats with ablations of the prefrontal cortex and subsequent lesions in amygdala. *J. Comp. Physiol. Psychol.*, 88: 36–39.

Warren, J.M. and Divac, I. (1972) Delayed response performance by rhesus monkeys with midprincipalis lesions. *Psychonomic Sci.*, 8: 146–147.

Wikmark, R.G.E. and Divac, I. (1973) Absence of effect of caudate lesions on delayed responses acquired after large frontal ablations in cats. *Israeli J. Med. Sci.*, 9: 92–97.

Wikmark, R.G.E., Divac, I. and Weiss, R. (1973) Retention of spatial delayed alternation in rats with lesions in the frontal lobes. Implications for a comparative neuropsychology of the prefrontal system. *Brain Behav. Evol.*, 8: 329–339.

Wörtwein, G., Mogensen, J. and Divac, I. (1993) Retention and relearning of spatial delayed alternation in rats after combined or sequental lesions of the prefrontal and parietal cortex. *Acta Neurobiol. Exp.*, 53: 357–366.

Wörtwein G, Mogensen J, Divac, I. (1994) Retention and relearning of spatial delayed alternation in rats after ablation of the prefrontal cortex or total nonprefrontal isocortex. *Behav. Brain Res.*, in press..

SECTION IV

Neuroplasticity

F. Bloom (Editor)
Progress in Brain Research, Vol. 100

CHAPTER 23

A comparison of the mechanistic relationships between development and learning in *Aplysia*

Emilie A. Marcus[1], Nigel J. Emptage[2], René Marois[3] and Thomas J. Carew[1-3]

Departments of [1]Biology, [2]Psychology and the [3]Interdepartmental Neuroscience Program, Yale University, New Haven, CT, USA

Introduction

At the turn of the century, Ramón y Cajal (1911) articulated the hypothesis that growth processes involved in the development of the nervous system persist into the adult where they subserve learning and memory. Since Cajal's seminal suggestion, considerable experimental attention has been aimed at elucidating the cellular and molecular mechanisms of both learning and neuronal development. Historically, however, preparations that have been favored by developmental neurobiologists have not been preferred for studying mechanisms of learning; likewise the systems that are best understood with respect to adult learning and memory have not been well-characterized developmentally. Consequently, to date no single experimental system has been extensively studied from both a developmental and a learning perspective. For this reason, comparisons between the mechanisms of learning and development often resort to analogies and inferences drawn across diverse systems. Ultimately, an assessment of the mechanistic commonalities between development and learning will require an analysis of similarities and differences within the same experimental system. Towards that end, in this review, we focus our attention on the marine mollusc *Aplysia californica* (with references to other systems where appropriate) with the goal of bringing together what is known about the relationship between development and learning in a single experimental system.

The principal advantage of *Aplysia* as a system in which to examine the relationship between developmental and adult plasticity stems from the fact that mechanisms of learning in the adult CNS have been extensively characterized at the cellular and molecular level (for review, see Kandel and Schwartz, 1982; Kandel and Hawkins, 1992). One general principle that has emerged from this work, as well as from related work in vertebrate systems, is that learning occurs through the modulation of synaptic efficacy. Perhaps the best understood form of learning in *Aplysia* is sensitization, which is defined as an increase in reflex response amplitude following the presentation of a noxious stimulus. By varying the number and magnitude of noxious stimuli, it is possible to produce sensitization of defensive withdrawal reflexes that lasts minutes to hours (short-term), or days to weeks (long-term). In both cases, the increase in reflex responsiveness is produced at least in part by presynaptic facilitation of a monosynaptic connection between the sensory and motor neurons that mediate the reflex. Presynaptic facilitation is in turn produced by release of the neuromodulatory transmitter serotonin (5HT) from interneurons which fire in response to the noxious stimulus. The short-term effects of 5HT on transmitter release are mediated by cAMP-dependent and protein kinase C-dependent reduction of specific potassium conductances; this results in a broadening of the sensory neuron action potential which permits an increase in intracellular calcium in the presynaptic terminals. The long-term effects on transmitter release involve modulation of the biophysical properties of sensory neurons described above, as well as morphological changes such as an increase in the number of synapses onto motor neurons, a process that requires gene transcription and protein synthesis (Bailey, 1991; Montarolo et al., 1986). Thus, in summary, considerable progress has been made in elucidating the mechanisms of short-term and long-term plasticity in adult *Aplysia*. In this review, we use this understanding of the cellular and molecular mechanisms of learning in *Aplysia* as a basis for comparison with the three principle stages of neuronal development: differentiation, neurite outgrowth and synapse formation.

Differentiation

The term differentiation refers to the process by which a cell acquires a specific identity. It is generally considered to be the first step in a developmental program during which a proliferative pleuripotent cell gives rise to a post-mitotic cell committed to a particular fate. In the development of the nervous system, the commitment to a neuronal cell fate typically is a prerequisite step for further elaboration, such as neurite outgrowth, axonal path finding and synapse formation. Thus, it might seem rather surprising to look for comparisons between this initial step of neuronal development and the cellular processes of learning and memory since the presumptive neurons have yet to become integrated in a functional neural circuit. However, the evidence described below suggests that both the cellular and molecular processes that lead to neuronal differentiation in *Aplysia* (as well as in the mammalian hippocampus) may have intriguing parallels to the mechanisms underlying the neural basis of learning and memory.

In *Aplysia* as in other systems, a cell generated from the proliferation of progenitor cells will follow a particular differentiation pathway in response to intrinsic and/or extrinsic factors. Whether intrinsic or extrinsic, the signal must ultimately be transduced in the nucleus where it initiates a cascade of molecular interactions that transform the phenotype of the cell. In this way, differentiation and long-term plasticity may share a fundamental common principle in that they both involve the transduction of a transient instructive biochemical signal into a self-sustained molecular process in order to retain cell-specific information. Below, we provide examples at both the cellular, transducing step and at the molecular, genomic level that illustrate the possible convergence of the mechanistic bases of differentiation and long-term plasticity.

Role of axosomatic contacts in differentiation and plasticity

Schacher et al. (1979) have shown that neuronal differentiation in the anlage of the abdominal ganglion of *Aplysia* appears to be triggered by an extrinsic signal in the form of a synaptic contact from a neuronal process that travels in the pleuro-abdominal connective. Following a transient synaptic contact by this fiber on their cell bodies, abdominal neurons appear to undergo major ultrastructural changes suggestive of a differentiative process: relaxation of the nuclear chromatin, proliferation of cytoplasmic constituents, and initiation of an axonal projection at precisely the site where the contact was previously established (Schacher et al., 1979). Although the identity of the pre-synaptic fiber is still unknown, recent work by Marois et al. (1992) has shown that serotonergic cells that have their cell bodies near the cerebral ganglion send processes in the pleuro-abdominal connective to the anlage of the abdominal ganglion where they make contact with presumptive neuronal cell bodies at a time that is consistent with their having a role in differentiation. These results raise the possibility that serotonergic neurons make appropriate synaptic contact with abdominal neurons, triggering their differentiation.

In parallel to the proposed role of axosomatic contacts in neuronal differentiation, there is also increasing evidence suggesting a role for serotonergic axosomatic contacts in the induction of long-term synaptic plasticity in *Aplysia*. For example, the somata of sensory neurons in the abdominal and pleural ganglia of adult *Aplysia* receive serotonergic contacts (Kistler et al., 1985; Zhang et al., 1991). Recently, Emptage and Carew (1993) have shown that exogenous application of 5HT to the cell body (and proximal synaptic region) of pleural sensory cells produces long-term synaptic facilitation of distant synapses that were never

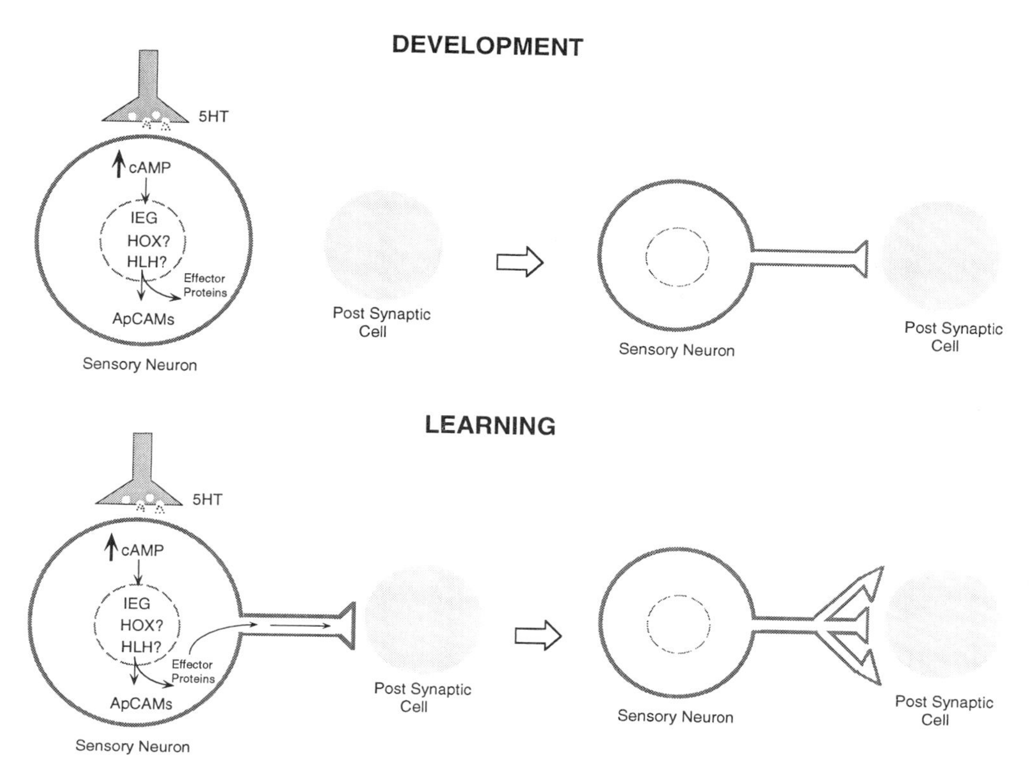

Fig. 1. A schematic representation of similarities in the cellular and molecular mechanisms of development and learning in *Aplysia*. The specific points of convergence include: (i) the role of serotonergic axosomatic contacts; (ii) the activation of transcription factors including immediate early genes and perhaps differentiation-inducing genes; (iii) the necessity of an appropriate post-synaptic target; (iv) the role of cAMP as a second messenger in the signal transduction cascade; and (v) the common role of cell adhesion molecules and other growth related proteins. During development, activation of these pathways leads to differentiation and neurite outgrowth; in adult learning, reactivation of these same pathways results in a growth-mediated increase in synaptic strength. See text for details.

exposed to 5HT (for similar findings in abdominal sensory neurons, see also Clark and Kandel, 1993). This raises the possibility that the somatic serotonergic input might be ideally situated to affect genomic (long-term) changes. If this were the case, then this system in the adult might correspond to that described above in development in which axosomatic contacts are involved in triggering the differentiation of neurons. Moreover, if the differentiation-inducing axosomatic contacts in developing *Aplysia* prove to be serotonergic, this may indicate that the serotonergic system impinges on the cell body region of neurons throughout the life of the animal to specifically activate genomic machinery; initially to trigger differentiation of a cell during development, and later to reactivate the genomic program in the context of long-term synaptic enhancement (Fig. 1).

Role of identified transcription factors

It is more difficult to pursue a comparison of differentiation and learning at the molecular level in *Aplysia* because, although the molecular events in long-term synaptic facilitation are being increasingly elucidated, the molecular basis of neuronal differentiation in *Aplysia* is virtually unexplored. However, the molecular basis of differentiation, as described in *Mus*, *Drosophila* and *C. elegans* shows a high degree of conservation, to the point where one can begin to consider a universal plan for differentiation of any cell, regardless of tissue type or animal species (see Anderson, 1992). The current model for understanding how information contained in the cytoplasm ultimately affects the nucleus centers on activation of two main sets of transcription factor genes. One set, the immediate-early genes (IEGs), is expressed rapidly and transiently, prior to or at the onset of differentiation. Based on the temporal pattern of expression, the IEGs have generally been regarded as serving as a rapid and amplifying nuclear signal to regulate and orchestrate the expression of downstream genes responsible for making the decision to either differentiate or re-enter the cell cycle (Sheng and Greenberg, 1990). The second class of transcription factor genes (which includes genes with homeobox or helix-loop-helix DNA-binding domains) are typically expressed over a long period of the differentiation process. This second class of genes tends to be seen as directing the cell's differentiation by orchestrating the expression of downstream effector genes that will endow the cell with its particular molecular identity (Gehring, 1987; Garrell and Campuzano, 1991).

Although still at an early stage, the study of the molecular basis of learning and memory already hints at interesting parallels with the molecular program of differentiation described above. For instance, changes in the expression of immediate-early proteins (IEPs) have been detected in paradigms involving the induction of long-term synaptic plasticity in *Aplysia* (Barzilai et al., 1989). The involvement of IEPs in the induction of long-term synaptic plasticity is further supported by the requirement for protein and RNA synthesis during and immediately following 5HT application, a time window that is consistent with the temporal expression of the IEPs (Montarolo et al., 1986; Barzilai et al., 1989; Castellucci et al., 1989). Specific IEGs are also differentially expressed in the mammalian hippocampus following the induction of long-term synaptic plasticity (Cole et al., 1989; Wisden et al., 1990; Nedivi et al., 1993; Qian et al., 1993). Although no genes belonging to the second class of nuclear proteins have yet been shown to be expressed differentially with long-term synaptic plasticity, the recent findings that genes of this class may regulate neurite outgrowth (LeRoux et al., 1993) and the expression of neural cell adhesion molecules (Jones et al., 1993; see below) suggest that such genes may also be implicated in neuronal plasticity. Thus, the activation of these two classes of transcription factors may prove to be central to the molecular basis of long-term synaptic changes accompanying long-term memory. These genes would be responsible for orchestrating the new levels of expression of an array of downstream effector genes that in turn would underlie the physiological and structural alterations associated with learning (Bailey, 1991; Madison et al., 1991; Bailey and Kandel, 1993). From this perspective, the molecular basis of long-term plasticity can be viewed as a reactivation of at least a portion of the molecular program for cell differentiation (Fig. 1).

Neurite outgrowth

Once a proliferative cell is committed to becoming a neuron, it enters the second stage in its development, the growth of neuronal processes. The nature of this growth is stringently regulated since it underlies the highly specific and reproducible patterns of neuronal connectivity that are characteristic of the adult nervous system. One principle that has emerged recently in the study of mechanisms of adult plasticity in both invertebrates and vertebrates, is that long-term changes in synaptic efficacy are associated with growth-dependent structural changes (Bailey and Kandel, 1993). For example, a detailed examination of learning-related morphological changes in neurons has been performed by Bailey and colleagues in *Aplysia* (Bailey, 1991). They found that behavioral training that produces long-term sensitization of the gill and siphon withdrawal reflex also produces long-lasting morphological changes in the sensory neurons that mediate this reflex. These changes include increases in the number and size of active zones, the number and distribution of synaptic vesicles, and the total number of presynaptic varicosities per sensory neuron. These findings raise the possibility that the same mechanisms which regulate outgrowth during development may be reutilized during learning to mediate long-term changes in neuronal morphology. Within this general framework of growth-related processes during development and learning, below we discuss several specific examples that demonstrate possible points of convergence between the mechanisms of neurite outgrowth and long-term plasticity.

Role of post-synaptic targets

Glanzman et al. (1989) analyzed *Aplysia* sensory neurons growing in culture and found that the presence of a postsynaptic cell appears to be critical for normal neuronal growth. Sensory neurons grown alone had neurites with few branches that tended to aggregate together into thick bundles. In contrast, sensory neurons grown in co-cultures with the appropriate post-synaptic cell had many finer neurites and numerous varicosities and thus appeared more comparable to sensory neurons observed in vivo. Moreover, if sensory neurons were co-cultured with inappropriate postsynaptic target cells (i.e. cells onto which they would not normally synapse) then the neuritic structure of the sensory neurons was similar to that seen for sensory neurons grown alone. It therefore appears that complex growth of the sensory neurons is not simply dependent on the presence of another neuron, but on the ability of the sensory neurons to recognize and form viable synaptic contacts with their appropriate post-synaptic target. However, it is important to note that these experiments have been carried out in regenerating neurons isolated from the mature CNS and grown in culture. In light of the as yet unclear relationship between development and regeneration, similar experiments will ultimately need to be carried out with embryonic neurons.

Glanzman et al. (1990) have also examined the role of the post-synaptic cell in mediating the effects of 5HT-induced growth in mature sensory neurons. 5HT repeatedly applied to co-cultures of sensory and motor neurons produced significant increases in the number of varicosities of the siphon sensory neurons. This cell growth was correlated with long-term synaptic enhancement, providing support for the idea that growth may be critical for production of long-term memory. In experiments where the postsynaptic target cell was not cultured with the sensory neuron, repeated presentations of 5HT did not produce morphological changes in the sensory neurons, suggesting that the presence of the postsynaptic cell is critical for these morphological changes to occur. Thus, the post-synaptic cell may be required not only for establishing appropriate sensory neuron morphology in development but also for the plasticity-associated growth induced by 5HT in the adult. It is therefore possible that the postsynaptic neuron subserves a similar mechanistic role for both types of growth processes (Fig. 1).

Role of second messengers

A second point of convergence that has come to light from work in developing and mature *Aplysia* neurons is the involvement of cAMP as a second messenger in both neuritic outgrowth and long-term synaptic plasticity. Direct support for a role for cAMP in development comes from studies examining the structural and biophysical consequences of manipulating cAMP concentrations in developing neurons. At the structural level, Forscher et al. (1987) have demonstrated that elevation of cAMP serves as the inductive agent in promoting growth cone differentiation into secretory terminals in cultured *Aplysia* bag cell neurons. In another mollusc, *Helisoma*, Mattson et al. (1988) have also shown that 5HT-induced elevation of cAMP can regulate neurite outgrowth. At the biophysical level, Belardetti et al. (1986) have demonstrated that 5HT-induced modulation of the 5HT-sensitive potassium current (an effect that is known to be mediated by activation of cAMP), is present in growth cones of cultured *Aplysia* sensory neurons. In addition, recent work examining the development of the biophysical properties of sensory neurons in vivo has shown that the effects of 5HT that are known to be mediated by cAMP in adults emerge early in development relative to other modulatory effects of 5HT (Marcus and Carew, 1990). Moreover, biochemical assays have revealed that 5HT is able to stimulate adenylate cyclase and elevate cAMP levels in the isolated central nervous system at the earliest stages of juvenile development (Chang et al., 1989). Taken collectively, these data demonstrate that a subcellular signaling pathway involving the elevation of cAMP as a second messenger is present and can be functionally triggered early in development and suggest that the role of this pathway may be to mediate neuritic outgrowth.

From the perspective of adult plasticity, there are now several lines of evidence implicating cAMP as an essential step in the intracellular cascade that mediates the long-term enhancement of synaptic efficacy associated with behavioral long-term sensitization in adult *Aplysia*. Recently, Nazif et al. (1991) have shown that cAMP can itself induce morphological changes (increases in the number of varicosities and the number of branch points) in *Aplysia* sensory neurons similar to those induced by 5HT and by sensitization training. Physiologically, elevation of cAMP in the sensory neurons produces a long-term decrease in the magnitude of the 5HT-sensitive potassium current (Scholz and Byrne, 1987) and an enhancement of the EPSP onto the motor neuron that persists for at least 24 h (Schacher et al., 1988). It has also recently been demonstrated that elevation of cAMP mimics the effects of long-term sensitization training on the expression and phosphorylation of specific proteins (Sweatt and Kandel, 1989; Noel et al., 1991). Finally, perhaps the most convincing demonstration of a role for cAMP in mediating long-term changes in synaptic efficacy in adult *Aplysia* comes from recent molecular analyses demonstrating a requirement for cAMP-dependent transcription factors in the induction of long-term synaptic plasticity (Dash et al., 1990). Thus, it appears that cAMP may be an important second messenger in the signal transduction cascades that regulate neurite outgrowth during development and mediate the reactivation of growth processes that accompany long-term plasticity (Fig. 1).

Regulation of specific proteins

Important advances have recently been made in the study of the molecular mechanisms that contribute to neurite outgrowth in *Aplysia*. Specifically, a family of proteins, designated ApCAMs, that are closely related to the cell adhesion molecules NCAM and fasciclin II, have recently been identified in the cells of developing *Aplysia* (Schacher et al., 1990). ApCAMs are found on the surface of all cells in the early blastula and with further development, their expression becomes restricted to the CNS (Schacher et al., 1990). That ApCAM proteins may be important in the growth of developing neurons has recently been demonstrated by Keller and Schacher (1990) who found that antibodies raised against ApCAM proteins disrupt fasciculation of growing axons in vitro, consequently altering neuronal cell morphology.

With regard to adult plasticity, Mayford et al. (1992) have shown that ApCAM expression decreases in the adult CNS in a protein synthesis-dependent manner following multiple applications of 5HT or cAMP. Since the level of cell adhesion molecules on the surface of the neuronal substrate has been shown to dramatically affect neurite outgrowth (Doherty et al., 1990), it has been suggested that a reduction in the level of ApCAM expression in adult *Aplysia* sensory neurons is the initial step in the process of neuritic outgrowth and the formation of new synaptic contacts between sensory and motor neurons that accompanies long-term plasticity. Thus, these results support the idea that a common set of adhesion molecules may be used during neural development and then re-utilized during learning-related growth at mature *Aplysia* synapses (Fig. 1).

Formation and modulation of synapses

A third critical step in the development of the nervous system is the establishment of functional synaptic connections between the individual elements of a neural circuit. In light of the current appreciation for the role of modulation of synaptic efficacy in mediating adult plasticity, it is reasonable to ask whether the cellular processes that underlie the modulation of synaptic strength in adult animals are the same processes that underlie the formation of synaptic connections during development. This question cannot yet be addressed in *Aplysia* because little is known about the mechanisms of synapse formation in this system.

Thus, although we have implicated mechanisms of long-term synaptic facilitation as playing a role in both differentiation and neurite outgrowth, it remains to be demonstrated whether these same mechanisms also play a role in synapse formation. There is reason, however, to be optimistic that such a comparison will be fruitful since it is known from work in vertebrate systems that there is considerable overlap between the mechanisms of synapse formation and long-term synaptic facilitation in the hippocampus (for review, see Kelly, 1992; Pffenninger et al., 1992)

One way in which to address the issue of whether mechanisms of synaptic facilitation contribute to synapse formation is to examine when different forms of synaptic plasticity first emerge developmentally. There are already several examples in *Aplysia* indicating that one form of synaptic modulation, short-term facilitation, emerges developmentally well after the formation of synaptic connections. It is therefore not likely that the mechanisms of short-term facilitation are involved in the initial development of synaptic connectivity.

Support for this view is provided by Ohmori (1982) who have examined the developmental expression of post-tetanic potentiation (a short-lasting form of homosynaptic facilitation) at identified inhibitory and excitatory synapses and have found that in both cases, short-term modulation of synaptic efficacy emerges at a distinct and later phase than the establishment of a functional synaptic connection.

In addition, Rayport and Camardo (1984) and Nolen and Carew (1988) have found a similar sequence for the developmental emergence of short-term heterosynaptic facilitation. Finally, Marcus and Carew (1990) have shown that spike broadening in *Aplysia* sensory neurons, an effect that has been strongly implicated in short-term synaptic facilitation in adult animals, emerges at the very end of the juvenile phase of development, well after functional synaptic connections have been made between sensory and motor neurons.

To date, the developmental expression of long-term synaptic facilitation has not been examined; however, based on our observations that elements of the mechanism of long-term facilitation play a role in differentiation and neurite outgrowth, we expect that some of these elements may also be involved in synapse formation. Thus, one further prediction of the early developmental role for the mechanisms of long-term synaptic facilitation is that long-term synaptic facilitation itself might develop earlier than short-term facilitation.

Concluding remarks

Our comparison of development and learning in *Aplysia* has revealed a striking number of mechanistic similarities between these two processes. These observations lend substantial support to Cajal's hypothesis that growth mechanisms involved in the development of the nervous system persist into the adult where they subserve learning and memory. It is particularly intriguing that growth processes were the focus of Cajal's attention in view of the finding that short-term synaptic plasticity, a process that does not involve growth but rather is dependent on the covalent modification of pre-existing proteins, emerges developmentally well after functional neuronal circuitry has been established. It therefore appears that the adaptive properties of the adult nervous system represent a combination of two classes of mechanisms, one that includes retained developmental processes, and a second class that appears to be specifically related to adult plasticity; the distinguishing feature between the two classes is the dependence on cell growth.

Where do we go from here? For studies in *Aplysia*, there are some clear avenues that warrant further investigation. For example, if we are correct in thinking that long-term plasticity involves the reactivation of at least part of the molecular program for differentiation, it will be important to look for regulation of differentiation genes such as homeobox and helix-loop-helix transcription factors in long-term synaptic plasticity. This is an avenue of research that remains virtually unexplored and may provide valuable insights into the molecular nature of long-term memory.

From the perspective of neurite outgrowth, important insights may come from an analysis of the role of ion channel modulation in the regulation of growth cone motility and axon guidance. In particular, the critical role that Ca^{2+} plays in these processes makes this an obvious focus for investigation. Specifically, it is known that in adult *Aplysia*, 5HT can modulate Ca^{2+} channels that do not contribute to normal synaptic transmitter release (Edmonds et al., 1990); perhaps it is through the modulation of these or related conductances in the growth cone that 5HT exerts an effect on neurite outgrowth.

Finally, with respect to synapse formation, recent advances have been made in vertebrate systems where synapse-specific proteins and their role in the development of connectivity and transmitter release are currently being elucidated (Kelly, 1992; DeCamilli, 1993). Thus, it may be useful to look for counterparts to these proteins in *Aplysia* where learning is well-characterized and determine whether they play comparable roles in synapse formation and modulation.

At a more general level, the real test of the relationship between development and learning will ultimately come from studies that ask whether the same process is *required* for both neuronal development and synaptic plasticity. Important advances in this direction have already been made with genetic manipulation in vertebrates and invertebrates. For example, it has been possible using homologous recombination technology in mice to demonstrate that knocking out the gene for a specific kinase can produce both developmental abnormalities in the CNS and a deficiency in long-term synaptic plasticity in the adult (Grant et al., 1992; Silva et al., 1992).

Likewise, in *Drosophila*, Drain et al. (1991) have used heat inducible promoters to control the temporal expression of inhibitors of the cAMP-dependent protein kinase and have shown that kinase activity in the adult fly is required for learning and memory. Therefore, the use of inducible promoters that allow selective spatial and temporal activation of downstream genes may provide a powerful strategy for specifying the developmental and adult requirement for the expression of specific processes and may in the future permit a means of directly testing the relationship between development and learning.

In summary, considerable progress has recently been made in the analysis of cellular and molecular mechanisms of both development and learning and there are now several points of convergence between these two fields. Thus, it appears that the ideas underlying the original development and formulation of Cajal's hypothesis have been retained through the intervening century and have helped to shape and guide our contemporary view of the mechanistic relationship between development and learning.

Acknowledgments

We thank Laura Stark for helpful comments on an earlier draft of the manuscript. Supported by NIMH post-doctoral training fellowship 5T332MH18397-07 to E.A.M., SERC-NATO post-doctoral fellowship to N.J.E., predoctoral NSERC and FCAR fellowships (Canada) to R.M. and NIH grant R01-MH-1083 and AFOSR award F49620-93-1-0273 to T.J.C.

References

Anderson, D.J. (1992) Molecular control of neural development. In: Z. Hall (Ed.), *Molecular Neurobiology*, Sinauer Associates, MA.

Bailey, C.H. (1991) Morphological basis of short- and long-term memory in *Aplysia*. In: R. Lister and H. Weingartner (Eds.), *Perspectives on Cognitive Neuroscience*, Oxford University Press, New York.

Bailey, C.H. and Kandel, E.R. (1993) Structural changes accompanying memory storage. *Annu. Rev. Physiol.*, 55: 397–426.

Barzilai, A., Kennedy, T.E., Sweatt, J.D. and Kandel, E.R. (1989) 5-HT modulates protein synthesis and the expression of specific proteins during long-term facilitation in *Aplysia* sensory neurons. *Neuron*, 2: 1577–1586.

Belardetti, F., Schacher, S., Kandel, E.R. and Siegelbaum, S.A. (1986) The growth cones of *Aplysia* sensory neurons: modulation by serotonin of action potential duration and single potassium currents. *Proc. Natl. Acad. Sci. USA*, 83: 7094–7098.

Cajal, S.R. (1911) Histologie du système nerveux de l'homme et des vertébrés.

Castellucci, V.F., Blumenfeld, H., Goelet, P. and Kandel, E.R. (1989) Inhibitor of protein synthesis blocks long-term behavioral sensitization in the isolated gill-withdrawal reflex of *Aplysia*. *J. Neurobiol.*, 20: 1–9.

Chang, T.N., Marcus, E.A., Dudai, Y. and Carew, T.J. (1989) Developmental analysis of adenylate cyclase activity and its modulation by 5-HT in the CNS of *Aplysia*. *Soc. Neurosci. Abstr.*, 15: 1019.

Clark, G.A. and Kandel, E.R. (1993) Induction of long-term facilitation in *Aplysia* sensory neurons by local application of 5HT to remote synapses. *Proc. Natl. Acad. Sci. USA*, 90: 11411–11415.

Cole, A.J., Saffen, D.W., Baraban, J.M. and Worley, P.F. (1989) Rapid increase of an immediate early gene messenger RNA in hippocampal neurons by synaptic NMDA receptor activation. *Nature*, 340: 474–476.

Dash, P.K., Hochner, B. and Kandel, E.R. (1990) Injection of the cAMP-responsive element into the nucleus of *Aplysia* sensory neurons blocks long-term facilitation. *Nature*, 345: 718–721.

DeCamilli, P. (1993) Exocytosis goes with a SNAP. *Nature*, 364: 387–388.

Doherty, P., Fruns, M., Seaton, P., Dickson, G., Barton, C.H., Sears, T.A. and Walsh, F.S. (1990) A threshold effect of the major isoforms of NCAM on neurite outgrowth. *Nature*, 343: 464–466.

Drain, P., Folkers, E. and Quinn, W.G. (1991) cAMP-dependent protein kinase and the disruption of learning in transgenic flies. *Neuron*, 6: 71–82.

Edmonds, B., Klein, M., Dale, N. and Kandel, E.R. (1990) Contributions of two types of calcium channels to synaptic transmission and plasticity. *Science*, 250: 1142–1147.

Emptage, N.J. and Carew T.J. (1993) Long-term synaptic facilitation in the absence of short-term synaptic facilitation in *Aplysia* neurons. *Science*, 262: 253–256.

Forscher, P., Kaczmarek, L.K., Buchanan, J. and Smith, S.J. (1987) Cyclic AMP induces changes in distribution and transport of organelles within growth cones of *Aplysia* bag cell neurons. *J. Neurosci.*, 7: 3600–3611.

Garrell, J. and Campuzano, S. (1991) The helix-loop-helix domain: a common motif for bristles, muscles and sex. *Bioessays*, 13: 493–498.

Gehring, W.J. (1987) Homeo boxes in the study of development. *Science*, 236: 1245–1252.

Glanzman, D.L., Kandel, E.R. and Schacher, S. (1989) Identified target motor neuron regulates neurite outgrowth and synapse formation of *Aplysia* sensory neurons *in vitro*. *Neuron*, 3: 441–450.

Glanzman, D.L., Kandel, E.R. and Schacher, S. (1990) Target dependent structural changes accompanying long-term synaptic facilitation in *Aplysia* neurons. *Science*, 249: 799–802.

Grant, S.G., O'Dell, T.J., Karl, K.A., Stein, P.L., Soriano, P. and Kandel, E.R. (1992) Impaired long-term potentiation. spatial learning, and hippocampal development in *fyn* mutant mice. *Science*, 258: 1903–1910.

Jones, F.S., Holst, B.D., Minowa, O.., DeRobertis, E.M. and Edelman, G.M. (1993) Binding and transcriptional activation of the promoter for the neural cell adhesion molecule by HoxC6 (Hox-3.3). *Proc. Natl. Acad. Sci. USA*, 90: 6557–6561.

Kandel, E.R. and Hawkins, R.D. (1992) The biological basis of learning and individuality. *Sci. Am.*, 267: 78–86.

Kandel, E.R. and Schwartz, J.H. (1982) Molecular biology of learning: Modulation of transmitter release. *Science*, 218: 433–443.

Keller, F. and Schacher, S. (1990) Neuron-specific membrane glycoproteins promoting neurite fasciculation in *Aplysia californica*. *J. Cell Biol.*, 111: 2637–2650.

Kelly, P.T. (1992) Calmodulin-dependent protein kinase II. *Mol. Neurobiol.*, 5: 153–177.

Kistler, H.B., Jr., Hawkins, R.D., Koester, J., Steinbusch, H.W., Kandel, E.R. and Schwartz, J.H. (1985) Distribution of serotonin-immunoreactive cell bodies and processes in the abdominal ganglion of mature *Aplysia*. *J. Neurosci.*, 5: 72–80.

LeRoux, I., Joliot, A.H., Bloch-Gallego, E., Prochiantz, A. and Volovitch, M. (1993) Neurotrophic activity of the Antennapedia homeodomain depends on its specific DNA-binding properties. *Proc. Natl. Acad. Sci.*, 90: 9120–9124.

Madison, D.V., Malenka, R.C. and Nicoll, R.A. (1991) Mechanisms underlying long-term potentiation of synaptic transmission. *Annu. Rev. Neurosci.*, 14: 379–397.

Marcus, E.A. and Carew, T.J. (1990) Differential modulation of excitability and spike duration in the tail sensory neurons of developing *Aplysia*. *Soc. Neurosci. Abstr.*, 16: 19.

Marois, R., Kelly, G.M., Hockfield, S. and Carew, T.J. (1992) An ultrastructural study of serotonergic cells in the CNS of embryonic and larval *Aplysia*. *Soc. Neursci. Abstr.*, 18: 1471.

Mattson, M.P., Taylor-Hunter, A. and Kater, S.B. (1988) Neurite outgrowth in individual neurons of a neuronal population is differentially regulated by calcium and cyclic AMP. *J. Neurosci.*, 8: 1704–1711.

Mayford, M., Barzilai, A., Keller, F., Schacher, S. and Kandel, E.R. (1992) Modulation of an NCAM-related adhesion molecule with long-term synaptic plasticity in *Aplysia*. *Science*, 256: 638–644.

Montarolo, P.G., Goelet, P., Castellucci, V.F., Morgan, J., Kandel, E.R. and Schacher, S. (1986) A critical period for macromolecular synthesis in long-term heterosynaptic facilitation in *Aplysia*. *Science*, 234: 1249–1254.

Nazif, F.A, Byrne, J.H and Cleary, L.J. (1991) cAMP induces long-term morphological changes in sensory neurons of *Aplysia*. *Brain Res.*, 539: 324–327.

Nedivi, E., Hevroni, D., Naot, D., Israeli, D. and Citri, Y. (1993) Numerous candidate plasticity-related genes revealed by differential cDNA cloning. *Nature*, 363: 718–721.

Noel, F., Scholz, K.P., Eskin, A. and Byrne, J.H. (1991) Common set of proteins in *Aplysia* sensory neurons affected by an in vitro analogue of long-term sensitization training, 5-HT and cAMP. *Brain Res.*, 568: 67–75.

Nolen T.G. and Carew, T.J. (1988) The cellular analog of sensitization emerges at the same time in development as behavioral sensitization in *Aplysia*. *J. Neurosci.*, 8: 212–222.

Ohmori, H. (1982) Development of post-tetanic potentiation at identified inhibitory and excitatory synapses in *Aplysia*. *J. Physiol.*, 322: 223–240.

Pfenninger, K.H., de la Houssaye, B.A., Helmke, S.M. and Quiroga, S. (1992) Growth-regulated proteins and neuronal plasticity. *Mol. Neurobiol.*, 5: 143–151.

Qian, Z., Gilbert, M.E., Colicos, M.A., Kandel, E.R. and Kuhl, D. (1993) Tissue-plasminogen activator is induced as an immediate-early gene during seizure, kindling, and long-term potentiation. *Nature*, 361: 453–457.

Rayport, S.G. and Camardo, J.S. (1984) Differential emergence of cellular mechanisms mediating habituation and sensitization in the developing *Aplysia* nervous system. *J. Neurosci.*, 4: 2528–2532.

Schacher, S., Kandel, E.R. and Wooley, R. (1979) Development of neurons in the abdominal ganglion of *Aplysia californica*: axosomatic synaptic contacts. *Dev. Biol.*, 71: 163–175.

Schacher, S., Castellucci, V.F. and Kandel, E.R. (1988) cAMP evokes long-term facilitation in *Aplysia* sensory neurons that requires new protein synthesis. *Science*, 240: 1667–1669.

Schacher, S., Glanzman, D., Barzilai, A., Dash, P., Grant, S.G.N., Keller, F., Mayford, M. and Kandel, E.R. (1990) Long-term facilitation in *Aplysia*: persistent phosphorylation and structural changes. *Cold Spring Harbor Symp. Quant. Biol.*, 55: 187–202.

Scholz, K.P. and Byrne, J.H. (1987) Intracellular injection of cAMP induces a long-term reduction of neuronal potassium currents. *Science*, 240: 1664–1666.

Sheng, M. and Greenberg, M.E. (1990) The regulation and function of c-*fos* and other immediate early genes in the nervous system. *Neuron*, 4: 477–485.

Silva, A.J., Stevens, C.F., Tonegawa, S. and Wang, Y. (1992) Deficient hippocampal long-term potentiation in α-calcium-calmodulin kinase II mutant mice. *Science*, 257: 210–206.

Sweatt, J.D. and Kandel, E.R. (1989) Persistent and transcripionally-dependent increase in protein phosphorylation in long-term facilitation of *Aplysia* sensory neurons. *Nature*, 339: 51–54.

Wisden, W., Errington, M.L., Williams, S., Dunnett, S.B., Waters, C., Hitchcock, D., Evan, G., Bliss, T.V.P. and Hunt, S.P. (1990) Differential expression of immediate early genes in the hippocampus and spinal cord. *Neuron*, 4: 603–614.

Zhang, Z.S., Fang, B., Marshak, D.W., Byrne, J.H. and Cleary, L.J. (1991) Serotoninergic varicosities make synaptic contacts with pleural sensory neurons of *Aplysia*. *J. Comp. Neurol.*, 311: 259–270.

F. Bloom (Editor)
Progress in Brain Research, Vol. 100

CHAPTER 24

Synapse-activated protein synthesis as a possible mechanism of plastic neural change

I.J. Weiler[1,2], X. Wang[3] and W.T. Greenough[1–4]

[1]Beckman Institute, [2]Department of Psychology, [3]Neuroscience Program and [4]Department of Cell and Structural Biology, University of Illinois, Urbana, IL, USA

There has been considerable interest for the last several decades in the extent of involvement of the metabolic machinery of the cell, and particularly de novo RNA and protein synthesis, in plastic neural change, including learning and memory. Much of current experimental work employs long-term potentiation (LTP) as a model (Bliss and Lomo, 1973); the term refers to long-lasting increased postsynaptic responses following presynaptic tetanization. Following LTP induction, both presynaptic alterations (e.g. modulation of transmitter release) and postsynaptic modifications (e.g. channel sensitivity or synaptic architecture changes) have been proposed to mediate the altered synaptic efficacy. In this review, we consider postsynaptic modifications.

There appear to be at least three sequential phases in the establishment of this altered response (Bliss and Collingridge, 1993): (a) a kinase-dependent, protein synthesis-independent stage, which is seen even in the presence of protein synthesis inhibitors such as anisomycin; (b) a protein synthesis-dependent stage, affected by anisomycin and other translation blockers; and (c) an mRNA transcription-dependent stage, blocked by actinomycin D. Although the first stage may last up to 3 h, it is followed by rapid decay if a protein translation inhibitor has been present during tetanization or during the first 15 min thereafter (Krug et al., 1984; Otani et al., 1989). Thus, during the first 15 min after tetanization, synthesis of key proteins which will prevent this rapid decay must be initiated. Finally, long-term effects are associated with the induction of immediate early genes such as *zif-268*, which are believed to induce a program of activation of specific genes which will be ultimately responsible for changes in cellular architecture (e.g. Worley et al., 1993).

There is evidence for involvement of dendritic protein synthesis in behaviorally driven neural plasticity. Rats reared in complex environments have, in the visual cortex, neurons with larger dendritic fields and more synapses than rats reared in pairs or individually in standard laboratory cages (e.g. Turner and Greenough, 1983). Greenough et al. (1985) reported that rats reared in a complex environment had a greater percentage of synapses with polyribosomal aggregates (PRA) in dendritic spines, relative to cage-housed groups.

An ambitious survey of protein changes in the dentate gyrus of the hippocampus, after LTP induction in anesthetized intact rats by tetanic stimulation, was made by Fazeli et al. (1993). Three hours after induction, eleven spots on two-dimensional protein gels were found to have altered densities; these included reductions as well as increases in spot intensity, suggesting that complex patterns of changes are occurring.

Since the transcriptional inhibitor actinomycin D does not initially affect LTP decay, the maintenance of LTP during at least the first 3–6 h depends on protein synthesis from pre-existing mRNA. Since a neuron may have over ten thousand synapses (Turner and Greenough, 1983; Harvey and Napper, 1991), a considerable degree of local control over rapid production of proteins would be required for specific, localized structural changes; and the mRNA must be targeted to specific cellular locations. Ca^{2+}-imaging techniques demonstrate that the Ca^{2+} elevation induced by tetanus is in fact localized to regions near dendritic spines (Mueller and Connor, 1991); there is increasing evidence for a complex interplay of calcium release from intracellular stores and influx of extracellular calcium (e.g. van den Pol et al., 1993).

Several laboratories have found evidence for cellular targeting of a few specific mRNAs to dendritic areas (e.g. Steward and Banker, 1992). Rao and Steward (1991) have shown that proteins can be synthesized within synaptosomes; and Torre and Steward (1992), employing an elegant tissue culture system in which dendrites can be definitively separated from cell somata, have demonstrated protein synthesis in isolated dendrites. In a recent report, Miyashiro et al. (1993), using poly-A primed cDNA amplification followed by PCR amplification of mRNA from single neurites of hippocampal neurons in tissue culture, presented evidence that in fact many species of mRNA may be present in neurites; identification of mRNA translated in or near dendritic spines will be harder to come by.

With regard to the more precise localization of mRNA in dendrites, Steward and Levy (1982) made the important observation that PRAs were more frequently found in postsynaptic spines during reactive synaptogenesis. Hwang and Greenough (1986) subsequently found, examining the repeatedly identifiable population of synapses on visual cortical layer V pyramidal neuron apical dendrites in layer IV, that PRAs were more frequently found postsynaptically during the period of peak synaptogenesis (days 13–20) than subsequently (≥25).

Because any attempt to examine synaptically associated PRA-related protein synthesis was likely to be masked by the vastly greater somatic protein synthesis, we have adapted a synaptoneurosome preparation (Hollingsworth et al, 1985) from the occipital-parietal cortex of 14–16-day-old Long-Evans rats, for further study of what we believe may be the same phenomenon. Synaptoneurosomes, depicted in Fig. 1, consist of pinched-off and resealed presynaptic terminals con-

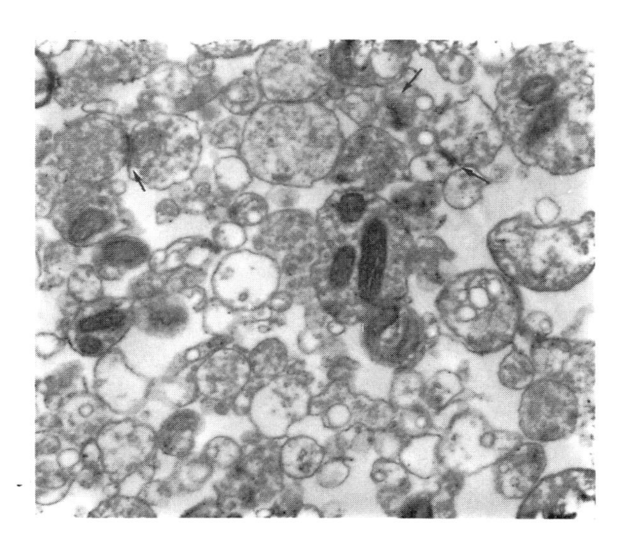

Fig. 1. Electron micrograph of synaptoneurosomal preparation, after filtration and pelleting, immersion fixation in 2.5% glutaraldehyde, followed by osmium tetroxide, sectioning and staining with uranyl acetate. Several complete synaptoneurosomes can be seen; some other profiles may also represent synaptoneurosomes in which the plane of section fails to include characteristic morphological components. Used with permission of Academic Press, copyright 1991.

nected to pinched-off and resealed postsynaptic compartments that appear to be larger, on average, than typical dendritic spines. They exhibit normal presynaptic neurotransmitter release in response to depolarization as well as apparently normal postsynaptic responses (Dunkley and Robinson, 1986; Recasens et al., 1987; Gorelick et al., 1988; Verhage et al., 1992). In a series of investigations of the biochemical events (see Fig. 2; numbers in bold refer to steps we have documented in this sequence) following in vitro depolarization by 40 mM K^+ (**1**), we have found a rapid aggregation of ribosomes with mRNA, which we have shown to be associated with accelerated synthesis of some new proteins (Weiler and Greenough, 1991). Furthermore, we found that the effect is mediated by glutamate release (**2**), which triggers protein synthesis via metabotropic glutamate receptors (mGluR) (Weiler and Greenough, 1993).

In this system we measured, at short intervals, the fraction of total synaptoneurosomal RNA trapped in the form of polyribosomal aggregates, selected by

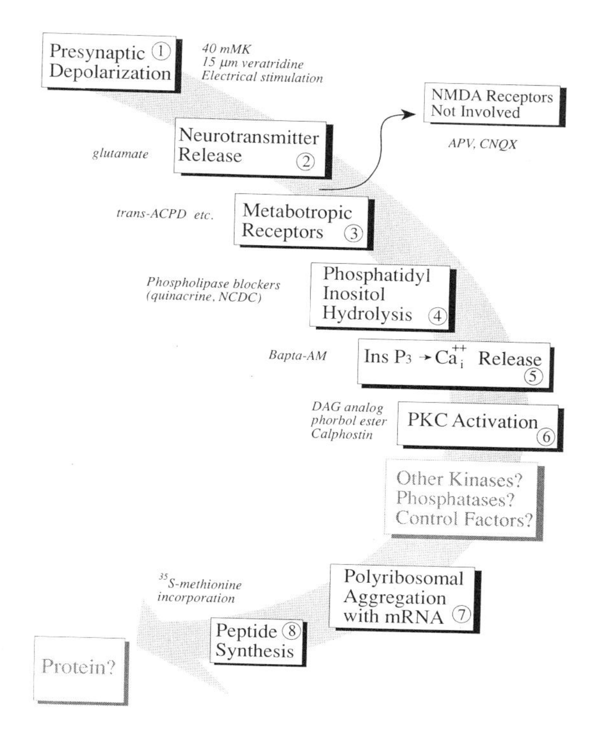

Fig. 2. Schematic description of probable cascade of intermediaries leading from presynaptic depolarization to postsynaptic protein synthesis. Numbers in circles correspond to references in text, where experimental documentation is available.

centrifugation through 1 M sucrose. We found that a rapid (1–2 min) loading of ribosomes onto mRNA was accompanied (**7**,**8**) by an accelerated rate of incorporation of [^{35}S]methionine into TCA-precipitable polypeptides. Entry of extracellular calcium appears not to be necessary to this process, since it was not prevented by calcium chelators in the buffer (0.04 mM BAPTA) and not weakened by co-application of 2-amino-5-phosphonovalerate (APV, a competitive NMDA receptor antagonist, **3**). An internal calcium buffer blocked the effect, however (**5**). Nifedipine, a blocker of voltage-sensitive calcium channels, also does not prevent the response. The susceptibility of the effect to a series of glutamate analogues (quisqualate, ibotenate and *trans*-aminocyclopentane-1*S*,3*R*-dicarboxylic acid, ACPD, **3**) matched that of mGluR receptors 1 and 5 (Schoepp et al., 1990).

Metabotropic receptors trigger the hydrolysis of membrane phosphatidyl inositol; in accord with this we found (**4**) that phospholipase blockers (quinacrine or 2-nitro-4-carboxyphenyl-*N*,*N*-dephenylcarbamate, NCDC) reduced the response. This hydrolysis leads to the production of inositol triphosphate (which releases Ca^{2+} from internal stores) and diacylglycerol which activates protein kinase C (PKC). We used phorbol ester and a diacylglycerol analogue (1-oleoyl-2-acetyl-glycerol) to activate PKC (**6**), and found a similar rapid increase of polyribosome aggregation. Calphostin C, a highly specific inhibitor which reacts with the regulatory domain of PKC, effectively blocked (80%) the response to ACPD. Within the same experiments, however, calphostin did not block the polyribosomal aggregation triggered by 40 mM K^+, suggesting that more than one activation system may be triggered by depolarization.

The induction of LTP involves both NMDA and metabotropic glutamate receptors. Activation of second messengers via metabotropic neurotransmitter receptors has been shown to play a crucial role in some forms of LTP (Anikstejn et al., 1992; Zheng and Gallagher, 1992; Bashir et al., 1993). Indeed, induction of LTP by application of the metabotropic agonist trans-ACPD alone has recently been reported (Bortolotto and Collingridge, 1993). Furthermore, protein synthesis has been demonstrated directly by Feig and Lipton (1993), pairing patterned electrical stimulation of hippocampal CA1 pyramidal cells, in an acute slice preparation, with exposure to 50 μM carbachol (a metabotropic cholinergic agonist). Neither electrical stimulation nor carbachol alone caused increased protein synthesis; but paired stimulation gave rise to a threefold increase of [^{3}H]leucine incorporation into dendrites, strongly suggesting de novo protein synthesis. This pairing has not been shown to produce LTP, although the metabotropic agonist *trans*-ACPD can induce LTP in hippocampal slices when it is coupled with a weak subthreshold presynaptic tetanus (Otani and Ben-Ari, 1991).

Initiation of protein synthesis is not likely to be a simple response; it is likely that several competing processes participate in regulation of the response to synaptic activation. In recent experiments we have found that Ca^{2+} entry (caused for example by stimulation with NMDA as well as by the calcium ionophore ionomycin) prior to mGluR activation markedly depresses the polyribosome aggregation (unpublished results). Thus, the threshold for polyribosomal aggregation in response to metabotropic glutamate receptor activation may be continuously adjusted according to the recent history of the postsynaptic Ca^{2+} level regulated by NMDA receptor activation or other sources. Similarly, the threshold for LTP induction is also closely regulated by the recent history of NMDA receptor activation (Huang et al., 1992). It is known that the activation of the NMDA receptors elicits a Ca^{2+} dependent release of nitric oxide (Garthwaite et al., 1988), which has been proposed as a retrograde messenger to affect presynaptic aspects of LTP, either in a positive or negative modulatory role (Izumi et al, 1992).

In recent work, we have been investigating the effects of electrical stimulation on synaptoneurosome suspensions. A polyribosome aggregation response is dependent on voltage (2 V, under these conditions about 1 mA), on burst length (20 ms), on frequency (100–200 Hz), and on the interval between bursts of stimuli (200 ms). These frequency parameters are similar to those defined as optimal for "prime burst" or "theta burst" potentiation (Larson and Lynch, 1986; Rose and Dunwiddie, 1986). Thus, there are, in addition to neurotransmitter responsiveness, a number of aspects in which experimental requirements for LTP parallel those we find for polyribosomal aggregation and initiation of protein synthesis. While there are tantalizing pharmacological links between the phospholipase-triggered second messenger effects and the initiation of protein synthesis in synaptoneurosomes, a direct link with LTP cannot yet be established.

There are several plausible candidates for alterations which would support the earliest phase of LTP, which can be maintained for 1–3 h without protein synthesis. The conformational state of NMDA and AMPA channels is known to be changed upon phosphorylation, such that Ca^{2+} entry is expedited (e.g. Reymann, 1993). Depending on which kinases are activated, either serine/threonine or tyrosine residues are phosphorylated. Some translational factors might require coordinate activation of both kinds of phosphorylation sites (e.g. Tuazon et al., 1989). Synaptic activation of protein phosphatases has also been shown to be an important element in regulation of synaptic strength (e.g. Mulkey et al., 1993). Calcium-calmodulin dependent kinase II (CaMKII) becomes autophosphorylated as a result of competing phosphorylating and dephosphorylating activities regulated in part by Ca^{2+}; the autophosphorylated form is then independent of calcium for some time. A biochemical model for storage of synaptic information based on this equilibrium in the state of phosphorylation of CaMKII has been suggested by Lisman (1989).

When we come to more permanent synaptic changes, requiring protein synthesis, there may simply be a shift in amount of proteins already present, but also some new proteins, such as PKM, the constitutively active form of PKC (Sacktor et al., 1993); this important question is still open to investigation. Geinisman (1993) and others have described a complex process of synaptic remodelling in hippocampal spines; a considerable array of proteases, structural proteins, and enzymatic complexes must be mobilized to implement these new structures, so that in addition to rapid synthesis of proteins from messages already present, we may expect delivery of new mRNA, and centrally synthesized proteins, also to play an important role.

Thus, an array of mechanisms is available which could maintain an altered postsynaptic state for minutes or hours on the basis of an altered phosphorylation state of pre-existing proteins; but altered phosphorylation may well also serve as a finely tuned regulatory device for stopping or starting the translation of proteins from localized RNA messages, held poised awaiting the proper combination of signals, which may be activated both by finely tuned Ca^{2+} oscillations and, as we have shown, by phospholipase-activated second messenger cascades which are triggered by metabotropic receptors. As in the case of growth factors, a complex interdigitation of interacting pathways could facilitate rapid information storage in a multifacetted way at the level of synapses. Given the number of intermediaries that the various pathways have in common, some sort of functional compartmentalization must exist to limit their interactions. Increasingly sophisticated techniques are being used to probe the finely tuned interplay between receptors, localized calcium ion flux, and a restricted, targeted set of messenger RNAs which might, under the appropriate conditions, be translated to effect a structural synaptic change. Which proteins these are, and how their translation is regulated, will be a challenging subject for future research.

Acknowledgement

The research reported here was supported by MH35321, NSF BNS 8821219, the Kiwanis Spastic Paralysis Foundation, and the Office of Naval Research.

References

Aniksztejn, L., Otani, S. and Ben-Ari, Y. (1992) Quisqualate metabotropic receptors modulate NMDA currents and facilitate induction of long-term potentiation through protein kinase C. *Eur. J. Neurosci*, 4: 500–505.

Bashir, Z.I., Bortolotto, Z.A., Davies, C.H., Berretta, N., Irving, A.J., Seal, A.J., Henley, J.M., Jane, D.E., Watkins, J.C. and Collingridge, G.L. (1993) Induction of LTP in the hippocampus needs synaptic activiation of glutamate metabotropic receptors. *Nature*, 363: 347–350.

Bliss, T.V.P. and Collingridge, G.L. (1993) A synaptic model of memory: long-term potentiation in the hippocampus. *Nature*, 361: 31–39.

Bliss, T.V.P. and Lomo, T. (1973) Long-lasting potentiation of synaptic transmission in the dentate area of the anesthetized rabbit following stimulation of the perforant path. *J. Physiol. (London)*, 232: 331–356.

Bortolotto, Z.A. and Collingridge, G.L. (1993) Characterisation of LTP induced by the activation of glutamate metabotropic receptors in area CA1 of the hippocampus. *Neuropharmacology*, 32: 1–9.

Dunkley, P.R. and Robinson P.J. (1986) Depolarization-dependent protein phosphorylation in synaptosomes: mechanisms and significance. In: W.H. Gispen and A. Routtenberg (Eds.), *Progress in Brain Research*, Vol. 69, Elsevier, Amsterdam, pp. 273–293.

Fazeli, M.S., Corbet, J., Dunn, M.J., Dolphin, A.C. and Bliss, T.V.P. (1993) Changes in protein synthesis accompanying long-term potentiation in the dentate gyrus *in vivo*. *J. Neurosci.*, 13: 1346–1353.

Feig, S. and Lipton, P. (1993) Pairing the cholinergic agonist carbachol with patterned Schaffer collateral stimulation initiates protein synthesis in hippocampal CA1 pyramidal cell dendrites via a muscarinic, NMDA-dependent mechanism. *J. Neurosci.*, 13: 1010–1021.

Garthwaite, J., Charles, S.L. and Chess-Williams, R. (1988) Endothelium-derived relaxing factor release relaxing factor release on activation of NMDA receptors suggests role as intercellular messenger in the brain. *Nature*, 336: 385–388.

Geinisman, Y. (1993) Perforated axospinous synapses with multiple, completely partitioned transmission zones: probable structural intermediates in synaptic plasticity. *Hippocampus*, 3: 417–434.

Gorelick, F.S., Wang, J.K.T., Lai, Y., Nairn, A.C. and Greengard, P. (1988) Autophosphorylation and activation of Ca^{2+}/calmodulin-dependent protein kinase II in intact nerve terminals. *J. Biol. Chem.*, 263: 17209–17212.

Greenough, W.T., Hwang, H.-M.F. and Gorman, C. (1985) Evidence for active synapse formation or altered postsynaptic metabolism in visual cortex of rats reared in complex environments. *Proc. Natl. Acad. Sci. USA*, 82: 4549–4552.

Harvey, R.J. and Napper, R.M.A. (1991) Quantitative studies on the mammalian cerebellum. *Prog. Neurobiol.*, 36: 437–463.

Hollingsworth, E.B., McNeal, E.T., Burton, J.L., Williams, R.J., Daly, J.W. and Creveling, C.R. (1985) Biochemical characterization of a filtered synaptoneurosome preparation from guinea pig cerebral cortex: cyclic adenosine 3′:5′-monophosphate-generating systems, receptors, and enzymes. *J. Neurosci.*, 5: 2240–2253.

Huang, Y-Y., Colino, A., Selig, D.K. and Malenka, R.C. (1992) The influence of prior synaptic activity on the induction of long-term potentiation. *Science*, 255: 730–733.

Hwang, H.M. and Greenough, W.T. (1984) Spine formation and synaptogenesis in rat visual cortex: a serial section developmental study. *Soc. Neurosci. Abstr.*, 10: 579.

Izumi, Y., Clifford, D.B. and Zorumski, C.F. (1992) Inhibition of long-term potentiation by NMDA-mediated nitric oxide release. *Science*, 257: 1273–1275.

Krug, M., Loessner, B. and Ott, T. (1984) Anisomycin blocks the late phase of long-term potentiation in the dentate gyrus of freely moving rats. *Brain Res. Bull.*, 13: 39–42.

Larson, J. and Lynch, G. (1986) Induction of synaptic potentiation in hippocampus by patterned stimulation involves two events. *Science*, 232: 985–988.

Lisman, J. (1989) A mechanism for the Hebb and the anti-Hebb processes underlying learning and memory. *Proc. Natl. Acad. Sci. USA*, 86: 9574–9578.

Miyashiro, K., Dichter, M. and Eberwine, J. (1993) Presence of multiple mRNA products in hippocampal neurites. *Soc. Neurosci. Abstr.*, 327: 16.

Mueller, W. and Connor, J.A. (1991) Dendritic spines as individual neuronal compartments for synaptic Ca^{2+} responses. *Nature*, 354: 73–76.

Mulkey, R.M., Herron, C.E. and Malenka, R.C. (1993) An essential role for protein phosphatases in hippocampal long-term depression. *Science*, 261: 1051–1055.

Otani, S. and Ben-Ari, Y. (1991) Metabotropic receptor-mediated long-term potentiation in rat hippocampal slices. *Eur. J. Pharmacol.*, 205: 325–326.

Otani, S., Marshall, C.J., Tate, W.P., Goddard, G.V. and Abraham, W.C. (1989) Maintenance of long-term potentiation in rat dentate gyrus requires protein synthesis but not messenger RNA synthesis immediately post-tetanization. *Neuroscience*, 28: 519–526.

Rao, A. and Steward, O. (1991) Evidence that protein constituents of postsynaptic membrane specializations are locally synthesized: analysis of proteins synthesized within synaptosomes. *J. Neurosci.*, 11: 2881–2895.

Recasens, M., Sassetti, I., Nourigat, A., Sladeczek, F. and Bockaert, J. (1987) Characterization of subtypes of excitatory amino acid receptors involved in the stimulation of inositol phosphate synthesis in rat brain synaptoneurosomes. *Eur. J. Pharmacol.*, 141: 87–93.

Reymann, K.G. (1993) Mechanisms underlying synaptic long-term potentiation in the hippocampus: focus on postsynaptic glutamate receptors and protein kinases. *Funct. Neurol.* 8 (Suppl. 5): 7–32.

Rose, G.M. and Dunwiddie, T.V.(1986) Induction of hippocampal long-term potentiation using physiologically patterned stimulation. *Neurosci. Lett.*, 69: 244–248.

Sacktor, T.C., Osten, P., Valsamis, H., Jiang, X., Naik, M.U. and Sublette, E. (1993) Persistent activation of the ζ isoform of protein kinase C in the maintenance of long-term potentiation. *Proc. Natl. Acad. Sci. USA*, 90: 8342–8346.

Schoepp, D., Bockaert, J. and Sladeczek, F. (1990) Pharmacological and functional characteristics of metabotropic excitatory amino acid receptors. *Trends Pharmacol. Sci.*, 111: 508–515.

Steward, O. and Banker, G.A. (1992) Getting the message from the gene to the synapse: sorting and intracellular transport of RNA in neurons. *Trends Neurosci.*, 15: 180–186.

Steward, O. and Levy, W. (1982) Preferential localization of polyribosomes under the base of dendritic spines in granule cells of the dentate gyrus. *J. Neurosci.*, 2: 284–291.

Torre, E.R. and Steward, O. (1992) Demonstration of local protein synthesis within dendrites using a new cell culture system that permits the isolation of living axons and dendrites from their cell bodies. *J. Neurosci.*, 12: 762–772.

Tuazon, P.T., Merrick, W.C. and Traugh, J.A. (1989) Comparative analysis of phosphorylation of translational initiation and elongation factors by seven protein kinases. *J. Biol. Chem.*, 264: 2773–2777.

Turner, A.M. and Greenough, W.T. (1983) Synapses per neuron and synaptic dimensions in occipital cortex of rats reared in complex, social or isolation housing. *ACTA Stereol.*, 2 (Suppl I): 239–244.

Van den Pol, A., Finkbeiner, S.M. and Cornell-Bell, A.H. (1993) Calcium excitability and oscillations in suprachiasmatic nucleus neurons and glia in vitro. *J. Neurosci.*, 12: 2648–2664.

Verhage, M., Sandman, H., Mosselveld, F., van de Velde, M., Hengst, P.A., Lopes da Silva, F.H. and Ghijsen, W.E.J.M. (1992) Perfusion of immobilized isolated nerve terminals as a model for the regulation of transmitter release: release of different, endogenous transmitters, repeated stimulation, and high time resolution. *J. Neurochem.*, 58: 1313–1320.

Weiler, I.J. and Greenough, W.T. (1991) Potassium ion stimulation triggers protein translation in synaptoneurosomal polyribosomes. *Mol. Cell. Neurosci.*, 2: 305–314.

Weiler, I.J. and Greenough, W.T. (1993) Metabotropic glutamate receptors trigger postsynaptic protein synthesis. *Proc. Natl. Acad. Sci. USA*, 90: 7168–7171.

Worley, P.F., Bhat, R.V., Baraban, J.M., Erickson, C.A., McNaughton, B.L. and Barnes, C.A. (1993) Thresholds for synaptic activation of transcription factors in hippocampus: correlation with long-term enhancement. *J. Neurosci.*, 13: 4776–4786.

Zheng, F. and Gallagher, J.P. (1992) Metabotropic glutamate receptors are required for the induction of long-term potentiation. *Neuron*, 9: 163–172.

F. Bloom (Editor)
Progress in Brain Research, Vol. 100

CHAPTER 25

Plasticity of adult central fibre tracts

G. Raisman

Norman and Sadie Lee Research Centre, Laboratory of Neurobiology, National Institute for Medical Research, The Ridgeway, Mill Hill, London NW7 1AA, UK

Introduction

When axons are severed in the adult brain or spinal cord they are unable to grow back to their original targets. Until we solve this problem, stroke and spinal cord injury will remain irreparable.

A number of observations suggest that the reason for the failure of regeneration of cut central axons lies in the white matter of the myelinated fibre tracts rather than in the grey matter of the neuropil:

(1) New short range synaptic connections readily form within denervated adult neuropil after partial deafferentation (Raisman, 1969, 1985),
(2) Embryonic transplants placed directly into neuropil both give and receive specific patterns of synaptic connections with an adult host brain (e.g. Zhou et al., 1985; Field et al., 1991; Björklund, 1992), and
(3) Peripheral nerve grafts induce cut central adult axons to regenerate for long distances and reinnervate their original targets (Villegas-Pérez et al., 1988; Carter et al., 1989; Thanos, 1992).

Thus, cut adult axons do have the power of regeneration, and the adult CNS retains the ability to mediate complex patterns of synaptic reconnection in neuropil after injury. Such observations sustain the hope that one day it may be possible to repair axon injuries in the brain and spinal cord, and they focus the next stages of the search for repair on the question of axon growth in central adult myelinated fibre tracts.

We have therefore directed our investigations towards the analysis of adult central fibre tracts: their glial structure and its development, their ability to incorporate additional transplanted glia and Schwann cells, and their capacity to permit the growth of embryonic and adult axons.

The glial organization of central fibre tracts

We have examined the arrangement of the glia in the adult rat fimbria, and found that the tract has a complex and regular organization (Fig. 1) (Suzuki and Raisman, 1992). The macroglial cell bodies are assembled in largely unicellular rows, which can be of great length, and which are aligned along the longitudinal, axonal axis of the tract. Although the regularity is not perfect, the arrangement of the glia can be described by a set of average dimensions. Sections along the longitudinal axis of the tract show that the cells consist of solitary astrocytes, interspersed singly at a repeat distance of about 40–80 μm between stretches of about 6–10 contiguous oligodendrocytes (Fig. 1a). In the transverse axis of the tract, the rows form an equidistant array, separated by about 10–20 μm. Each row is surrounded by about 1000–1400 axons, of which 600–1000 are myelinated.

Both oligodendrocytes and astrocytes each give rise to two different types of processes which lie in the radial and longitudinal planes of the tract. The oligodendrocytes (Fig. 1a) each have about 20–40 fine radial processes which cross the axons, and each process gives rise to a myelin internode which ensheathes a stretch of about 150–250 μm of axon in the longitudinal axis of the tract. The astrocytic cell bodies (Fig.

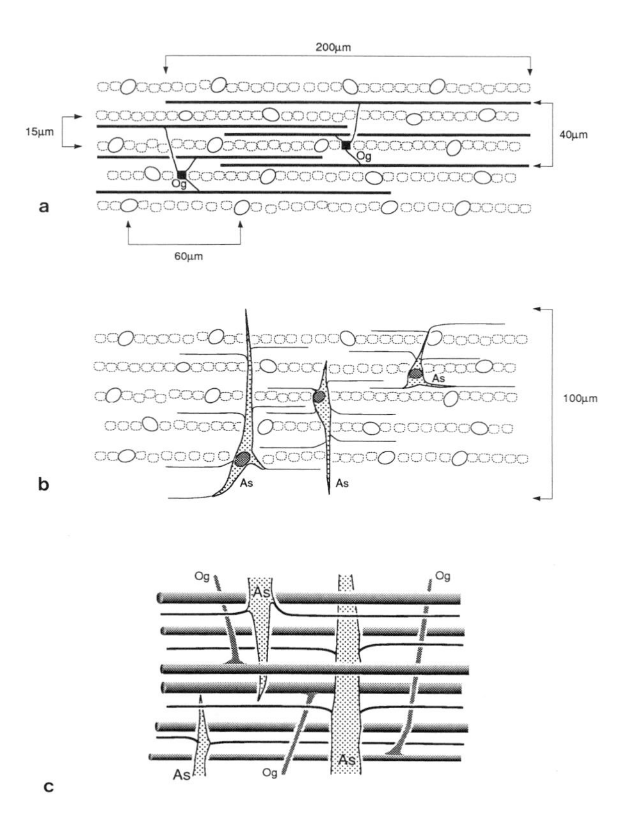

Fig. 1. (a) Radial (stem) and longitudinal (myelinating) processes of two oligodendrocytes (Og, black squares) shown against a background of the interfascicular glial nuclear rows in a horizontal section of the adult rat fimbria. (b) Three astrocytes (As) with radial and longitudinal processes. (c) Higher magnification of the meshwork of oligodendrocytic (Og) and astrocytic (As) processes. Typical dimensions: core-to-core distance between interfascicular glial rows, 15 μm; interastrocytic distance within a row, 60 μm; oligodendrocytic radial span, 40 μm; length of internode, 200 μm; astrocytic radial span, 100 μm.

1b) are prolonged into tapering radial processes which cross the axons and generate large numbers of fine, untapering longitudinal processes which run parallel to and among the axons (Fig. 2a). Thus, the radial and longitudinal processes of the oligodendrocytes and the astrocytes are interwoven to form a rectilinear network through which the tract axons run (Fig. 1c).

We have found similar glial arrangements in the descending corticospinal tract and in the ascending sensory columns of the dorsal horn of the cervical spinal cord (Y. Ajayi et al., unpublished observations; see also Matthews and Duncan, 1971). Thus, despite their very different fibre compositions and times of formation (the sensory tracts very early in embryonic CNS development, the fimbria after the 15th day of embryonic life in the rat, and the corticospinal tract postnatally), the glial frameworks of each of these tracts are qualitatively alike, differing only in quantitative parameters such as the numbers of axons per interfascicular glial row, the dimensions of the glial cells, their spacing, or the lengths of their processes.

Compared with the ability of peripheral axons to regenerate in the Schwann cell environment of peripheral nerves, the complexity of the glial framework may be a reason why axons do not regenerate in adult central tracts. During the developmental period when axons are growing, the future central fibre tracts have a

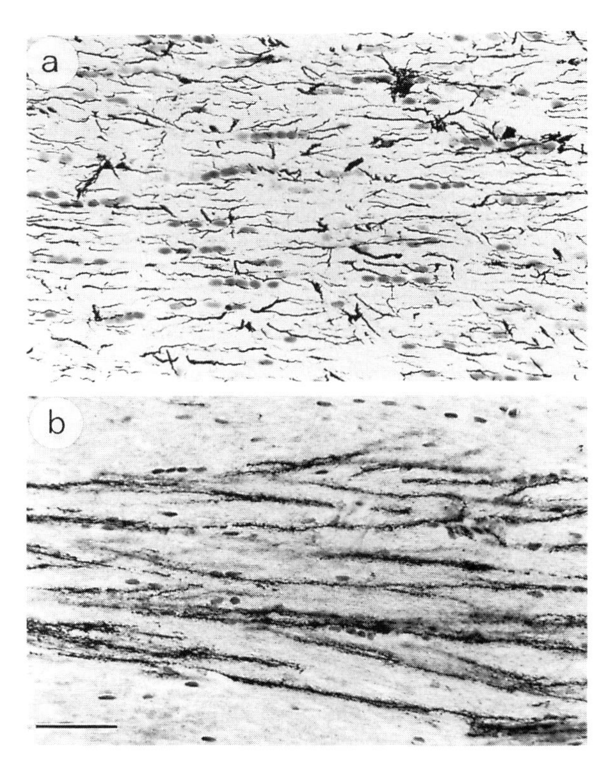

Fig. 2. (a) Section of the fimbria cut along its horizontal axis to show the parallel array of longitudinal processes of the astrocytes. GFAP immunohistochemistry. (b) An adjacent serial section showing a loose fascicle of axons running parallel to the longitudinal axis of the host tract. The axons belong to neurons from a suspension of E18 embryonic mouse neocortical cells micro-transplanted into the adult rat fimbria. M6 immunohistochemistry, 6 days after transplantation. Scale bar (for both), 50 μm.

simple structure consisting of radial glia whose cell bodies are confined to the ventricular layer, so that the developing axons grow through a region occupied by radial glial processes which lie at right angles to the axons and which run from the ventricular zone to the pial surface (Rickmann et al., 1987; Sievers et al., 1992).

At birth, the fimbria is still at the radial glial stage, and the postnatal formation of the definitive glial skeleton (M. Suzuki and G. Raisman, unpublished observations) is associated with the migration of glial cells out of the ventricular layer (finally leaving only the inert, adult unicellular lining ependyma). The first interfascicular glia are scattered, solitary cells. Over the first 10 days of postnatal life, cell division (and accompanying cell death) leads to a continuous increase in the numbers of glial cells, which become assembled into longer and longer rows, which themselves gradually become aligned into the regular adult pattern of evenly spaced unicellular interfascicular glial rows. The oligodendrocytes are formed entirely postnatally. Myelination is a multifocal process, starting by simultaneous differentiation of regularly scattered individual cells, and taking several months to reach adult density.

Axons are present in the fimbria from the 15th day of embryonic life (Valentino and Jones, 1982), when the glial skeleton still consists only of the radial processes of cells in the ventricular layer. It is not known when axon growth ceases in the fimbria, but it certainly continues for some time into the postnatal period, since some of the hippocampal pyramidal cells which will subsequently send axons into the fimbria do not have their final birthdays until around the time of birth (Schlessinger et al., 1978). It is also possible that, as has been demonstrated for the corpus callosum (Innocenti and Clarke, 1984), there may be considerable remodelling of axonal pathways in the postnatal period.

Growth of embryonic axons in adult fibre tracts

That the adult glial tract skeleton is not universally inhibitory to axon growth has recently been shown by the use of embryonic transplants (e.g. Wictorin et al., 1990). We have developed a minimally traumatic approach (Emmett et al., 1990) to transplant small volumes of suspensions of late embryonic hippocampal neurons directly into the adult fimbria (Davies et al., 1993). The effect of this form of transplantation was that the host tract astrocytes did not form reactive scars, and the longitudinal processes of the tract astrocytes continued undeflected into the transplants, which themselves developed an internal glial skeleton whose astrocytic processes flowed out in smooth continuity with the longitudinal astrocytic processes of the host tract.

From as little as 2 days after operation, the neurons extended a beam of abundant axons through the fimbria (Fig. 2b) at a rate of at least 1 mm per day. This is equivalent to the rate of regeneration of peripheral nerve axons. The axons grew for at least 10 mm in the tract, and reached the contralateral hippocampus by 10 days. Thus, adult myelinated central tracts retain the capacity to direct profuse and rapid growth of axons from embryonic donor neurons.

The donor axons formed a loose bundle, but did not fasciculate with each other. The donor projections travelled in both directions in the host fimbria, and did not seem to require any appreciable degree of destruction of host axons. Both for the host and the donor axons, a characteristic mode of exit from the tract was by the formation of collaterals at right angles to the parent axonal stem. The donor intrafimbrial axons projected to the septal nuclei, and the ipsilateral and contralateral hippocampi, where they diffusely invaded terminal fields appropriate to hippocampal projection fibres.

We have compared the pattern of axon growth after transplanting donor cell suspensions taken from three different brain areas (hippocampus, neocortex and superior colliculus) into two different host tracts: the fimbria and the corpus callosum (Davies et al., 1994). In both tracts the speed and density of the intrafascicular axon projection was the same for all three types of donor tissue. The pattern of distribution of the donor projections was determined not by the type of donor tissue, but by its position in the host tract. This formation of inappropriate types of projection is in marked contrast to the situation in neuropil, where embryonic transplants usually only form correct specific patterns of projection (e.g. Björklund and Stenevi, 1984; Zhou et al., 1990). We do not know whether the projections we have observed after transplantation of incorrectly matched donor tissue into fibre tracts would be permanent. Our observations were based on a mouse-specific axonal marker, M6, which was only suitable for following projections for a few weeks, so that it is possible that what we were detecting was an initial "exuberant" formation of temporary projections which might, if inappropriate to the postsynaptic targets, later be retracted (e.g. Innocenti and Clarke, 1984; O'Leary et al., 1990).

Regardless of the donor cell type, the intrafascicular projections followed rigidly controlled routes through the host tracts. The common feature of these routes was that they were parallel to the host tract axons and to the longitudinal processes of the host tract astrocytes, and they respected the boundaries between fibre tracts. When a transplant occupied only part of a tract, the projection was also restricted to a band which did not occupy the full width of the tract, and whose distribution was restricted to the part of the terminal field reached by that band.

That adult myelinated central fibre tracts are facilitatory for axon growth from embryonic transplants was confirmed in a series of experiments in which we micro-transplanted embryonic hippocampal cells into the dorsal columns of the adult rat spinal cord (Li and Raisman, 1993). The donor axons grew at a similar rate to those in the fimbria, and also strictly in alignment with the longitudinal axis of the host tracts. The donor axons grew equally in the rostral and in the caudal directions in both the descending corticospinal tract and the ascending sensory dorsal column tracts. The axons showed a marked preference for white (rather than grey) matter, and they were neither stopped nor deflected by an adjacent transplant of hippocampal tissue which would have contained their normal postsynaptic targets.

Because of the transformation of the glial skeleton of central fibre tracts from radial glia (the structure present during the developmental stage when most axon growth occurs) to the complex astro/oligodendrocytic meshwork of adult tracts, the growth of embryonic axons along adult fibre tracts involves a series of axon/glial interactions which cannot be regarded as a simple recapitulation of the events of normal development. It is possible that the ability of the embryonic axons to grow along adult tracts is due to some special property of immature axons (e.g. that they do not have a receptor to respond to an axon growth-inhibitory molecule expressed on oligodendrocytes (Schwab et al., 1993). Another possible factor preventing regeneration could be the response of the glia in the adult tracts: thus, the axons from our minimally traumatic intra-tract embryonic transplants were growing in a tract whose glial orientation was relatively undisturbed, and which was continuous with the glial orientation in the transplants. In contrast, adult axons attempting to regenerate after a traumatic tract lesion are faced with reactive tract astrocytes which become rearranged into "scar" formations (Reier, 1986) which may have the effect of distorting what otherwise might be growth permissive astrocytic surfaces into an impenetrable tangle.

Migration of glia and Schwann cells into central tracts

Surprisingly, although the glial processes form a dense meshwork in adult tracts, the glial array remains permissive to the ingrowth of additional glia from embryonic transplants, and also transplanted cultured Schwann cells.

When either solid embryonic grafts (Lindsay and Raisman, 1984) or suspensions of purified astrocytes (Emmett et al., 1991) were injected into the hippocampus in the vicinity of the fimbria or alveus, donor astroglial cells migrated for considerable distances out of the transplants. While these migrating glial cells were capable of traversing neuropil, and also leaving the brain and entering the subarachnoid space, their preferred routes of migration were along blood vessels or fibre tracts. In some cases, we found that the donor astrocytes had inserted themselves into the interfascicular glial rows of the host at some distance from the contact with the graft and in regions of the tract that showed no detectable damage.

The use of peripheral nerve grafts to bring about regenerative growth of cut adult central axons (Villegas-

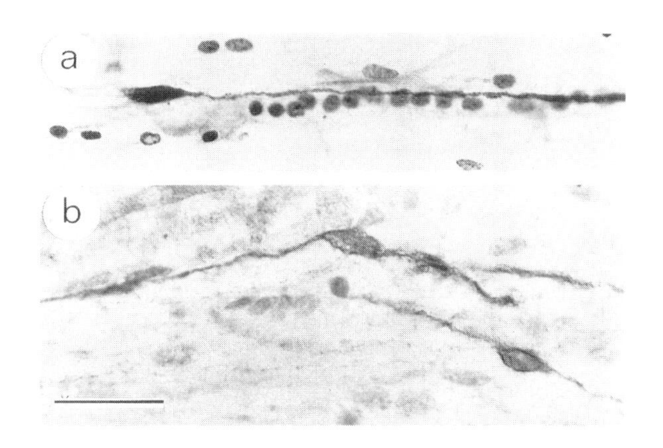

Fig. 3. (a),(b) Ribbon-like elongated Schwann cells from a microtransplanted suspension of cultured adult peripheral nerve Schwann cells migrating along the longitudinal axis of the adult rat fimbria. Low affinity NGF receptor immunohistochemistry (MC192 monoclonal antibody), (a) 10 and (b) 22 days after transplantation. Scale bar (for both), 50 μm.

Pérez et al., 1988; Carter et al., 1989; Thanos, 1992), and the demonstration that this effect depends on the presence of viable Schwann cells (Berry et al., 1988; Smith and Stevenson, 1988) has led to experiments designed to explore whether injections of suspensions of purified cultured Schwann cells can also be used to induce regenerative axon growth (Neuberger et al., 1992; Brook et al., 1994).

We have found (Brook et al., 1993) that transplanted cultured Schwann cells were able to migrate rapidly through the adult brain. Within 1–2 days of injecting a small volume of a suspension of cultured Schwann cells into the adult fimbria, the donor cells started to disperse from the injection site. Initially the cells migrated as cuffs along the blood vessels of the transplant region. By 4–5 days, however, individual cells had detached themselves from the perivascular cuffs, and had become transformed into ribbon-like, bipolar cells (Fig. 3) which entered the host fibre tract and migrated as dispersed single cells along the longitudinal axis of the tract.

Schwann cells and adult central axons

When suspensions of cultured Schwann cells were injected into the long tracts of the dorsal columns of the spinal cord, they induced sprouting of both cut and uncut axons in both the corticospinal tract and the ascending sensory dorsal columns (Li and Raisman, 1994). This sprouting occurred both at the cut ends of the axons and also at intervals (probably at myelin internodes) along the axonal shafts of both cut and uncut axons. In these experiments, the sprouts formed branching terminal plexuses with presynaptic-like varicosities.

In order to convey axon sprouts across a lesion site and into distant deafferented terminal territories, it would be necessary to induce elongative growth. In the case of peripheral nerve grafts, a directional factor is provided by the basal lamina and the fibroblastic/collagenous framework of the nerve, as well as by the aligned bands of Bungner formed by the reactive dividing Schwann cells. We have now found (Brook et al., 1994) that a similar, highly effective directional

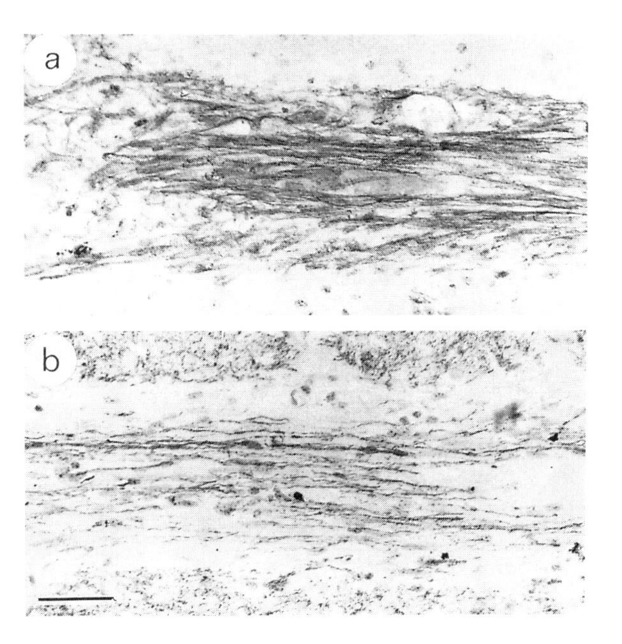

Fig. 4. (a) Parallel fascicles of cultured adult rat peripheral nerve Schwann cells transplanted as a column by continuous extrusion from a micropipette which was gradually withdrawn along a vertical track through the thalamus. Low affinity NGF receptor immunohistochemistry, 11 days after transplantation. (b) Adjacent serial section showing parallel fascicles of adult host axons induced to grow along the Schwann cell column. Neurofilament immunohistochemistry (RT97 monoclonal antibody). Scale bar (for both), 50 μm.

effect can be produced by "extrusion grafting" of suspensions of Schwann cells.

A suspension of donor Schwann cells was continuously extruded from a micropipette tip which was placed in the ventral thalamus/subthalamic region and gradually withdrawn dorsally through the thalamus and overlying forebrain during expulsion. Within 1–2 days, the donor Schwann cells became aligned to form a columnar track along the axis of retraction (Fig. 4a). Host axons progressively invaded the Schwann cell column, and became aligned along its vertical axis (Fig. 4b), where they extended for considerable distances (up to 4 mm). The extrusion grafts were able to carry axons through a perforation of the pial surface, so that regenerating fibres were carried out of the dorsal surface of the thalamus, across the choroid fissure and back into the brain at the ventral surface of the hippocampus.

Conclusions

Although adult central myelinated fibre tracts have a complex and regular glial structure, with densely interwoven processes, the tracts show considerable plasticity. Thus:

(1) embryonic axons grow readily in adult fibre tracts, and follow the orientation of the elements of the glial skeleton, and
(2) adult central fibre tracts allow the incorporation of additional cells, both embryonic glia and also cultured Schwann cells.

The incorporation of Schwann and/or other cells into adult tracts provides a possible future method to encourage the regeneration of cut adult fibres along these tracts.

References

Berry, M., Rees, L., Hall, S., Yiu, P. and Sievers, J. (1988) Optic axons regenerate into sciatic nerve isografts only in the presence of Schwann cells. *Brain Res. Bull.*, 20: 223–231.

Björklund, A. (1992) Dopaminergic transplants in experimental parkinsonism: cellular mechanisms of graft-induced functional recovery. *Curr. Opinions Neurobiol.*, 2: 683–689.

Björklund, A. and Stenevi, U. (1984) Intracerebral neural implants: neuronal replacement and reconstruction of damaged circuitries. *Annu. Rev. Neurosci.*, 7: 279–308.

Brook, G.A., Lawrence, J.M. and Raisman, G. (1993) Morphology and migration of cultured Schwann cells transplanted into the fimbria and hippocampus in adult rats. *Glia*, 9: 292–304.

Brook, G.A., Lawrence, J.M., Shah, B. and Raisman, G. (1994) Extrusion transplantation of Schwann cells into the adult rat thalamus induces directional host axon growth. *Exp. Neurol.*, in press.

Carter, D.A., Bray, G.M. and Aguayo, A.J. (1989) Regenerated retinal ganglion cell axons can form well-differentiated synapses in the superior colliculus of adult hamsters. *J. Neurosci.*, 9: 4042–4050.

Davies, S.J.A., Field, P.M. and Raisman, G. (1993) Long fibre growth by axons of embryonic mouse hippocampal neurons micro-transplanted into the adult rat fimbria. *Eur. J. Neurosci.*, 5: 95–106.

Davies, S.J.A., Field, P.M. and Raisman, G. (1994) Long interfascicular axon growth from embryonic neurons transplanted into adult myelinated tracts. *J. Neurosci.*, 14: 1596–1612.

Emmett, C.J., Jaques-Berg, W. and Seeley, P.J. (1990) Microtransplantation of neural cells into adult rat brain. *Neuroscience*, 38: 213–222.

Emmett, C.J., Lawrence, J.M., Raisman, G. and Seeley, P.J. (1991) Cultured epithelioid astrocytes migrate after transplantation into the adult rat brain. *J. Comp. Neurol.*, 310: 330–341.

Field, P.M., Seeley, P.J., Frotscher, M. and Raisman, G. (1991) Selective innervation of embryonic hippocampal transplants by adult host dentate granule cell axons. *Neuroscience*, 41: 713–727.

Innocenti, G.M. and Clarke, S. (1984) The organization of immature callosal connections. *J. Comp. Neurol.*, 230: 287–309.

Li, Y. and Raisman, G. (1993) Long interfascicular axon growth from embryonic mouse hippocampal neurons transplanted into the myelinated corticospinal tracts and dorsal columns of immunosuppressed adult rat hosts. *Brain Res.*, 629: 115–127.

Li, Y. and Raisman, G. (1994) Schwann cells induce sprouting in motor and sensory axons in the adult rat spinal cord. *J. Neurosci.*, in press.

Lindsay, R.M. and Raisman, G. (1984) An autoradiographic study of neuronal development, vascularization and glial cell migration from hippocampal transplants labelled in intermediate explant culture. *Neuroscience*, 12: 513–530.

Matthews, M.A. and Duncan, D. (1971) A quantitative study of morphological changes accompanying the initiation and progress of myelin production in the dorsal funiculus of the rat spinal cord. *J. Comp. Neurol.*, 142: 1–22.

Neuberger, T.J., Cornbrooks, C.J. and Kromer, L.F. (1992) Effects of delayed transplantation of cultured Schwann cells on axonal regeneration from central nervous system cholinergic neurons. *J. Comp. Neurol.*, 315: 16–33.

O'Leary, D.D.M., Bicknese, A.R., De Carlos, J.A., Heffner, C.D., Koester, S.I., Kutka, L.J. and Terashima, T. (1990) Target selection by cortical axons: alternative mechanisms to establish axonal connections in the developing brain. *Cold Spring Harbor Symp. Quant. Biol.*, 55: 453–468.

Raisman, G. (1969) Neuronal plasticity in the septal nuclei of the adult rat. *Brain Res.*, 14: 25–48.

Raisman, G. (1985) Synapse formation in the septal nuclei of adult rats. In: C.W. Cotman (Ed.), *Synaptic Plasticity*, Guilford Press, New York, pp. 13–38.

Reier, P.J. (1986) Gliosis following CNS injury: the anatomy of astrocytic scars and their influences on axonal elongation. In: S. Fedoroff and A. Vernadakis (Eds.), *Astrocytes,* Vol. 3, Raven Press, New York, pp. 163–196.

Rickmann, M., Amaral, D.G. and Cowan, W.M. (1987) Organization of radial glial cells during the development of the rat dentate gyrus. *J. Comp. Neurol.*, 264: 449–479.

Schlessinger, A.R., Cowan, W.M. and Swanson, L.W. (1978) The time of origin of neurons in Ammon's horn and the associated retrohippocampal fields. *Anat. Embryol.*, 154: 153–173.

Schwab, M.E., Kapfhammer, J.P. and Bandtlow, C.E. (1993) Inhibitors of neurite growth. *Annu. Rev. Neurosci.*, 16: 565–595.

Sievers, J., Hartmann, D., Pehlemann, F.W. and Berry, M. (1992) Development of astroglial cells in the proliferative matrices, the dentate granule cell layer, and the hippocampal fissure of the hamster dentate gyrus. *J. Comp. Neurol.*, 320: 1–32.

Smith, G.V. and Stevenson, J.A. (1988) Peripheral nerve grafts lacking viable Schwann cells fail to support central nervous system axonal regeneration. *Exp. Brain Res.*, 69: 299–306.

Suzuki, M. and Raisman, G. (1992) The glial framework of central white matter tracts: segmented rows of contiguous interfascicular oligodendrocytes and solitary astrocytes give rise to a continuous meshwork of transverse and longitudinal processes in the adult rat fimbria. *Glia*, 6: 222–235.

Thanos, S. (1992) Adult retinofugal axons regenerating through peripheral nerve grafts can restore the light-induced pupilloconstriction reflex. *Eur. J. Neurosci.*, 4: 691–699.

Valentino, K.L. and Jones, E.G. (1982) The early formation of the corpus callosum: a light and electron microscopic study in foetal and neonatal rats. *J. Neurocytol.*, 11: 583–609.

Villegas-Pérez, M.P., Vidal-Sanz, M., Bray, G.M. and Aguayo, A.J. (1988) Influences of peripheral nerve grafts on the survival and regrowth of axotomized retinal ganglion cells in adult rats. *J. Neurosci.*, 8: 265–280.

Wictorin, K., Brundin, P., Gustavii, B., Lindvall, O. and Björklund, A. (1990) Reformation of long axon pathways in adult rat central nervous system by human forebrain neuroblasts. *Nature*, 347: 556–558.

Zhou, C.F., Raisman, G. and Morris, R.J. (1985) Specific patterns of fibre outgrowth from transplants to host mice hippocampi, shown immunohistochemically by the use of allelic forms of Thy-1. *Neuroscience*, 16: 819–833.

Zhou, C.F., Li, Y., Morris, R.J. and Raisman, G. (1990) Accurate reconstruction of three complementary laminar afferents to the adult hippocampus by embryonic neural grafts. *Neurosci. Res.*, 13(Suppl.): S43–S53.

F. Bloom (Editor)
Progress in Brain Research, Vol. 100

CHAPTER 26

Brain damage and recovery

Donald G. Stein

Institute of Animal Behavior, Rutgers, The State University of New Jersey, Newark, NJ 07102, USA

Introduction

Although clinical neurologists and neuropsychologists have known for a long time that some recovery after damage to the brain and spinal cord is possible, it has only been within the last two decades that a serious and consistent effort has been made to understand its physiological bases. Now there are many dozens of laboratories interested in studying the mechanisms of recovery from brain and spinal cord damage, but back in the mid-1960s, when I began my career, there were only a handful of people working on the problem.

Rather than provide yet another traditional review of the pharmacology of recovery mechanisms (there are now many excellent and comprehensive papers available), the editors of this commemorative volume asked me to give a personal retrospective. So, in this essay I would like to start with a very brief background of the area and then highlight some of the conceptual and practical issues that I suggest require further thought and study. However, given what I have to say, I would guess that those of you with no interest in behavioral neuroscience would probably be happier skipping to one of the other articles in this volume.

Research on functional recovery may actually have begun in the early part of the 19th century. In the 1840s, the physiologist, Pierre Flourens, was probably one of the first experimentalists to manipulate brain damage and observe recovery of function in laboratory animals. He mutilated the brains of pigeons and despite the massive damage, Flourens noted that the birds gradually re-acquired all of their sensory and motor functions (Finger and Stein, 1982). Although there were other reports on recovery from brain damage, it was Karl Lashley's pioneering work (1929, see also Finger, 1994 for more) from the 1920s to the 1950s, which provided the most convincing data demonstrating that laboratory rats could suffer massive brain injuries and yet show dramatic sparing of function on a variety of behavioral tasks.

Lashley was less interested in studying recovery of function for its clinical implications than in demonstrating that the idea of discrete localization of associative and cognitive functions to specific brain regions, was not the best way to describe cerebral organization (Finger, 1994). Except for a few devoted followers, most of Lashley's work was not welcomed because it did not fit the prevailing paradigm in which behavioral deficits following specific lesions were used to infer the functions of the damaged tissue. At the time, cortical cartography was capturing the imagination and talents of many investigators and great amounts of empirical data could be generated with little concern for theoretical implications. Defining cortical and subcortical regions of the brain also fit very well with the idea that all behaviors could be represented or "mediated" by a specific cerebral organ. By the 1960s, the focus on map-making and localization of function had become so strong that the leading neuroanatomist, Walle Nauta, had to express his concern as follows:

> "It seems that if we try to discover the ways in which any part of the brain functions, it is only logical to try to find out in what way it acts within the brain as a whole. There is evidence that whatever the frontal lobe does must in some way affect the reticular formation, hypothalamus,

> limbic system, and a number of other structures about which we do not know very much. I think the point should be made that no part of the brain functions on its own, but only through the other parts of the brain with which it is connected." (Nauta, 1964)

The overemphasis on cartography that disturbed Walle Nauta is not one that now can simply be consigned to the history textbooks because it is still being taught in many places. For example, as recently as 1981, the editors (Kandel and Schwartz, 1981) of a major textbook in neuroscience stated the argument as follows:

> "Clinical studies and their counterparts in experimental animals suggest that all behavior, including higher (cognitive as well as affective) mental functions, is localizable to specific regions or constellations of regions within the brain." p. 11

The case for inferring function from damage to brain regions was made more strongly as a result of testing the large numbers of head-injured veterans returning from World War II, and a few years later, the Korean war. In the clinical literature, much of what was written about brain–behavior relationships and cerebral organization derived from studies in patients with bullet and shrapnel wounds to the head. Yet, very little of this extensive literature concerned itself with how, or whether, the victims of head injury would ever recover.

Head-injured veterans were entitled to receive comprehensive rehabilitation but there was nothing available in the way of early intervention or treatment to promote cognitive or functional recovery. Many patients simply languished in VA hospitals or were discharged to a life of dependency. For the most part, neuropsychologists focused on defining and precisely measuring the nature and severity of the deficits caused by the trauma rather than in providing "therapy" to the patients.

During the post-war period and up until the 1960s, one of the few people writing on recovery in humans was Alexander Luria. Luria was one of the Soviet Union's most distinguished neurologists and he wrote extensively on recovery and cerebral organization (Luria, 1966). He thought that direct localization of complex cerebral functions to circumscribed areas of the brain did not make much sense. In that regard, he antedated the notions of serial and parallel processing. He believed that the "performance of a given function necessitates the integrity of far more extensive and far more structurally varied zones of the cortex than was assumed by classical neurology." (p. 13)

Indeed, it would not have been possible for Luria to stress recovery of function after massive trauma, if the functions were permanently lost as a result of damage to specific regions thought to mediate them. Although Luria clearly had a few strong followers here and abroad, most of his theories of cerebral organization and approaches to treatment were generally ignored because they also did not fit with the prevailing localization paradigm.

Hans-Lukas Teuber, who helped introduce Luria's work to American neuropsychologists, was one of the first post-World War II neuropsychologists to examine the residual capacities of brain-damaged patients and investigate the role of cerebral shock (diaschisis) in blocking recovery of function. Teuber (1975) did long-term follow-up studies with young, brain-damaged patients and was able to show remarkable recovery that required years to develop . He was one of the first to highlight clinical research on recovery and the role of early versus late brain damage. His studies gave credence to the idea that recovery of function in human patients merited further study and reflection.

Working independently in the late 1950s and 1960s, the physiological psychologists, Donald and Patricia Meyer (Meyer et al., 1963) Walter Isaac (1964) and Louis Petrinovich (Petrinovich and Carew, 1969), began to examine some of the factors contributing to functional recovery from cortical injuries in laboratory animals. These investigators looked at how specific types of training and control of environmental conditions before and after injury, could promote functional recovery. They were also interested in studying whether stimulant drugs like amphetamine could enhance recovery from cortical damage (Meyer, 1972). In their studies, the benefits conferred by drug treatments, seemed to last only while the animals were intoxicated. Once the drug wore off, the deficits returned so that recovery was not sustained.

My own work in the area of brain injury began as a straightforward examination of the role of the hippocampus in learning and the consolidation of memory traces (Stein and Kimble, 1966). In the course of my experiments, I created almost total, bilateral aspiration hippocampectomies in rats and then tested the animals on a variety of maze tasks. Although I obtained the expected deficits, I also noticed that some of the animals in the lesion groups performed almost as well as normal controls. Some might attribute the sparing to the typical "variability" that one sees in all living systems. However, when I did the lesion reconstructions and found that some animals had no behavioral deficits despite the fact that they also had no hippocampus, it became more difficult for me to think that the hippocampus was "necessary" for short-term memory functions. I was very puzzled by these data and by the few reports on recovery that were appearing at the time. However, I really wanted to get my PhD and get on with my life so I did not try to pursue the question of sparing of function and its theoretical implications until I began my first job at Clark University. At Clark, I was able to return to the work of John Adametz (1959), a neurosurgeon who showed that if one slowly damaged the rostral reticular formation in cats by doing staged surgeries, spaced 3 weeks apart, the animals would show normal sleep-wake cycles, groom normally and maintain themselves quite well. In contrast, if the same kind of lesions were made in a single sitting, all of the cats lapsed into coma and died within a few weeks. Adametz did not perform any systematic, cognitive testing of his animals, so my students and I decided to perform staged lesions in a number of different brain areas and then test our animals on a variety of behavioral tasks.

To make a long story short, we were able to show that, if complete bilateral injuries to different brain structures were done in stages (i.e. with about 2 weeks between each surgical operation), the animals were able to perform as well as intact controls on all of our behavioral tests. In contrast, animals with the same bilateral lesions inflicted in one sitting, demonstrated all of the deficits typically associated with that type of injury. In a number of different experiments, we examined what we then called the "serial lesion effect" in the frontal cortex (Stein et al., 1969), hippocampus (Isseroff et al., 1976), amygdala (McIntyre and Stein, 1973), motor cortex (Gentile et al., 1978), caudate nucleus (Schultze and Stein, 1975), lateral hypothalamus (Fass et al., 1975), and superior colliculus (Weinberg and Stein, 1978). In each case, we obtained essentially the same results. One-stage injuries produced the "expected" deficits, whereas the same damage inflicted in stages permitted the animals to escape the impairments and perform as well as controls.

In collaboration with Nelson Butters and Jeffrey Rosen (Rosen et al., 1971, 1975; Butters et al., 1974; Stein et al., 1977), we did a series of studies to replicate our findings in adult monkeys. We were able to demonstrate that multiple stage removals (i.e. in four operations removing one bank of the *sulcus principalis* at a time) of the frontal cortex led to spared performance on spatial learning performance. Monkeys with two stage lesions, spaced 4 weeks apart, were in turn, significantly better than those with one-stage lesions. What was interesting to note was that the recovered monkeys with four-stage lesions actually had more extensive damage and more gliosis than their one-stage counterparts who were very impaired. At the time we had no idea that glia might actually have contributed to the recovery process by serving as a source of trophic factors that could support the survival of remaining, damaged or dysfunctional neurons (Ermakova et al., 1993).

Around the same time, Stanley Finger was reproducing similar results in tactile discrimination after bilateral removal of the somatosensory cortex (Finger et al., 1971). Similar recovery was also being observed after staged damage to other sensory systems. For example, Dru et al. (1975) and Spear and Barbas (1975) were showing that, under appropriate conditions, serial lesions of the visual cortex would lead to functional sparing of visual discrimination performance. These and many other additional findings made it clear that the serial lesion effect was very robust and generalizable to different areas of the brain, different types of injuries and different species. The only problem was that no one really had a good explanation of why staged lesions almost always led to better outcomes that when the same damage was inflicted in a single setting (but see reference to Bulsara et al., 1992 for attempts to address this question).

To this day there is no completely satisfactory answer to what is inherently a very fundamental problem: why should the context in which an injury occurs, that is, "how" you remove a structure determine whether its "function" is lost or preserved? Also, the implications of this work for theories of cerebral organization have not been given much serious consideration in the neuroscience community. This is due to the fact that the serial lesion results simply do not fit easily with reductionist/localizationist theories implying that each and every structure has a specific function and that damage to the structure should result in the permanent loss of the specific function.

One way to dismiss the problem of what type of reorganization in response to bilateral injury can mediate recovery, is to propose that there are both "serial and parallel systems" involved in behavior and any one component can "take over" for the lost parts through some type of vicariation (see Slavin et al., 1988 for more discussion of this issue). This idea has become very tempting, especially in the light of newer imaging techniques which can show the brain areas that become metabolically active by measuring blood flow or glucose utilization. How much predictive power do we gain from talking about serial and parallel processing in recovery? Not a lot. Recently, Douglas and Martin (1991) discussed how map making based on the neuron doctrine needs to be reexamined. They presented a figure showing that, in 1983, the wiring diagram for cortical visual areas consisted of ten structures with a number of reciprocal connections. By 1990, the number of structures just "devoted" to processing cortical visual information had grown to 43 with reciprocal connections numbering in the hundreds. In applying the same logic to complex associative functions like learning and memory, which involve the processing of motor, sensory, kinesthetic and proprioceptive cues, one can easily begin to understand why the concept of mapping the structures in "parallel processing" begins to lose much of its explanatory value.

This is why working with isolated neurons or in vitro preparations, is so much more captivating; the issues are easier to deal with and can be explained with much more mechanistic concepts. In times of serious fiscal constraint, no study section wants to award research dollars to examine old theoretical constructs that do not fit well with established paradigms. So, to be more practical, our laboratory put aside the serial lesion puzzle and turned to the investigation of using newly developed pharmacological agents to promote functional recovery. We never did get back to the serial lesion effect.

By the mid-1980s things changed for the better. Policy makers, patients and their families, and pharmaceutical companies grew to recognize that head injury is a significant medical problem that needs more focused attention and better funding. Extramural support for work on neural plasticity became available and a whole new field of research developed. As a result, research on traumatic brain injury entered a new stage as changes in both conceptual thinking about the brain and exciting discoveries about mechanisms, began to appear almost on a daily basis.

The traditional notion of the brain as a static, genetically determined, collection of structures, each with a specific function, has (grudgingly?) given way to a much more dynamic view based on principles of plasticity, and capacity for rapid change in response to trauma. In addition, there is a much more comprehensive understanding of the immediate and long-lasting biochemical and anatomic changes that occur in response to cerebral trauma (for detailed reviews, see for example, Feeney and Sutton, 1988; Ziven and Choi, 1991; Stein and Glasier, 1992; Brodkey et al., 1993; Povlishock, 1993).

These last ten years have witnessed a "revolutionary" paradigm shift in thinking about brain injury repair. It is now accepted that functional recovery after traumatic injury to the brain can be obtained once the cascade of injury processes that accompanies trauma is properly controlled. We know much more about the mechanisms of injury-induced, neuronal sprouting; the role of glial cells and trophic factors in promoting the rescue of neurons; the effects of blocking high levels of excitatory amino acids; the role of free radicals in the injury process, and a host of other mechanistic factors that can determine the success or failure of the organism to survive and recover from brain damage. The problem is to figure out what do and when to do it at the biochemical, morphological and behavioral levels of analysis.

Like most other laboratories concerned with head trauma and recovery, we have focused our efforts on examining the specific molecular and anatomic variables that we can manipulate to enhance recovery. But this reductionistic approach provides only one part of the answer. A problem that must also be addressed if we are to move from laboratory research on recovery using animals as subjects, to the clinical rehabilitation of patients, is: what do we mean by the term "recovery"? Why is recovery so clearly seen in some laboratories and clinics and not others? Are the tests of clinical recovery relevant to what the patient really needs to accomplish? Why do certain drugs seem to have such dramatic and beneficial effects in some cases and fail just as dramatically in others? Are there surgical alternatives to promoting recovery? Why do some groups report highly successful instances of fetal brain tissue grafts in the clinical treatment of Parkinson's disease while others using the same surgical techniques fail to get benefits? Should grafts also be used to aid victims of head trauma?

One reason for the discrepancy is that scientists and clinicians do not often agree about how they would define "recovery". Different groups often have their own definitions for recovery of function, rehabilitation outcomes and clinical restoration. For example, in discussing recovery from brain injury, Laurence and Stein (1978) stressed that the phrase, "recovery of function" can refer to the straightforward accomplishment of certain goals by the subject (achieving an "end") or it can refer to the strategies (or "means") by which organisms accomplish their goals.

On the one hand, in the laboratory, recovery may mean only that after a specific injury, a particular drug or treatment causes neurotransmitter levels in one area of the brain to return or exceed pre-injury levels. On the other hand, for therapists and their patients, the more important issue is to improve quality of life; there is less concern about the particular mechanisms or underlying, physiological substrates. Thus, the question of "how" may be less important than "whether" a clinical goal is obtained. Yet, if each laboratory or rehabilitation clinic sets its own arbitrary definition of "recovery", it is not surprising that there are so many differences of opinion concerning whether or not a given function is said to be "recovered"; not to mention what might satisfy the patients or their families.

A parallel question to that of defining what is meant by recovery, is what is meant by the term, "function". Do we really know how is the function of a given brain area is defined, or even what the "function" is in the first place? In the past, functions of brain regions were defined by observing and describing the loss or impairments of behaviors that accompanied damage to the structure in question.

Ernst Poppel (1989) recently defined (CNS) functions "as those that, in principle, can be lost after circumscribed injuries of the brain." (p. 6). But, as I have discussed above, there are numerous cases in which "circumscribed injuries" do not lead to significant, functional impairments. As I mentioned earlier, there is a substantial literature demonstrating that slow growing, bilateral injuries to the brain (e.g. serial lesions), often spare the complex behaviors thought to be mediated by the injured structures (Finger et al., 1973; Finger and Stein, 1982). In these cases involving rats, cats monkeys and people, slowly occurring bilateral damage does not always lead to behavioral symptomatology or to the same severity of impairments.

Recently, Anderson et al. (1990) carefully examined patients with neuropsychological deficits associated with lesions caused by tumors or stroke. They carefully matched the patients on the locus and the extent of their brain injuries. The neuropsychologists found that all of the patients with stroke in the left hemisphere (i.e. rapid onset of injury) have more severe language impairments than did their counterparts with slowly forming tumors (which had larger areas of damage). In fact, most of the tumor patients were normal on all of the language tests. What constitutes an adequate definition of an injury and a "deficit" in the context I have just described? What is the appropriate behavioral test to identify when and how a function is lost? For example, although Poppel (1989) has argued that "the integrity of local neuronal structures is essential for the availability of specific psychological functions," (p. 222) the serial lesion data and the tumor/stroke data argue against that view.

In addition, in the clinic, Poppel et al. (1973) and others (e.g. Zihl, 1981; Pommerenke and Markowitsch, 1989) have demonstrated that patients with cerebral "blindness" following posterior cortical injuries, have residual vision and can make limited visual discriminations if they are properly trained; a phenomenon they refer to as "blindsight". In this case are visual functions lost or are they suppressed or 'blocked' by the neural insult? Animal research on this topic as well as the blindsight data, can be taken to indicate that "inhibition" of function rather than loss, may be one result of "localized" brain injury. Under the circumstances, the problem becomes how to "unlock" or "unmask" the repressed function.

I mentioned above that about 20 years ago, Donald Meyer (1972) and his colleagues showed that cats with massive, bilateral, posterior cortical lesions could be rendered almost totally blind. However, following an injection of amphetamine, a potent neuroleptic, the "blind" cats were able to make visually guided placing responses as long as they were under the effects of the drug. Meyer argued that the lesions blocked access to the visual "engrams" which could be "derepressed" by the activating properties of the amphetamine. More recently, Feeney and Sutton (1988), showed that cats and rats could recover almost normal limb movements and gait following bilateral sensorimotor cortex lesions if they were given a single injection of amphetamine at the time of the injury. Without recourse to any drugs, Held et al. (1985) were able to show that enriched sensory activity and opportunity to explore the environment, could lead to better (although not complete) recovery of gait in adult rats which had received extensive injury to the sensorimotor cortex. In combination, these studies reveal that the "functions" of a given area may be altered (or determined?) by the organism's experience and post-traumatic environment and not just by its genetically "fixed" neuronal connections (or their disruption).

William Jenkins' research most directly speaks to this important issue. He, Michael Merzenich and their colleagues (Jenkins et al., 1990) have repeatedly demonstrated that the classical view suggesting that maturation of neuronal connections determines their "functions" is simply not true. Using precise, electrophysiological recording of single neurons in the sensorimotor cortex to map their receptive fields, Jenkins and his colleagues have shown that such fields can be modeled and remodeled by experience and training. More importantly for this discussion, following loss of sensory inputs, these same neurons may change their receptive fields dramatically (i.e. skin surface representation shifts). In some cases after cortical damage, the representation of skin sensory areas may shift to undamaged, previously unused portions of the remaining cortex. Likewise, as animals are trained on a sensory discrimination task, there are continuous shifts in the locus of the field as in their size and shape characteristics. This means that environment and training can model and alter neuronal properties in an adaptive manner.

The functional reorganization seen in Jenkins' work occurs far more rapidly than could be explained by the physical regeneration of nerve pathways themselves. The findings suggest that there are "latent" or "silent" pathways that occur in the normal development of the CNS. Such pathways can be unmasked and modified by experience or injury and later play a role in the recovery process. Although there are many questions that could be addressed in more detail, two with particular clinical relevance will close my review. The first is concerned with how soon after injury, the treatment should begin. Thanks to a large body of recent laboratory research, it is now quite clear that early and aggressive intervention and therapy may lead to the best prognosis for recovery. It has to be emphasized, here again, that brain injury and the concomitant neuronal loss is not a single, unitary event. Rather, a cascade of processes, some of which may last for years, is the complex result of neural damage. For example, in the first stage(s) of injury, substances in the brain itself are produced which are toxic to neurons and which diffuse beyond the immediate area of the injury to kill or damage vulnerable cells (Nieto-Sampedro, 1988; McIntosh, 1993). Toxic substances can also enter through the blood supply or through a compromised blood–brain barrier, leading to inflammation and immune reactions. As a result, neurons initially spared by the trauma could begin to die in massive numbers resulting in cell loss more devastating to the patient than the initial injury itself.

This is one reason why research on how much time should be devoted to rehabilitation or other forms of post-injury "therapy" is another area that needs careful investigation. Norman Geschwind (1985), felt that most clinicians not only did not pay sufficient attention to "seeking the right maneuvers" to elicit recovery, they also did not give enough attention to the time necessary for the processes leading to recovery to become manifest. Referring to the aphasias, He said that:

> "Most neurologists are gloomy about the prognosis of severe adult aphasias after a few weeks and pessimism is reinforced by a lack of prolonged follow-up in most cases. I have, however, seen patients severely aphasic for over one year, who then made excellent recovery. One patient returned to work as a salesman, the other as a psychiatrist. Furthermore, there are patients who continue to improve over many years, for example, the patient whose aphasia is still quite evident six years after onset, cleared up substantially by 18 years." (p. 3)

I take Geschwind's observations to suggest that perhaps we are not doing enough to provide long-term therapy and rehabilitation to patients whose progress through the spectrum of recovery may be arduous and long. There is no principle of neuronal and behavioral organization that demands that functional and behavioral recovery occur immediately or not at all. The decision to withhold or terminate either pharmacological treatments or rehabilitation therapy is probably based more on economic (i.e. it would cost too much to continue) and social factors (let's get on to something with a more rapid, personal payoff; i.e. seeing a quick recovery) than it is on purely physiologic grounds; i.e. what we know about how the nervous system actually works.

The long time-course of neuronal recovery also means that the specifics of the therapy itself should change as the brain's endogenous response to the trauma progresses. As the underlying brain chemistry and morphology change over time, so should the course and type of therapy. Thus, for example, in the earliest stages, it might be appropriate to give drugs which block edema and reduce excitotoxicity, then shortly follow this regimen with injection of neurotrophic factors that could enhance new neuronal growth or reduce neuronal loss by repairing damaged cell membranes. It is important to emphasize that pharmacologic therapy alone may not likely be sufficient to ensure complete and long-lasting functional recovery in patients with serious brain damage. Systematic, physical rehabilitation and cognitive therapy will continue to play a critical role in returning the patient to full health. Psychological and emotional aspects of recovery should be considered and manipulated in a combined program of medical and psychosocial management.

Much more research needs to be done to determine what specific types of cognitive and behavioral rehabilitation should be provided. Paul Bach-y-Rita, a well known physiatrist, has often stated that:

> "there is little or no demonstrated scientific basis for most of the (rehabilitation) procedures, and few of these procedures have been validated by controlled studies. The importance of the intervention of the various therapies has not been demonstrated."

For rehabilitation to be fully accepted by the medical and scientific community, rigorous controlled studies will have to demonstrate that the procedures employed are those that are, indeed, beneficial and appropriate to the patient's complete recovery.

Although a contextual approach to "recovery of function" is a new paradigm in the process of its own development, it is worthy of serious consideration and study. This approach puts emphasis on a more holistic approach to patients and their potential therapy. It is a view which leads to specifically testable hypotheses and proposes new alternatives for therapy and rehabilitation. Given the rapid progress being made these last few years, it is certain that brain-injured patients, who only too recently had nothing to hope for, may soon expect to participate more fully in all of the range of activities that are part of daily life.

By now, everyone working in the field of neural plasticity knows that Santiago Ramon y Cajal (1928), the Nobel prize-winning neuroanatomist, expressed initial pessimism about the possibility for regenerative growth in the adult central nervous system. He stated that:

> "Once development was ended, the fonts of growth and regeneration of the axon and dendrites dried up irrevocably. In adult centers, the nerve paths are something fixed and immutable; everything may die, nothing may be regenerated." (p. 750)

However, Cajal was not ready to give up on the question of how one might go about unlocking or promoting neural plasticity. He also stated that:

> "It is the duty of future scientists ... with the inspiration of high ideals ... to continue working to avoid or modulate the gradual and continuous decay of neurons, the almost invincible rigidity of their connections and, finally, to obtain the establishment of new neuronal pathways when diseases untie the intimately associated populations of neurons." (cited in Portera-Sanchez, 1987)

In looking back on the last 25 years of research and teaching, I believe that we are well on the way to realizing Cajal's dreams and aspirations.

References

Adametz, J. (1959) Rate of recovery of functioning in cats with rostral reticular lesions. *J. Neurosurg.*, 16: 85–98.

Anderson, S.W., Damasio, H. and Tranel, D. (1990) Neuropsychological impairments associated with lesions caused by tumor or stroke. *Arch. Neurol.*, 47: 397–405.

Bach-y-Rita, P. (1994) Applications of principles of brain plasticity and training to restore function. In: R.R. Young and P.J. Delwaide (Eds.), *Principles of Restorative Neurology*, Butterworths, London, in press.

Brodkey, J.A., Gates, M.A., Laywell, E.D. and Steindler, D.A. (1993) The complex nature of interactive neuroregeneration-related molecules. *Exp. Neurol.*, 123: 251–270.

Bulsara, K.R., Manibo, J.F. and Ramirez, J.J. (1992) Progressive entorhinal lesions accelerate hippocampal sprouting in rats. *Soc. Neurosci. Abstr.*, 18: 345.

Butters, N., Rosen, J.J. and Stein, D.G. (1974) Recovery of behavioral functions after sequential ablations of the frontal lobes of monkeys. In: D.G. Stein et al. (Eds.), *Plasticity and Recovery of Function in the Central Nervous System*, Academic Press, New York, pp. 429–466.

Douglas, R.J. and Martin, K.A.C. (1991) Opening the grey box. *Trends Neurosci.*, 14: 286–293.

Dru, D.O., Walker, J.P. and Walker, J.B. (1975) Recovery of CNS function:sSelf produced locomotion restores visual capacity after striate lesions. *Science*, 187: 265–266.

Ermakova, L., Fulop, Z., Geller, H.M., Chachaj, J., Mody, N. and Stein, D.G. (1993) Cultured astrocytes implanted into damaged NBM 48 hours after ibotenic acid lesions facilitate passive avoidance performance in rats. *Soc. Neurosci. Abstr.*, 19: 447.

Fass, B., Jordan, H., Rubman, A., Seibel, S. and Stein, D.G. (1975) Recovery of function after serial or one-stage lesions of the lateral hypothalamic area in rats. *Behav. Biol.*, 14: 283–294.

Feeney, D.M. and Sutton, R.L. (1988), Catecholamines and recovery of function after brain damage. In: D.G. Stein and B.A. Sabel (Eds.), *Pharmacological Approaches to the Treatment of Brain and Spinal Cord Injury*, Plenum Press, New York, pp. 121–142.

Finger, S. (1994) Origins of Neuroscience, New York, Oxford University Press.

Finger, S. and Stein, D.G. (1982) Brain Damage and Recovery: Research and Clinical Perspectives, Academic Press, New York.

Finger, S., Marshak, R.A., Cohen, M., Scheff, S. Trace, R. and Niemand, D. (1971) Effects of successive and simultaneous lesions of somatosensory cortex on tactile discrimination in the rat. *J. Comp. Physiol. Psychol.*, 77: 221–227.

Finger, S., Walbran, B., and Stein, D.G. (1973) Brain damage and behavioral recovery: serial lesion phenomena. *Brain Res.*, 63: 1–18.

Gentile, A.M., Green, S., Nieburgs, A., Schmelzer, W. and Stein, D.G. (19??) Disruption and recovery of locomotor and manipulatory behavior following cortical lesions in rats. *Behav. Biol.*, 22: 417–455.

Geschwind, N. (1985) Mechanisms of change after brain lesions. In: F. Nottebohm. (Ed.), Hope for a new neurology. *Ann. N. Y. Acad. Sci.*, 457: 1–11.

Held, J.M., Gordon, J. and Gentile, A.M. (1985) Environmental influences on locomotor recovery following cortical lesions in rats. *Behav. Neurosci.*, 99: 678–690.

Isaac, W. (1964) Role of stimulation and time in the effects of spaced occipital ablations. *Psychol. Rep.*, 14: 151.

Isseroff, A., Leveton, L., Freeman, G., Lewis, M.E. and Stein, D.G. (1976) Differences in the behavioral effects of single stage and serial lesions of the hippocampus. *Exp. Neurol.*, 53: 339–354.

Jenkins, W.M., Merzenich, M.M. and Recanzone, G. (1990) Neocortical representational dynamics in adult primates: implications for neuropsychology. *Neuropsychologia*, 23: 573–584.

Kandel, E.R. and Schwartz, J.H. (1981) *Principles of Neural Science*, Elsevier, Amsterdam, p. 11.

Lashley, K. (1929) *Brain Mechanisms and Intelligence*, University of Chicago Press, Chicago, IL.

Laurence, S., and Stein, D.G. (1978), Recovery after brain damage and the concept of localization of function. In: S. Finger (Ed.), *Recovery from Brain Damage: Research and Theory*, Plenum Press, New York, pp. 369–407.

Luria, A.R. (1966) Human Brain and Psychological Processes, Harper and Row, New York, p. 13.

McIntosh, T.K (1993) Novel pharmacologic therapies in the treatment of experimental traumatic brain injury: a review. *J. Neurotrauma*, 10: 215–262.

McIntyre, M. and Stein, D.G. (1973) Differential effects of one versus two-stage amygdaloid lesions on activity, exploratory and avoidance behavior in the albino rat. *Behav. Biol.*, 9: 451-465.

Meyer, D.R. (1972) Access to engrams. *Am. Psychol.*, 27: 124–33.

Meyer, P.M., Horel, J.A. and Meyer, D.R. (1963) Effects of dl-amphetamine upon placing responses in neodecorticate cats. *J. Comp. Physiol. Psychol.*, 56: 402–404.

Nauta, W.J.H. (1964) Discussion of Stamm, J.S., Retardation and facilitation in learning by stimulation of frontal cortex in monkeys. In: M. Warren and K.A. Akert (Eds.), *The Frontal Granular Cortex and Behavior*, McGraw-Hill, New York.

Nieto-Sampedro, M. (1988) Growth factor induction and order of events in CNS repair. In: D.G. Stein and B.A. Sabel (Eds.), *Pharmacological Approaches to the Treatment of Brain and Spinal Cord Injury*. Plenum Press, New York, pp. 301–338.

Petrinovich, L. and Carew, T.J. (1969) Interaction of neocortical lesion size and interoperative experience in retention of a learned brightness discrimination. *J. Comp. Physiol. Psychol.*, 68: 451–454.

Pommerenke, K. and Markowitsch, H.J. (1989) Rehabilitation training of homonymous visual field defects in patients with postgeniculate damage of the visual system. *Restor. Neurol. Neurosci.*, 1: 47–64.

Poppel, E. (1989) Taxonomy of the subjective: an evolutionary perspective. In: J.W. Brown (Ed.), *Neuropsychology of Visual Perception*, Lawrence Earlbaum, New Jersey, pp. 219–232.

Poppel, E., Held, R. and Frost, D. (1973) Residual visual function after brain wounds involving the central visual pathways in man. Nature (London), 243: 295–296.

Portera-Sanchez, A. (1987) Cajal's concepts on plasticity in the central nervous system revisited: a perspective. In: R.L. Masland, A. Portera-Sanchez and G. Toffano (Eds.), *Neuroplasticity: A New Therapeutic Tool in the CNS Pathology*, Springer-Verlag, Berlin, pp. 9–30.

Povlishock, J.T. (1993) Traumatic brain injury: the pathobiology of injury and repair. In: A. Gorio (Ed.), *Neuroregeneration*, New York, Raven Press, pp. 185–216.

Ramon y Cajal, S. (1928) *Degeneration and Regeneration of the Nervous System*, translated by R.M. May, Oxford University Press, London.

Rosen, J.J., Stein, D.G. and Butters, N. (1971) Recovery of function after serial ablation of prefrontal cortex in the monkey. *Science*, 173: 353–355.

Rosen, J., Butters, N., Soeldner, C. and Stein, D.G. (1975) Effects of one stage and serial ablations of the middle third of sulcus principalis on delayed alternation performance in monkeys. *J. Comp. Physiol. Psychol.*, 89: 1077–1082.

Schultze, M.J. and Stein, D.G. (1975) Recovery of function in the albino rat following either simultaneous or seriatim lesions of the caudate nucleus. *Exp. Neurol.*, 46, 291–301.

Slavin, M.D., Laurence, S. and Stein, D.G. (1988) Another look at vicariation. In: S. Finger, T.E. LeVere, C.R. Almli and D.G. Stein (Eds.), *Brain Injury and Recovery: Theoretical and Controversial Issues*, Plenum Press, New York, pp. 165–168.

Spear, P.D. and Barbas, H. (1975) Recovery of pattern discrimination ability in rats receiving serial or one-stage visual cortex lesions. *Brain Res.*, 94: 337–346.

Stein, D.G. and Glasier, M. (1992) An overview of developments in research on recovery from brain injury. *Adv. Exp. Med. Biol.*, 325: 1–22.

Stein, D.G. and Kimble, D.P. (1966) Effects of hippocampal lesions and post-trial strychnine administration on maze behavior in the rat. *J. Comp. Physiol. Psychol.*, 62: 243–249.

Stein, D.G., Rosen, J.J., Graziadei, J., Mishkin, D. and Brink, J.J. (1969) Central nervous system: recovery of function. *Science*, 166: 528–530.

Stein, D.G., Butters, N. and Rosen, J.J. (1977) A comparison of two and four-stage ablations of sulcus principalis on recovery of spatial performance in the monkey. *Neuropsychologia*, 15: 179–182.

Teuber, H.-L. (1975) Recovery of function after brain injury in man. In: Outcome of Severe Damage to the CNS, Ciba Foundation Symposium, Elsevier, Amsterdam, pp. 159–186.

Weinberg, D. and Stein, D.G. (1978) Impairment and recovery of visual functions after bilateral lesions of superior colliculus. Physiol. Behav., 20: 323–329.

Zihl, J. (1981) Recovery of visual field associated with specific training in patients with cerebral damage. In: M.W. van Hof and G. Mohn (Eds.), *Functional Recovery from Brain Damage*, Elsevier, Amsterdam, pp. 189–202.

Ziven, J. and Choi, D.W. (1991) Stroke therapy. *Sci. Am.*, 265: 56–63.

F. Bloom (Editor)
Progress in Brain Research, Vol. 100

CHAPTER 27

Trophic factor therapy in the adult CNS: remodelling of injured basalo-cortical neurons

A. Claudio Cuello

McGill University, Department of Pharmacology and Therapeutics, McIntyre Medical Building, 3655 Drummond Street, Suite 1325, Montreal, Quebec, Canada H3G 1Y6

Introduction

It is a great pleasure to contribute to the 100th volume of *Progress in Brain Research* with some thoughts on one of the most exciting developments in the neurosciences. For me, this is a great satisfaction as this collection has accompanied my career since one of the first volumes was lent to me in the early 1960s, in Buenos Aires, by Professor Eduardo De Robertis. Much has changed in neurobiology since those early years.

The concept that future neurology might resort to pharmacological intervention, to attenuate or even reverse neurological processes leading to neuronal cell death or loss of synaptic connections, would have been an untenable proposition just over a decade ago. However, scientific advances have provided strong foundations for this proposition. Presently, not only academic units are investigating potential avenues for the pharmacological treatment of neurodegenerative processes, but also well established pharmaceutical industries, as well as newly born biotechnological enterprises. Indeed, the neurosciences are becoming one of the most exciting scientific areas with challenging advances comparable to those of developmental biology and cancer research.

Two important factors have contributed significantly to this. One is the demonstration that the neurons of the CNS possess a capacity of recovery well beyond that assumed by the traditional "medical culture". The successful neuronal grafting within the CNS and the offering of adequate "milieu" to central neurons for the growth of axonal processes have done much to question traditional concepts on the immutability of the CNS. Another crucial aspect which contributed to the generation of this new scenario has been the realization that neurotrophic factors (NTFs) known to elicit fundamental phenotypic changes in neurons during development were also capable of profoundly affecting these cells in mature, fully differentiated animals (for reviews see Hefti et al., 1989; Thoenen, 1991; Cuello, 1993).

A concept which is emerging from such experimentation is that NTFs (or their derivatives) can be used as *drugs* to elicit trophic responses in damaged neurons of the CNS. Figure 1 represents in schematic fashion the circumstances in which exogenously applied NTFs to the adult and fully differentiated CNS might elicit trophic responses leading to the re-establishment or compensation of neural function lost by damage of cell bodies or processes. The hypothetical scheme assumes that diverse NTFs are produced in relatively large amount during development, securing the differentiation and connectivity of neurones in a programmed temporal-spatial fashion. During the developmental period, distinct sets of neurons are probably dependent on these factors for their survival; a situation that can closely be reproduced in tissue culture conditions. The levels of NTFs decrease considerably during early post-natal stages to "basal" levels during adulthood. Although this is an issue of some controversy, we can tentatively assume that the main conse-

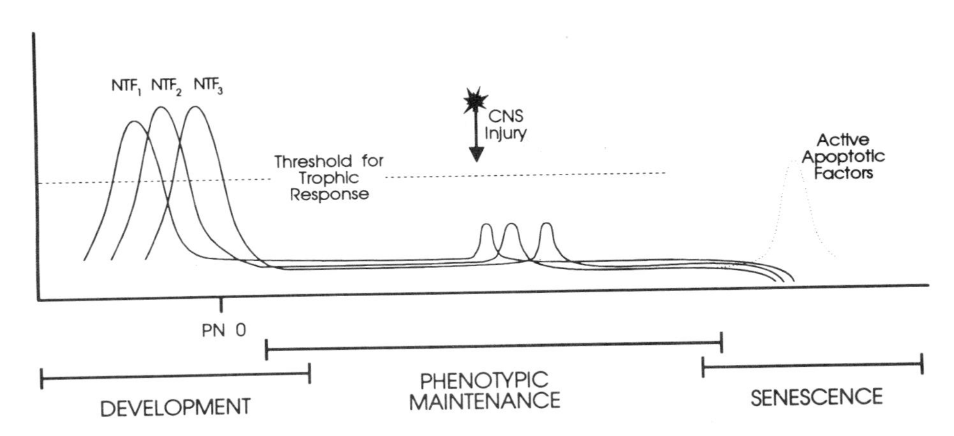

Fig. 1. Hypothetical circumstances in which the production of neurotrophic factors (NTFs) might control the survival, development and specific connections of distinct neuronal populations. During development, relatively high levels of NTFs are produced at various stages to secure differentiation of specific sets of neurons. During adulthood, basal levels of NTFs are produced presumably to maintain the differential phenotypic characteristics of mature CNS neurons and probably their pattern of synaptic connections. The enhanced production of NTFs consequent to neural injury during this ontogenic phase is apparently insufficient to reach the threshold for trophic responses, which can be evoked by exogenous NTFs. It is as yet unclear if CNS senescence with the loss of neurons and connections is due, to some extent, either to a diminution in the expression of NTFs and their receptors or to de novo expression of presumptive apoptotic factors (from Cuello et al., 1993).

quence of this basal production of NTFs during adulthood is the maintenance of phenotypic characteristics of mature neurons, their arborization and synaptic connectivity.

Experimental injury of the CNS in adult mammals results in upregulation of the synthesis of some NTFs. Is this an attempt of the mature injured CNS to recapitulate the embryonic growth of neural cells? If so, this would fulfil Ramon y Cajal's assertion that "...the central axons, and in a less measure the dendrites, possess, *ab initio*, the capacity inherent in the constitution itself of the nervous protoplasm, of increasing their mass intrinsically and extrinsically, that is, of rebuilding their structure and of expanding into new projections" (Ramon y Cajal, 1928).

It is clear, however, that these surges in the production of NTFs are insufficient in adulthood to recover the ab initio properties of neurons. The episodic increase in endogenous NTFs is unable to re-establishing neuronal phenotypic characteristics and connectivity following a major injury of the CNS. The experimental exogenous application of NTFs is proving to be capable of doing precisely that for certain subsets of CNS neurons. Much of this knowledge and current excitement derives from studies in which the application of neurotrophins (nerve growth factor (NGF), brain derived neurotrophic factor (BDNF), neurotrophin 3 (NT3)) importantly redresses the neuronal cell loss or atrophy which follows from lesions of the CNS of adult animals. Some of these aspects, centering on our experience with cholinergic neurons of the nucleus basalis magnocellularis, are discussed below.

The rescue of cholinergic neurons of the nucleus basalis magnocellularis

In the past decade we have been interested in the capacity of recovery of the cholinergic neurons of nucleus basalis magnocellularis (nbm). These neurons are particularly interesting in that they provide the bulk of the cholinergic innervation to neocortex. They constitute the "basalo-cortical" cholinergic pathway. Their transmitter (acetylcholine: ACh) has been associated with higher functions such as memory, attention and learning. Some of the symptoms of Alzheimer's disease (AD) have been attributed to the failure of cortical cholinergic function. Indeed, this transmitter and its biosynthetic enzyme (choline acetyltransferase: ChAT) are depleted to an important extent (albeit not exclusively) in this disease.

The cortical pathology of AD is accompanied in the basal forebrain by severe loss and atrophy of the cells of the nbm (nucleus magnocellularis of Meynert in the human species). From the experimental viewpoint, these cells provide an excellent opportunity to investigate CNS plasticity as they are clearly well endowed (as is also the case for cholinergic medial septum neurons) with both the low ($p75^{LNGFR}$) and the high ($p140^{trk}$, the proto-oncogen product: tyrosine kinase A) affinity receptor to NGF (for review, see Bothwell, 1991).

The abundance of NGF receptor (low and high affinity) mRNAs in the nbm of the rat is illustrated in Fig. 2. The lesion model in which we have explored extensively the responsiveness of nbm neurons consists of partial, unilateral cortical infarctions involving mainly the parietal cortex and adjacent portions of the neocortex (see Fig. 3). This ischemic injury deprives a large portion of the nbm of their normal target sites and the resulting lesion engulfs the *terminal* portion of these forebrain cholinergic neurons. The experimental procedure is such that it spares the main axonal shaft of these neurons and, in consequence, it is not an *axotomy* lesion model.

The outcome of this lesion is a gradual retrograde degeneration of nbm cholinergic neurons, ending with a marked and prolonged atrophy of these cells. Ostensibly, cell death does not occur, since the ChAT immunoreactive neurons persist unchanged in number. However, they appear considerably shrunken (Sofroniew et al., 1983), with important retraction of their dendritic processes in the nbm area and axonal networks in the remaining cortex of the lesioned side (Garofalo et al., 1992). These changes are accompanied by a marked depletion in ChAT enzymatic activity in the microdissected region of the nbm (Stephens et al., 1985). The intracerebro-ventricular administration of relatively low amounts of NGF completely redresses the biochemical and morphological signs for

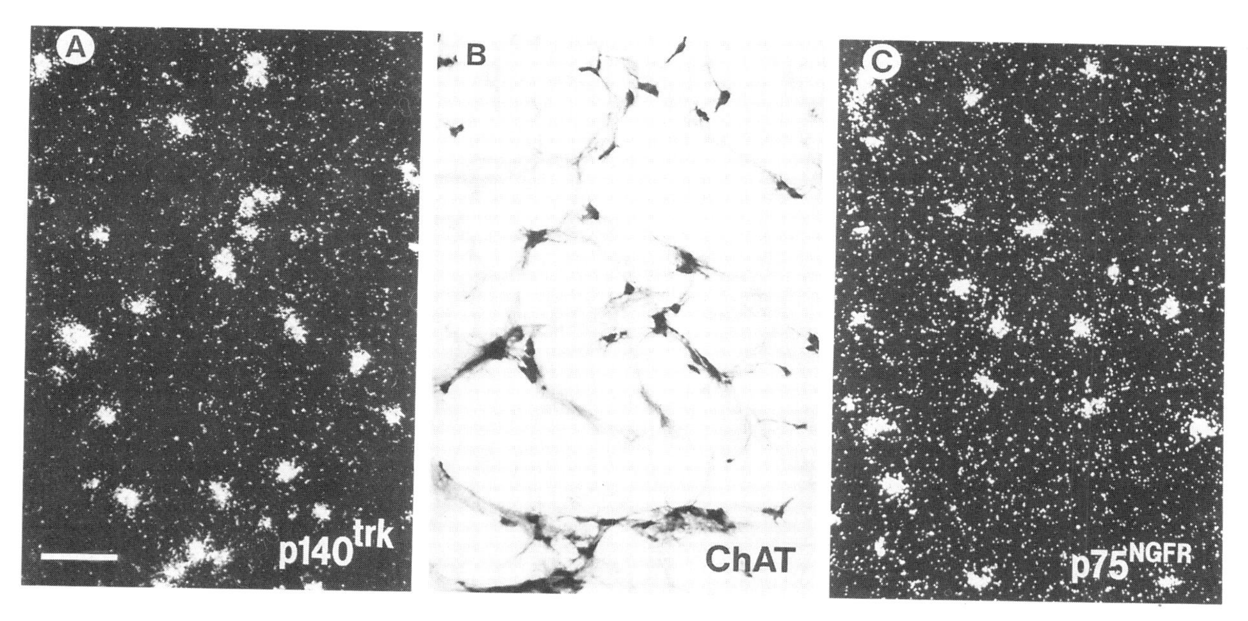

Fig. 2. Typical distribution of neurons at the level of the middle portion of the nucleus basalis magnocellularis of the rat. (*A*) and (*C*) Dark-field photomicrographs of neurons expressing $p140^{trk}$ and $p75^{NGFR}$ mRNAs, respectively. In situ hybridization experiments were performed with 10-μm-thick coronal sections using radioactive antisense DNA oligoprobes. Note the close correspondence in the pattern of distribution of the low and high affinity of NGF receptors with the nbm cholinergic neurons. (*B*) Photomicrograph of 50-μm thick coronal section at the same level as *A* and *B*, displaying ChAT-immunopositive neurons. Bar, 100 μm (*A–C*) (from Figueiredo, Skup, Bedard, Tetzlaff and Cuello, unpublished).

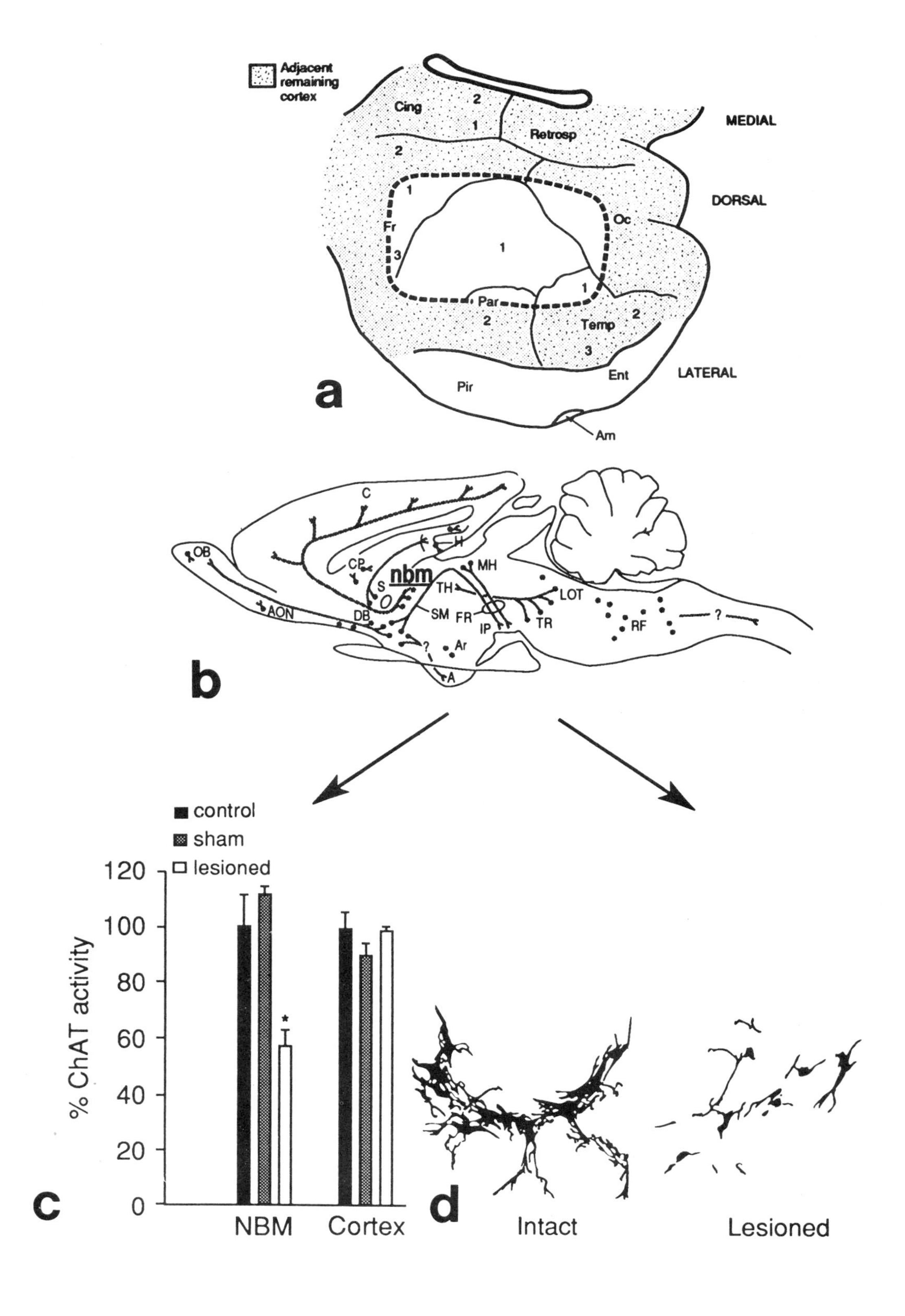

Adjacent remaining cortex
MEDIAL
DORSAL
LATERAL
Cing
Retrosp
Fr
Oc
Par
Temp
Ent
Pir
Am
a
C
H
OB
CP
nbm
MH
S
TH
LOT
AON
DB
SM
FR
IP
TR
RF
Ar
A
b
control
sham
lesioned
% ChAT activity
NBM
Cortex
c
d
Intact
Lesioned

the retrograde degeneration in the nbm (Cuello et al., 1989).

Other NTFs such as acidic fibroblast growth factor can achieve the same effect (Figueiredo et al., 1993) while the neurotrophins BDNF and NT3 are seemingly ineffective in the same experimental circumstances (Skup et al., 1994). Interestingly, the application of NGF in this experimental basalo-cortical lesion model not only protects cholinergic cell somata but also results in supranormal levels of ChAT activity in the terminal field, i.e. in the remaining cortex ipsilateral to the lesion (Cuello et al., 1989).

Does this imply that NTFs are capable of provoking changes compensatory to the deficits provoked by the infarction? Indeed, the neurotrophic therapy results in a marked increase in high affinity choline uptake sites in synaptosomes obtained from rat cortices bearing lesions and treated with neurotrophins (Garofalo and Cuello, 1993) and in a facilitated spontaneous output of endogenous ACh, recorded in vivo with micro-dialysis probes implanted in the remaining cortex of lesioned, neurotrophin-treated animals (Maysinger et al., 1992).

Neurotrophin-induced synaptic remodelling of the cerebral cortex

Are the structural cholinergic terminal elements also altered by the trophic factor therapy? The immunocytochemical investigation combined with computerized image analysis of these fine cholinergic cortical fibres revealed important modifications in the terminal network of the basalo-cortical pathway induced by the experimental neurotrophic therapy (Garofalo et al., 1992). Thus, a significant retraction of cholinergic fibers terminating in the neocortex occurs in the ipsilateral cortex to an ischemic lesion.

Administration of small amounts of NGF for 7 days suffice to correct this retrenchment of the cholinergic network and increase above normal the number of identifiable varicosities along these axons. These varicosities were further investigated at electron microscopic level concentrating the analysis in lamina 5, which is a prominent area of termination of basolo-cortical cholinergic fibers. At the ultrastructural level, it was seen that, concomitantly with the retraction of the fiber network and the diminution of the number of cholinergic varicosities, a contraction of the size of ChAT immunoreactive boutons occurs in lesioned animals. This contraction in the size of the individual ChAT immunoreactive boutons is not only corrected by the NGF administration but this neurotrophin provokes a considerable hypertrophy of these presynaptic cholinergic elements in lesioned animals.

These consequences of neurotrophic factor therapy are illustrated in Fig. 4. Furthermore (also illustrated in Fig. 4), the NGF application provoked a remarkable incidence in the number of synaptic differentiations as revealed by high resolution immunocytochemistry. Interestingly, none of these changes were noticed in unoperated animals receiving NGF. These observations demonstrate that NGF is capable of provoking profound remodelling of the cortical synaptic circuitry and, furthermore, by increasing the number of varicosities and synaptic differentiations of generating additional synapses, i.e. "synaptogenesis". Synaptic remodelling and synaptogenesis have in the past been identified in the adult CNS. For example, synaptic changes occur as a consequence of considerable deafferentation of the affected nuclei while pre- and post-synaptic alterations have been shown to develop in the hippocampal formation after long-term potentiation (for review, see Cotman, 1985). However, the NGF-induced synaptogenesis (Garofalo et al., 1992) is the first direct demonstration that a *drug* (a NTF in this case), is capable of inducing remodelling of the terminal network and synapses in mature, fully differentiated animals.

Fig. 3. Schematic representation of the cortical devascularizing lesion model. (*a*) Planimetric representation of the rat cerebral cortex from the medial aspects to the piriform (Pir) cortex; Fr, frontal; Oc, occipital; temp, temporal cortices. The removing of pial vessels in the framed, empty area causes an infarct of the frontal-parietal rat cortex. The shaded areas depict the remaining neocortex after infraction. This lesion leads to retrograde and anterograde degeneration of cholinergic neurons of the nbm. The basalo-cortical cholinergic pathway is indicated in (*b*). The depletion of ChAT enzymatic activity in the microdissected nbm is indicated in (*c*), and the extent of cell shrinkage and retraction of neurites is illustrated in (*d*) (Camera lucida drawing) (from Cuello, 1993).

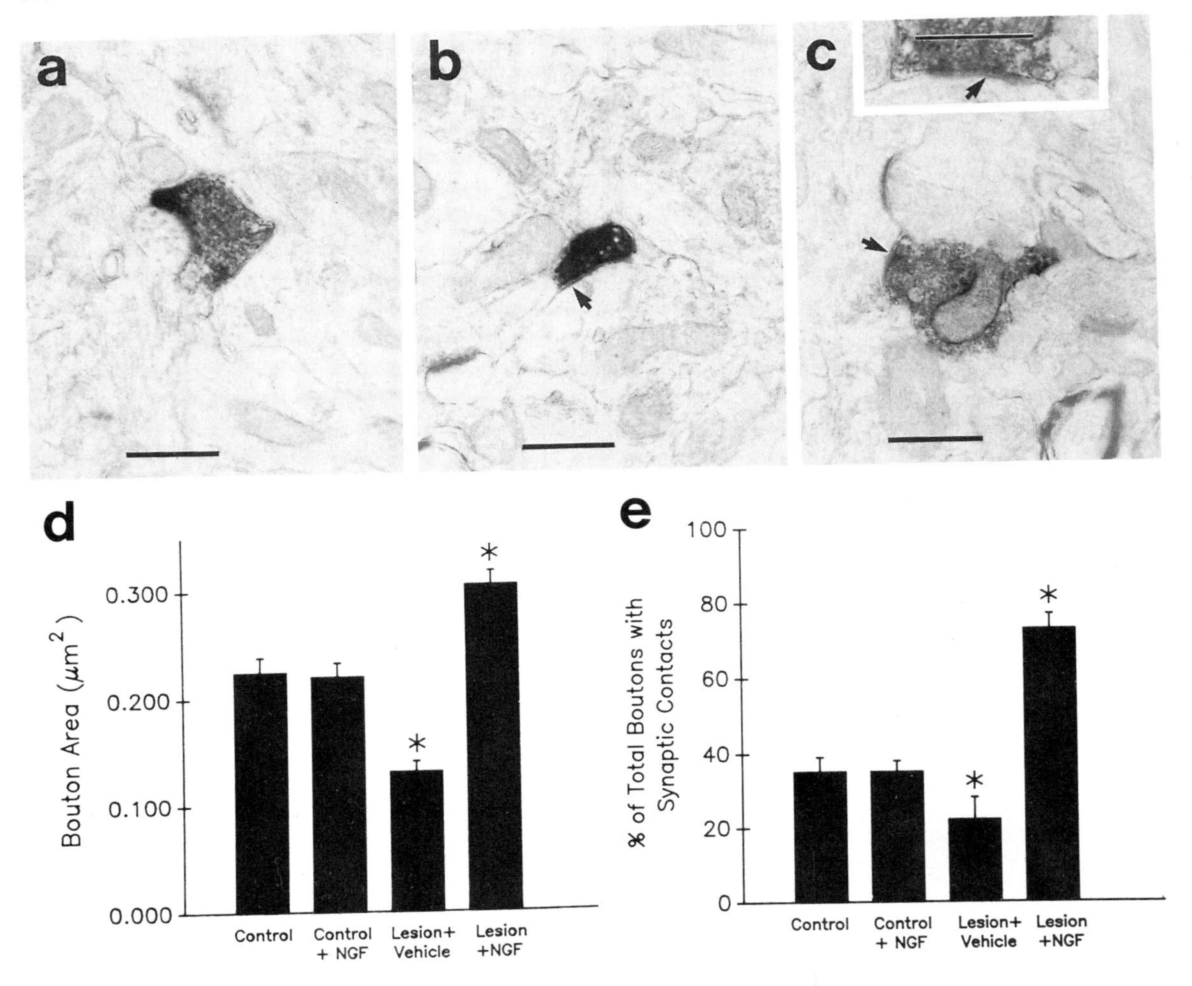

Fig. 4. Electron micrographs of cortical ChAT-IR boutons from layer V of rat somatosensory cortex. Representative profiles of cholinergic presynaptic terminal boutons of control (*a*), lesioned vehicle-treated (*b*), and lesioned NGF-treated (*c*) rats. Scale bars = 0.5 μm.) (*d*) Cross-sectional area of ChAT immunoreactive boutons in cortical layer V. (*e*) Percentage of varicosity profiles quantified with visible synaptic contacts. $^*P < 0.01$ from control (analysis of variance post hoc Tukey test) (from Garofalo et al., 1992).

The trophic factor therapy in primates

It is obviously essential that before embarking on neurotrophic therapies in the clinical scenario, that these agents are exhaustively investigated in subhuman primates. The thorough knowledge of the organization of the two main basal forebrain cholinergic projections (the septo-hippocampal and the basalo-cortical pathways) have greatly facilitated this task. The therapeutic capabilities of NGF in the *axotomy* lesion model in primates (unilateral fimbria-fornix sections resulting in "loss" of cholinergic neurons of the medial septum) have proven its efficacy. Thus, Tuszynski et al. (1990, 1991) and Koliatsos et al. (1990, 1991) have shown in *Macaca fascicularis*, utilizing either mouse or human recombinant NGF, that they were able to prevent cholinergic cell loss in the medial septum. They have even demonstrated a NGF-induced hypertrophy of forebrain cholinergic cell somata. The survival time of these lesioned-treated animals was 2–4 weeks.

We were able to reproduce in *Cercopithecus aethiops* the rodent model of retrograde degeneration of basolo-cortical cholinergic neurons. This was done by producing unilateral, ischemic, infarctions of the most superficial gyri in a region of the neocortex analogous to the rodent lesion model (Pioro et al., 1993). The ischemic lesion comprised portions of posterior frontal, superior temporal, parietal and anterior occipital cortices while sparing Brodmann area 4. Human-recombinant NGF (Genentech) was applied to these animals in the form of a gelatin implant in the area of the affected cortex in amounts comparable to those previously applied in rodents.

These investigations demonstrated that a retrograde cholinergic degeneration of cells of the nucleus basalis of Meynert (nbM) occurred, exclusively restricted to its intermediate region, that is the Ch4i group according to the classification of Mesulam et al. (1986). This region of the nbM is the sole region responsible for the cholinergic innervation of the affected areas of the neocortex. Shrunken cholinergic neurons in the intermediate nbM (Ch4i) were apparent some 6 months after surgical intervention (Liberini et al., 1993; Pioro et al., 1993). The morphology and dimensions of these cholinergic nbM cells appeared normal after 6 months survival time and more than 5 months after the complete dissolution of the gel containing the human-recombinant NGF (Liberini et al., 1993).

In addition, the deficits in ChAT enzymatic activity observed in the microdissected nbM were corrected by the NGF application (Liberini et al., 1993). Furthermore, a supranormal enzymatic activity of ChAT was found in samples of cortical tissue adjacent to the infarcted area. This upregulation of the activity of the acetyl choline synthesising enzyme is reminiscent of the biochemical changes observed in rodents (Cuello et al., 1989) which occur concomitantly with the remodelling of cortical synaptic elements (Garofalo et al., 1992).

It is therefore, very possible that profound plasticity of forebrain cholinergic neurons can be achieved in subhuman primates with the therapeutic administration of NTFs. Table I summarizes the results obtained with the application of neurotrophins in primates in two different lesion models after short and long survival times.

TABLE I

Effects of trophic agent treatment in different primate lesion models affecting the cholinergic system (from Cuello et al., 1993)

Lesion model	Species	Treatment	Survival time	Effects	Ref.[b]
FF transection	*Macaca fascicularis*		**Short survival**		
		Mouse-NGF (360 μg i.c.v.)	4 weeks	Cell loss prevention Neuronal hypertrophy	1
		Mouse-NGF (5 mg i.c.v.)	2 weeks	Cell loss prevention	2
		Human-NGF (360 μg i.c.v.)	4 weeks	Cell loss prevention	3
		Human-NGF (5 mg i.c.v.)	2 weeks	Cell loss prevention Neuronal hypertrophy	4
Cortical devascularization	*Cercopithecus aethiops*		**Long survival**		
		GM1 (175 mg/gel)	6 months	Rescue from cell atrophy ChAT decrease prevention[a]	5
		Human NGF (2.8 mg/gel)	6 months	Rescue from cell atrophy ChAT decrease prevention[a]	6
		GM1 + human NGF (175 mg + 2.8 mg/gel)	6 months	Rescue from cell atrophy Dendritic expansion Supranormal ChAT increase[a]	6

[a]ChAT assay was performed in the nbM ipsilateral to the lesion side and in cortex surrounding the devascularized area.
[b]1, Tuszynski et al. (1990); 2, Koliatsos et al. (1990); 3, Tuszynski et al. (1991); 4, Koliatsos et al. (1991); 5, Pioro et al. (1993); 6, Liberini et al. (1993).

Significance of CNS trophic factor therapy and future possibilities

Does trophic factor therapy have a clinical significance or it is merely an approach to acquire further knowledge on brain plasticity? The two quests, individually, are each important enough to concentrate our attention in the working hypothesis that most CNS neurons can be rescued from atrophy, and further that this therapy might lead to compensatory remodelling of synaptic circuitry. The minimal outcome from these studies will be a better understanding of brain function and dysfunction. The optimal outcome will be the designing of novel neurological therapies directly based on experience gathered with the experimental application of diverse NTFs (or their derivatives) in a variety of animal models, reproducing as closely as possible traumatic and neurodegenerative conditions.

The outstanding questions largely exceed the number of answers already assembled. The spectrum of NTFs' specificity, the receptors responsible for their biological actions as well as their intermediate intracellular messengers are yet to be established in most cases. This information, and that of the responsiveness of distinct CNS neuronal sets to the application of NTFs in a variety of experimental circumstances, will be gathered in the coming years. By then, the question of the right indication in the clinical setting will be paramount. At this point in time it is tempting to speculate that the cortical synaptic remodelling of cholinergic fibers might be a desirable therapeutic goal in AD.

Our confidence in bridging the experimental to clinical gap will be, among other factors, based on the fidelity with which animal models reproduce neuropathological conditions. The path to clinical application is full of potential drawbacks. One of the most immediate is the thought that NTFs might additionally activate undesired genes. Thus, the possibility has been raised of the potential overproduction of beta-amyloid or the generation of neuritic plaques in the case of Alzheimer's disease (for review, see Butcher and Woolf, 1989).

Age might also be a factor for consideration. It is clear from present information that the CNS of younger animals responds better to experimental trophic therapy. However, sustained application of NGF has shown to improve cholinergic markers and some behavioral responses in aged, impaired animals (for review, see Williams et al., 1993). Would the NTF-induced neuronal remodelling provoke disorganized connectivity? It is possible. We do not have enough information in this regard. It is, nonetheless, rewarding to note that this therapy in the lesioned basalo-cortical model resulted in correction of the deficient behaviors (Garofalo and Cuello, 1993).

Assuming that trophic factor therapy gathers sufficient momentum for clinical application, pharmaceutical issues will become crucial. For instance, should NTFs be administered directly into the CNS and, if so, how? Experimentally both permanent cannulae linked to mini-pumps and the grafting of microcapsule containing NTFs have proven successful. Adaptation of these procedures are possible but also cumbersome.

The possibility that future pharmacology might resort to the utilization of other active molecules capable of crossing the blood–brain barrier is on the horizon. These can be agents which either act cooperatively with endogenous NTFs (as might be the case for the sialogangliosides), substances which interfere with the cellular mechanisms involved in the trophic response (secondary, tertiary cellular messengers' responses) or tailored peptide "mimetics" able to activate specific NTF receptors. In the not so distant future it is conceivable that neurologists will consider the grafting of genetically transformed cells producing specific NTFs. This is a path that has been successfully pioneered by Gage et al. (1991) and which we have found of value in the basalo-cortical lesion model (Piccardo et al., 1992). Beyond that, the in vivo somatic cell transgenesis in the CNS could be an exciting prospect if the means for directing DNA material coding for NTFs is specifically incorporated by the desired cells and in the desired region of the CNS. All these options are full of unknowns but will certainly be explored in years to come.

Acknowledgements

The author gratefully acknowledges the support from

the MRC (Canada), the fruitful interactions with present and past collaborators, the secretarial assistance of Ms. Dawn Torsein and Ms. Oralia Mackprang and Dr. Paul Clarke and Dr. Bonald Figueiredo for their critical reading of the manuscript.

References

Bothwell, M. (1991) Keeping track of neuropeptide receptors. *Cell*, 65: 915–918.

Butcher, L.L. and Woolf, N.J. (1989) Neurotrophic agents may exacerbate the pathologic cascade of Alzheimer's disease. *Neurobiol. Aging*, 10: 557–570.

Cotman, C.W. (1985) Growth factor induction and temporal order in central nervous system repair. In: C.W. Cotman (Ed.), *Synaptic Plasticity*, Guilford Press, New York, pp. 407–456.

Cuello, A.C. (1993) Trophic responses of forebrain cholinergic neurons: a discussion. In: A.C. Cuello (Ed.), *Cholinergic Function and Dysfunction, Progress in Brain Research*, Vol. 98, Elsevier, Amsterdam, pp. 265–277.

Cuello, A.C., Garofalo, L., Kenigsberg, R.L. and Maysinger, D. (1989) Gangliosides potentiate in vivo and in vitro effects of nerve growth factor on central cholinergic neurons. *Proc. Natl. Acad. Sci. USA*, 86: 2056–2060.

Cuello, A.C., Liberini, P. and Piccardo, P. (1993) Atrophy and regrowth of CNS forebrain neurons. Models of study and clinical significance. In: A.C. Cuello (Ed.), *Neuronal Cell Death and Repair*, Elsevier, Amsterdam, pp. 173–191.

Figueiredo, B.C., Piccardo, P., Maysinger, D., Clarke, P.B.S. and Cuello, A.C. (1993) Effects of acidic fibroblast growth factor on cholinergic neurons of nucleus basalis magnocellularis and in a spatial memory task following cortical devascularization. *Neuroscience*, 56: 955–963.

Gage, F.H., Kawaja, M.D. and Fisher, L.J. (1991) Genetically modified cells: applications for intracerebral grafting. *Trends Neurosci.*, 14: 328–333.

Garofalo, L, and Cuello, A.C. (1993) Nerve growth factor and the monosialoganglioside GM1: analogous and different in vivo effects on biochemical, morphological and behavioral parameters of adult cortically lesioned rats. *Exp. Neurol.*, 125: 195–217.

Garofalo, L., Ribeiro-da-Silva, A. and Cuello, A.C. (1992) Nerve growth factor-induced synaptogenesis and hypertrophy of cortical cholinergic terminals. *Proc. Natl. Acad. Sci. USA*, 89: 2639–2643.

Hefti, F., Hartikka, J. and Knusel, B. (1989) Function of neurotrophic factors in the adult and aging brain and their possible use in the treatment of neurodegenerative diseases. *Neurobiol. Aging*, 10: 515–533.

Koliatsos, V.E, Martin, L.J., Walker, L.D., Richardson, R.T., DeLong, M.R. and Price, D.L. (1990) Mouse nerve growth factor prevents degeneration of axotomized basal forebrain cholinergic neurons in the monkey. *J. Neurosci.*, 10: 3801–3813.

Koliatsos, V.E., Clatterbuck, R.E., Nauta, H.J.W., Knüsel, B., Burton, L.E., Hefti, F.F., Mobley, W.C. and Price, D.L. (1991) Human nerve growth factor prevents degeneration of basal forebrain cholinergic neurons in primates. *Ann. Neurol.*, 30: 831–840.

Liberini, P., Pioro, E.P., Maysinger, D., Ervin, F.R. and Cuello, A.C. (1993) Long-term protective effects of human recombinant nerve growth factor and monosialoganglioside GM1 treatment on primate nucleus basalis cholinergic neurons after neocortical infarction. *Neuroscience*, 53: 625–637.

Maysinger, D., Herrera-Marschitz, M., Goiny, M., Ungerstedt, U. and Cuello, A.C. (1992) Effects of nerve growth factor on cortical and striatal acetylcholine and dopamine release in rats with cortical devascularizing lesions. *Brain Res.*, 577: 300–305.

Mesulam, M.M., Mufson, E.J. and Wainer, B.H. (1986) Three-dimensional representation and cortical projection topography of the nucleus basalis (CH4) in the macque: current demonstration of choline acetyltransferase and retrograde transport with a stabilized tetramethybenzidine method for horse radish peroxidase. *Brain Res.*, 367: 301–308.

Piccardo, P., Maysinger, D. and Cuello, A.C. (1992) Recovery of nucleus basalis cholinergic neurons by grafting NGF secretor fibroblasts. *Neurosci. Rep.*, 3: 353–356.

Pioro, E.P., Maysinger, D., Ervin, F.R., Desypris, G. and Cuello, A.C. (1993) Primate nucleus basalis of Meynert $p75^{NGFR}$-containing cholinergic neurons are protected from retrograde degeneration by the ganglioside GM_1. *Neuroscience*, 53: 49–56.

Ramon y Cajal, S. (1928) *Degeneration and Regeneration of the Nervous System*, translated by Raoul M. May, Oxford University Press, London.

Skup, M., Figueiredo, B.C. and Cuello, A.C. (1994) Intraventricular application of BDNF and NT-3 failed to protect NBM cholinergic neurons. *NeuroReport*, in press.

Sofroniew, M.V., Pearson, R.C., Eckenstein, F., Cuello, A.C. and Powell T.P. (1983) Retrograde changes in cholinergic neurons in the basal forebrain of the rat following cortical damage. *Brain Res.*, 289: 370–374.

Stephens, P.H., Cuello, A.C., Sofroniew, M.V., Pearson, R.C. and Tagari, P. (1985) Effect of unilateral decortication on choline acetyltransferase activity in the nucleus basalis and other areas of the rat brain. *J. Neurochem.*, 45: 1021–1026.

Thoenen, H. (1991) The changing scene in neurotrophic factors. *Trends Neurosci.*, 14: 165–170.

Tuszynski, M.H., Sang, U.H., Amaral, D.G. and Gage, F.H. (1990) Nerve growth factor infusion in the primate brain reduces lesion-induced neural degeneration. *J. Neurosci.*, 10: 3604–3614.

Tuszynski, M.H., Sang, U.H., Yoshida, K. and Gage, F.H. (1991) Recombinant human nerve growth factor infusions prevent cholinergic neuronal degeneration in the adult primate brain. *Ann. Neurol.*, 30 625–636.

Williams, L.R., Rylett, R.J., Ingram, D.K., Joseph, J.A., Moises, H.C., Tang, A.H. and Mervis, R.F. (1993) NGF affects the cholinergic neurochemistry and behavior of aged rats. In: A.C. Cuello (Ed.), *Cholinergic Function and Dysfunction, Progress in Brain Research*, Vol. 98, Elsevier, Amsterdam, pp. 251–256.

F. Bloom (Editor)
Progress in Brain Research, Vol. 100

CHAPTER 28

ACTH/MSH-derived peptides and peripheral nerve plasticity: neuropathies, neuroprotection and repair

Willem Hendrik Gispen[1], Joost Verhaagen[1] and Dop Bär[2]

[1]*Department of Medical Pharmacology and* [2]*Department of Neurology, Rudolf Magnus Institute for Neurosciences, Utrecht University, Universiteitsweg 100, 3584 CG Utrecht, The Netherlands*

Introduction

The regenerative capacity of the peripheral nervous system is limited. Since mature neurons do not have the ability to divide, the recovery of function depends completely on the potential of injured neurons to repair damaged nerve fibres. Numerous humoral and structural factors of neuronal, glial or target cell origin, most of which are active during development, appear to facilitate nerve repair by governing different stages of the repair process. Moreover, such factors may be of significance under conditions when neuronal function is compromised as a result of metabolic disturbance or neurointoxication. Restoring the balance between regenerative and degenerative forces may aid the neuron to cope with noxious stimuli. In this chapter, we review some of the evidence concerning the putative efficacy of neurotrophic peptides related to ACTH and MSH in animal and human disorders of the peripheral nervous system. As their neurotrophic or protective efficacy is manifest irrespective of the nature of the threatening conditions, eventual application may involve many disorders that have the dysfunction of neurons in common (for more extensive reviews, see Strand et al. (1991) and Bär et al. (1990)).

Mechanical trauma

Various laboratories have shown in animal experiments that treatment with peptides related to ACTH and MSH improves post-lesion repair of the sciatic nerve as illustrated by morphological, neurophysiological and functional evidence. Structure-activity studies pointed to the significance of the amino acid sequence ACTH/MSH-(4–10) (Met-Glu-His-Phe-Arg-Trp-Gly). This sequence lacks the classical corticotrophic activity and thus its neurotrophic property is not due to adrenal steroid release but acts rather by a direct effect on neural tissue. Although steroids are often used in the treatment of central or peripheral nervous system diseases, the rationale behind that treatment refers to their anti inflammatory or immunosuppressive action and seldom to their presumed neurotrophic or protective effect. Peptide dose response curves display an inverted U-shape. Effective routes of administration are subcutaneous injection (repeated), implantation (osmotic micropumps, biodegradable microspheres) or local administration at the site of injury whereas oral or nasal administration so far have been unsuccessful. Treatment must commence shortly following the injury of the nerve and should last at least 8 days in order to initiate an improvement in the recovery process (Bär et al., 1990).

Diabetic neuropathy

Diabetic neuropathy is defined by the American Diabetics Association as a demonstrable disorder, evident either clinically or subclinically, that occurs in the setting of diabetes mellitus in the absence of other possible causes for the peripheral neuropathy. Prevalence of the neuropathy varies between 5 and 80% and increases with both age and duration of diabetes mellitus. There is considerable uncertainty about the pathogenesis of the neural disorder. Roughly three working hypotheses can be distinguished. Firstly, the "metabolic" hypothesis suggests that chronic hyperglycaemia causes activation of the polyol pathway at the cost of myoinositol, resulting in a reduced neural Na^+/K^+ ATP-ase activity and consequently in a reduced nerve conduction velocity. Secondly, high blood glucose levels would lead to glycation of neural proteins and hence disturbing neuronal function. Lastly, chronic hyperglycaemia is known to lead to pathology of the endoneurial vessels resulting in decreased blood flow in the peripheral nerve. The subsequent anoxia is thus taken as the primary cause in nerve degeneration. However, it may well be that a combination of these pathological factors underlies the development of diabetic neuropathy (Dyck et al., 1987).

In two different experimental models of diabetic neuropathy, the efficacy of chronic treatment with the ACTH-(4–9) analog was determined. The first model involved the single administration of the cytostatic drug streptozotocin, which is known to kill the β-cells in the islets of Langerhans and thus to increase blood glucose. No insulin therapy was applied. The second model employed the BB/Wor rat, a strain of rats that genetically develop an autoimmune disease resulting in hyperglycaemia. Insulin therapy is required to maintain the rats that develop this disease. Longitudinal measurement of nerve conduction velocities was used to assess the onset and development of the peripheral neuropathy. In the streptozotocin model subcutaneous treatment with 75 μg/kg ACTH-(4–9) analog every 48 h improved peripheral nerve function both in a preventive and in an interventive protocol (Van der Zee et al., 1989; Bravenboer et al., 1993). Histological examination of sciatic nerves revealed a normal fiber size distribution in peptide-treated diabetic rats. In the BB/Wor rat, peptide treatment of animals with an existing neuropathy resulted in a marked improvement in neuronal function without a major demonstrable effect on histological parameters (Bravenboer et al., 1992b). The peptide treatment did not affect parameters indicative of the diabetic disease as such in any of the experiments . Preliminary studies revealed that peptide treatment may also interfere with the development of autonomic neuropathies as evidenced by improved presynaptic regulation of peripheral blood pressure (Van der Zee et al., 1990). Based on these promising animal data, a first clinical double blind study on the efficacy of chronic treatment with the ACTH-(4–9) analog in diabetic neuropathy was performed involving 62 patients suffering from Type 1 diabetes mellitus randomly divided over a placebo and a peptide treatment group. During the treatment period of 12 months, a number of parameters for peripheral and autonomic neuropathy were assessed. The peptide treatment was effective in lowering the vibration threshold in patients with diabetic neuropathy. There were no differences in any other parameter studied (Bravenboer et al., 1992a). Currently a dose finding study is being performed to optimize the efficacy of peptide treatment in diabetic neuropathy.

Chemotherapy related neuropathies

Peripheral nerve intoxication due to oncolytic treatment is among the most prominent side effects encountered in the treatment of various cancers. Peripheral neurotoxicity as a dose-limiting factor and discontinuation of the oncolytic treatment may profoundly interfere with adequate treatment of the malignancy. Therefore, prevention or diminishing of the neurotoxic side effects is of the utmost significance in the chemotherapy of cancer.

Cisplatin

The cytotoxic drug cisplatin (*cis*-diamine dichloroplatinum) has shown particular efficacy against bladder, testicular and ovarian cancer. Its neurotoxicity is manifested by a primarily sensory neuropathy leading to diminished proprioception and sensory ataxia. This selective neurotoxicity points to a particular sensitivity of dorsal root ganglia, although the neurotoxic action of cisplatin in these sensory neurons is poorly understood (Hamers et al., 1991a).

In rats, as in humans, repeated injections with cisplatin result in a selective decrease in the sensory nerve conduction velocity of the sciatic nerve. Various studies have demonstrated that co-treatment with the ACTH-(4–9) analog nearly completely counteracted this cisplatin-induced decrease in sensory conduction velocity. The protective effect was evident in both young adult and fullgrown rats and during relatively mild and severe cisplatin treatment regimens (Hamers et al., 1991b, 1993a). Notably, in two different models to evaluate antitumor efficacy of cisplatin, the peptide treatment did not interfere with the oncolytic action of the drug. In view of the close resemblance between cisplatin neurotoxicity in rats and humans, a randomized, double blind, placebo-controlled study was initiated to assess the efficacy of peptide treatment in the prevention of cisplatin neurotoxicity in women suffering from ovarian cancer. Fifty-four women were treated with a cisplatin based therapy and received either placebo, a low (0.25 mg/m^2) or a high (1 mg/m^2) dose of ACTH-(4–9). The principle measure of neurotoxicity was the vibration perception threshold, although a number of neurological signs and symptoms were also recorded. After six cycles of chemotherapy, the perception threshold in cisplatin-placebo treated women showed an increase over 8-fold from the baseline whereas the patients treated with the high dose peptide regimen displayed a non-significant increase of less than 2-fold. Similarly, the peptide-treated group showed significantly less clinical signs and symptoms than the placebo-treated patients (Gerritsen van der Hoop et al., 1990). Although patients from all groups showed a progression of the neuropathy following discontinuation of all treatment, such progression was smaller in patients originally belonging to the high peptide dose regimen group (Hovestadt et al., 1992). These results are a first indication that treatment with a neurotrophic peptide can prevent or diminish cisplatin neuropathy in the clinic.

Vincristine and taxol

Both *Vinca* alkaloids and taxol are known to interact with microtubules. Undoubtedly, the interference with microtubular kinetics is responsible for their oncolytic action and perhaps also for their neurotoxicity as the cytoskeleton is essential for proper neuronal function. Again, clinical use is threatened by their neurotoxic side effects. Taxol is a novel oncolytic drug with a major indication in cisplatin refractory carcinomas. Recently, it was demonstrated that, in rats, taxol-induced sensory peripheral neuropathy can be counteracted by chronic co-treatment with the ACTH-(4–9) analog (Hamers et al., 1993b) (Fig. 1).

Vinca alkaloids such as vincristine are widely used in the treatment of leukemia and malignant lymphomas. Using neurons of the snail *Lymnea stagnalis* as a model to study neurotoxic effects of cytostatic compounds, it was reported that the ACTH-(4–9) analog greatly diminishes the severity of vincristine neurotoxicity (Müller et al., 1992). In a randomized, double-blind placebo-controlled pilot study, the effect of the peptide was studied on neurotoxicity in 28 patients with lymphoma, who were treated with vincristine and vinblastine. Thirteen patients received the ACTH-(4–9) analog and 15 patients received placebo treatment. The placebo-treated patients had more autonomic complaints, motor deficits and sensory disturbances than the peptide-treated patients. The authors warrant some caution, however, as the average age of the peptide-treated patients was significantly lower than that of the placebo-treated patients (Van Kooten et al., 1992).

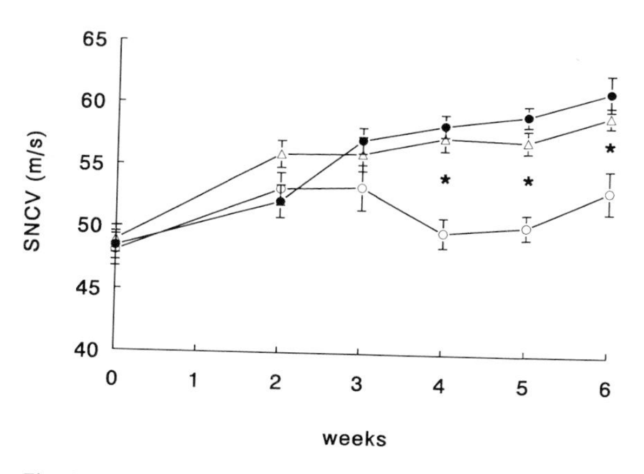

Fig. 1. Effects of taxol treatment on the sensory nerve conduction velocity (SNCV). (Δ) Age controls, (○) taxol/saline, (●) taxol/Org 2766. Values are given as means ± SEM. Taxol was injected weekly i.p. in a dose of 9 mg/kg. Org 2766 was injected s.c. every 48 h in a dose of 75 μg/kg during the whole experimental period. (*$p < 0.02$; age controls and taxol/Org 2766 versus taxol/saline).

The Guillain-Barré syndrome

Demyelinating diseases form an important group of life-threatening neurological disorders that await effective pharmacotherapy. The underlying mechanism appears to be a cell-mediated immune response directed against myelin components. Current therapeutic strategies are based on the use of anti-inflammatory and immunosuppressive drugs. As under demyelinating conditions neurons are in despair and their function compromised, it was hypothesized that neurotrophic or neuroprotective agents might be of benefit in improving neuronal function in these disorders. An established model of the human demyelinating syndrome of the peripheral nervous system (Guillain-Barré syndrome) is experimental allergic neuritis induced in the Lewis rat by peripheral myelin components. Recently, we were able to demonstrate that chronic treatment of such rats with the ACTH-(4–9) analog markedly suppresses the clinical symptoms (Fig. 2), protects against loss of motor coordination and prevents the degeneration of myelinated axons in the affected peripheral nerve. Subsequently, it was demonstrated that peptide treatment was also effective when the treatment commenced at the first appearance of clinical symptoms indicative of experimental allergic neuritis. These data were taken to illustrate that treatment with a neurotrophic peptide may provide a new approach in the therapy of peripheral demyelinating polyneuropathies (Duckers et al. in Vaudry and Eberle, 1993).

Mechanism of action of melanocortins

The mechanism by which melanocortins exert their beneficial effect is still largely unknown. However, several observations concerning the mode of action of melanocortins have been made.

(1) In vitro experiments show that melanocortins stimulate nerve outgrowth of cultured primary sensory and motoneurons, suggesting that melanocortins exert a direct trophic effect on one or more of the cell types present in these cultures (Van der Neut et al., 1992). Peptide signal transduction involves the activation of adenylate cyclase and c-fos (Hol et al., 1993). Tissue culture experiments also revealed putative binding sites for melanocortins in non-neuronal cells (Dyer et al. in Vaudry and Eberle, 1993) in the nervous system suggesting the significance of neuron–glia interaction in peptide induced nerve repair and protect. The neuroprotective effect of melanocortins in cisplatin neuropathy was confirmed in cultured dorsal root ganglia (Bär et al. in Vaudry and Eberle, 1993; Hol et al., 1994).

(2) Treatment schedules have demonstrated that melanocortins stimulate peripheral nerve regeneration only during a restricted period early following the lesion.

(3) Experiments in which different routes of melanocortin administration were tested demonstrated that local application at the site of injury is effective in facilitating nerve regeneration, implying that melanocortins act at the damaged nerve site (Bär et al., 1990).

Studies demonstrating an increase in POMC-derived peptides at the damaged nerve site following nerve injury are in line with the idea of a physiological role of endogenous ACTH/MSH-like peptides in the process of nerve regeneration (Bär et al., 1990). Firstly, it has been demonstrated with an in vitro bioassay for α-MSH that degenerating nerve portions contain agents with MSH-like activity, whereas intact nerves do not (Edwards and Gispen, 1985). Secondly, following nerve transection, increased α-MSH and β-endorphin immunoreactivity has been detected in the damaged nerve (Hughes and Smith, 1988). Increased immunoreactivity for ACTH/MSH-like peptides has also been demonstrated in rodents with neuromuscular disorders, such as inherited motoneuron disease in the wobbler mouse and inherited muscular dystrophy in C57BL/6J or REF/129 mice (Haynes and Smith, 1985). Rats treated with β,β'-iminodiproprionitrile (IDPN), a neurotoxin that affects motoneurons, and mice with streptozotocin-induced diabetes have been shown to express increased levels of POMC-derived peptides in motoneurons (Hughes et al., 1992). These results collectively suggest that increased immunoreactivity for ACTH/MSH-like peptides is present whenever neuropathic changes occur in motor nerves, whether this is due to a neurological disorder, to neurotoxicity or to mechanical damage. We have sug-

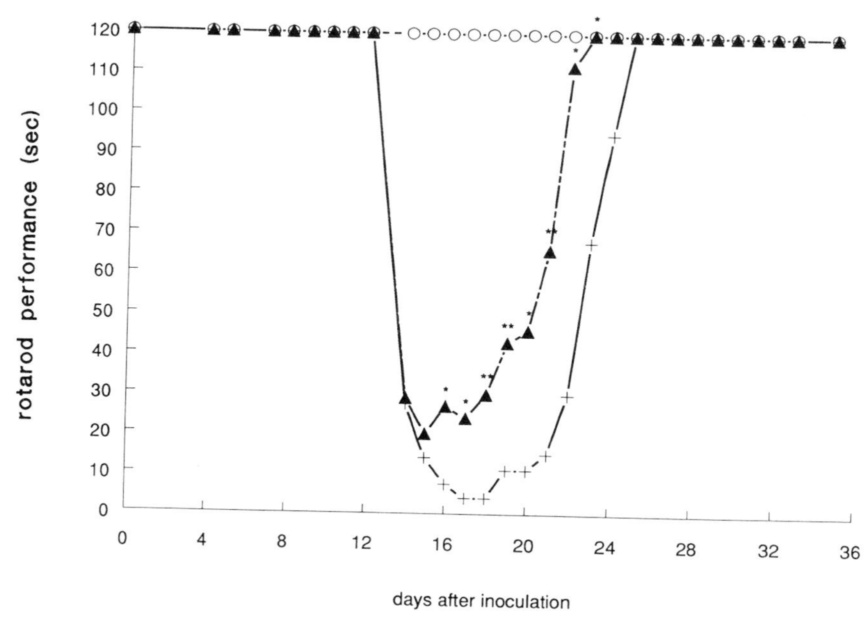

Fig. 2. Rotarod performance test. Rotarod performance scores are presented as medians. Solid line ($n = 10$) represents Lewis rats with EAN treated with 0.5 ml saline; dashed line (closed triangle, $n = 10$) represents Lewis rats with EAN treated with 75 μg Org 2766/kg body wt. in 0.5 ml saline; dashed line with open circles represents age-matched, control rats treated with 0.5 ml saline ($n = 10$). Saline and the peptide were administered by subcutaneous injections in the neck every 48 h. Statistics: Kruskal–Wallis test followed by Mann–Whitney U-tests (EAN-saline group versus EAN-Org 2766 group): $*p < 0.05$, $**p < 0.01$.

gested that exogenous melanocortins mimic or amplify a naturally occurring ACTH/MSH-like peptide signal early in the repair process that is part of the regenerative repertoire.

Two different working hypotheses have been put forward to explain the source and the nature of the naturally occurring endogenous melanocortins after nerve damage (Edwards and Gispen, 1985). Firstly, mature neurons might re-express proopiomelanocortin (POMC), the large precursor peptide from which ACTH and α-MSH derive, in their cell bodies after peripheral nerve damage. POMC mRNA is expressed in the developing spinal cord of rat embryos and is subsequently downregulated, although immunoreactivity for the POMC-derived peptides β-endorphin, α-MSH and ACTH has been demonstrated in a small portion of the motor nerves of the normal adult mouse and rat (Hughes and Smith, 1988). However, no increase in the expression of POMC mRNA has been detected in dorsal root ganglia, spinal cord or in the damaged nerve after sciatic nerve crush (Plantinga et al., 1992). Therefore, at this point it is not clear whether POMC mRNA expression in the nerve cell bodies and in the nerve contribute to the physiological stimulus leading to regeneration following nerve injury.

The second hypothesis on the nature and source of endogenous melanocortin production following nerve injury suggests that ACTH/MSH-like peptides are derived from the degenerating distal nerve stump (Edwards and Gispen, 1985). Immunoblotting and immunohistochemistry have demonstrated epitopes shared by α-MSH and the 150 kDa neurofilament protein (NF150). Since NF150 in the distal nerve stump is rapidly degraded after injury, we suggested that breakdown of NF150 results in the formation of an α-MSH-like peptide substance. Current research is aimed at further characterizing that peptide.

The work on the mechanism of action of melanocortins has been hampered by the fact that no receptor for Org 2766 has been identified. Putative binding sites of [^{3}H]Org 2766 have been demonstrated in spinal cord sections, but the binding of radiolabeled peptide was poorly displaceable by unlabeled Org 2766 (Dekker and Tonnaer, 1989). Recently, three different members of the ACTH/MSH receptor family have been cloned (Mountjoy et al., 1992; Gantz et al., 1993a,b). The human melanocortin-1 receptor has been identified as an α-MSH receptor and is expressed in melanoma cells, whereas the melanocortin-2 receptor comprises an ACTH receptor and is localized in the adrenal tissue. The third member of this group, the melanocortin-3 receptor, appears to be an MSH receptor that also recognizes the ACTH-(4–10) heptapeptide core common to ACTH, α-, β- and γ-MSH. It is expressed in brain, placenta, and gut but not in melanocytes or in the adrenal gland. Low stringency hybridization of human genomic DNA has indicated the existence of additional melanocortin receptors. Identification of the receptor for melanocortins involved in the neurotrophic action of these peptides will have a significant impact on advances in understanding the molecular mechanisms underlying the stimulatory effect of ACTH/MSH-like peptides on nerve regeneration.

References

Bär, P.R.D., Schrama, L.H. and Gispen, W.H. (1990) Neurotrophic effects of the ACTH/MSH-like peptides in the peripheral nervous system. In: D. de Wied (Ed.), *Neuropeptide Concept*, Elsevier, Amsterdam, pp. 175–211.

Bravenboer, B., Hendriksen, P.H., Oey, P.L., Van Huffelen, A.C., Gispen, W.H. and Erkelens, D.W. (1992a) $ACTH_{4-9}$ analogue in a randomized, double-blind, placebo-controlled trial in diabetic patients with neuropathy. *Diabetologia*, 35 (suppl. 1): 50.

Bravenboer, B., Kappelle, A.C., Van Buren, T., Erkelens, D.W. and Gispen, W.H. (1992b) ACTH4–9-analogue ORG 2766 improves existing diabetic neuropathy in the BB/Wor model. *Diabetes*, 41 (suppl 1): 491.

Bravenboer, B., Kappelle, A.C., Buren van, T., Erkelens, D.W. and Gispen, W.H. (1993) $ACTH_{4-9}$ analogue ORG 2766 can improve existing neuropathy in streptozocin-induced diabetic rats. *Acta Diabetol.*, 30: 21–24.

Dekker, A.J.A.M. and Tonnaer, J.A.D.M. (1989) Binding of the neurotrophic peptide Org 2766 to rat spinal cord sections is affected by a sciatic nerve crush. *Brain Res.*, 477: 327–331.

Dyck, P.J., Thomas, P.K., Asbury, A.K., Winegrad, A.I. and Porte, D. (1987) *Diabetic Neuropathy*, W.B. Saunders, Philadelphia.

Edwards, P.M. and Gispen, W.H. (1985) Melanocortin peptides and neural plasticity. In: J. Traber and W.H. Gispen (Eds.), *Senile Dementia of the Alzheimer's Type: Early Diagnosis, Neuropathology and Animal Models*, Springer-Verlag, Heidelberg, pp. 231–240.

Gantz, I., Konda, Y., Tashiro, T., Shimoto, H., Miwa, H., Munzert, G., Watson, S.J., DelValle, J. and Yamada, T. (1993a) Molecular cloning of a novel melanocortin receptor. *J. Biol. Chem.*, 268: 8246–8250.

Gantz, I., Miwa, H., Konda, Y., Shimoto, Y., Tashiro, T., Watson, S.J., DelValle, J. and Yamada, T. (1993b) Molecular cloning, expression, and gene localization of a fourth melanocortin receptor. *J. Biol. Chem.*, 268: 15174–15179.

Gerritsen van der Hoop, R., Vecht, C.J., Van der Burg, M.E.L., Elderson, A., Boogerd, W., Heimans, J.J., Els, D., Vries, P., Van Houwelingen, J.C., Jennekens, F.G.I., Gispen, W.H. and Neijt, J.P. (1990) Prevention of cisplatin neurotoxicity with an ACTH(4-9) analogue in patients with ovarian cancer. *N. Engl. J. Med.*, 322: 89–94.

Hamers, F.P.T., Gispen, W.H. and Neijt, J.P. (1991a) Neurotoxic side-effects of cisplatin. *Eur. J. Cancer*, 27: 372–376.

Hamers, F.P.T., Gerritsen van der Hoop, R., Steerenburg, P.A., Neijt, J.P. and Gispen, W.H. (1991b) Putative neurotrophic factors in the protection of cisplatin-induced peripheral neuropathy in rats. *Toxicol. Appl. Pharmacol.*, 111: 514–522.

Hamers, F.P.T., Pette, C., Bravenboer, B., Vecht, C.J., Neijt, J.P. and Gispen, W.H. (1993a) Cisplatin-induced neuropathy in mature rats. Effects of the melanocortin-like peptide ORG 2766. *Cancer Chemother. Pharmacol.*, 32: 162–166.

Hamers, F.P.T., Pette, C., Neijt, J.P. and Gispen, W.H. (1993b) The ACTH(4–9) analog ORG 2766, prevents taxol-induced neuropathy in rats. *Eur. J. Pharmacol.*, 233: 177–178.

Haynes, L.W. and Smith, M.E. (1985) Presence of immunoreactive α-melanotropin and β-endorphin in spinal motor neurons of the dystrophic mouse. *Neurosci. Lett.*, 53: 13–18.

Hol, E.M., Hermens, W.T.J.M.C., Verhaagen, J., Gispen, W.H. and Bär, P.R. (1993) α-MSH but not ORG 2766 induces expression of c-fos in cultured rat spinal cord cells. *NeuroRep.*, 4: 651–654.

Hol, E.M., Mandys, V., Sodaar, P., Gispen, W.H. and Bär, P.K. (1994) Protection by an $ACTH_{4-9}$ analogue against the toxic effects of cisplatin and taxol on sensory neurons and glial cells in vitro. *J. Neurosci. Res.*, in press.

Hovestadt, A., Van der Burg, M.E.L., Verbiest, H.B.C., Van Putten, W.L.J. and Vecht, Ch.J. (1992) The course of neuropathy after cessation of cisplatin treatment, combined with Org 2766 or placebo. *J. Neurol.*, 239: 143–146.

Hughes, S. and Smith, M.E. (1988) Effect of nerve transection on β-endorphin and α-melanotropin immunoreactivity in motor nerves of normal and dystrophic mice. *Neurosci. Lett.*, 92: 1–7.

Hughes, S., Smith, M.E., Simpson, M.G. and Allen, S.L. (1992) Effect of EDPN on the expression of POMC-derived peptides in rat motoneurons. *Peptides*, 13: 1021–1023.

Mountjoy, K.G., Robbins, L.S., Mortrud, M.T. and Cone, R.D. (1992) The cloning of a family of genes that encode the melanocortin receptors. *Science*, 257: 1248–1251.

Müller, L.J., Moorer-van Delft, C.M. and Boer, H.H. (1992) The ACTH(4–9) analog ORG 2766 stimulates microtubule formation in axons of the central nervous system of the snail *Lymnea stagnalis*. *Peptides*, 13: 769–774.

Plantinga, L.C., Verhaagen, J., Edwards, P.M., Schrama L.H., Burbach, J.P.H. and Gispen, W.H. (1992) Expression of the proopiomelanocortin gene in dorsal root ganglia, spinal cord and sciatic nerve after sciatic nerve crush in the rat. *Mol. Brain Res.*, 16: 135–142.

Strand, F.L., Rose, K.J., Zuccarelli, L.A., Kume, J., Alves, S.E., Antonawich, F.J. and Garrett, L.Y. (1991) Neuropeptide hormones as neurotrophic factors. *Physiol. Rev.*, 71: 1017–1046.

Van der Neut, R., Hol, E.M., Gispen, W.H. and Bär, P.R. (1992) Stimulation by melanocortins of neurite outgrowth from spinal and sensory neurons in vitro. *Peptides,* 13: 1109–1115.

Van der Zee, C.E.E.M., Gerritsen van der Hoop, R. and Gispen, W.H. (1989) Beneficial effect of Org 277 in the treatment of peripheral neuropathy in streptozocin-induced diabetic rats. *Diabetes,* 38: 225–230.

Van der Zee, C.E.E.M., Van den Buuse, M. and Gispen, W.H. (1990) Beneficial effect of an ACTH-(4–9) analog on peripheral neuropathy and blood pressure response to tyramine in streptozocin diabetic rats. *Eur. J. Pharmacol.,* 117: 211–213.

Van Kooten, B., Van Diemen, H.A.M., Groenhout, K.M., Huijgens, P.C., Ossenkoppele, G.I., Nauta, J.J.P. and Heimans, J.J. (1992) A pilot study on the influence of a corticotropin (4–9) analogue on *Vinca* alkaloid-induced neuropathy. *Arch. Neurol.,* 49: 1027–1031.

Vaudry, H. and Eberle, A.N. (1993) *The Melanotropic Peptides*, The New York Academy of Sciences, New York.

SECTION V

Neuro-Psychiatric Conditions

F. Bloom (Editor)
Progress in Brain Research, Vol. 100

CHAPTER 29

Enhancement of action potential conduction following demyelination: experimental approaches to restoration of function in multiple sclerosis and spinal cord injury

S.G. Waxman, D.A. Utzschneider and J.D. Kocsis

Department of Neurology, Yale University School of Medicine, New Haven, CT 06510; and PVA/EPVA Neuroscience Research Center, VA Hospital, West Haven, CT 06516, USA

Introduction

Restoration of function is a major goal of current research on demyelinating diseases. Multiple sclerosis (MS) represents a prototype demyelinating disorder of the CNS, in which axons usually maintain continuity through the lesions, but myelin is damaged. Researchers have been especially interested in restoration of function in MS since its course can include partial or complete remissions characterized by clinical recovery; moreover, subclinical demyelinated plaques (i.e. demyelinated lesions that are not accompanied by an appropriate clinical deficit) are well documented in MS, and may account for more than 50% of the lesions in many patients with this disorder (for review, see Waxman, 1988).

Non-penetrating spinal cord injury and spinal cord compression provide other examples of disorders in which there can be significant demyelination. Histological and electron microscopic studies have demonstrated demyelination in spinal cord white matter in experimental compressive and contusive spinal cord injury (Gledhill et al., 1973; Harrison and McDonald, 1977; Griffiths and McCulloch, 1983). Moreover, electron microscopic studies have provided evidence suggesting that myelin may be stripped from spinal cord axons following contusive injury, as a result of delayed invasion by inflammatory cells (Blight, 1985). There is also evidence for demyelination in spinal cord white matter following spinal cord injury and spinal cord compression in man (Byrne and Waxman, 1990; Bunge et al., 1993). These observations provide a pathological correlate for the presence, in some patients in whom spinal cord injury has been judged to be clinically complete, of residual descending influences on spinal reflex activity, a situation that has led to the concept of "dyscomplete" injury (Sherwood et al., 1992). Although a relapsing-remitting course is not usually observed after spinal cord injury, in some patients who were initially judged to have "complete" spinal cord injury, careful serial examinations have demonstrated improvements in neurological status that can occur over time periods extending at least as long as 1 year (Young, 1989). This delayed clinical improvement, together with the evidence for demyelination cited above, suggests the possibility that restoration of conduction in previously demyelinated axons may provide a basis for functional recovery following spinal cord injury.

This chapter briefly reviews recent studies which have explored two approaches aimed at restoration of function following demyelination of central axons.

Pharmacologic manipulation of ionic conductances in demyelinated axons

The distribution of voltage-sensitive ion channels within the axon membrane of myelinated axons is dis-

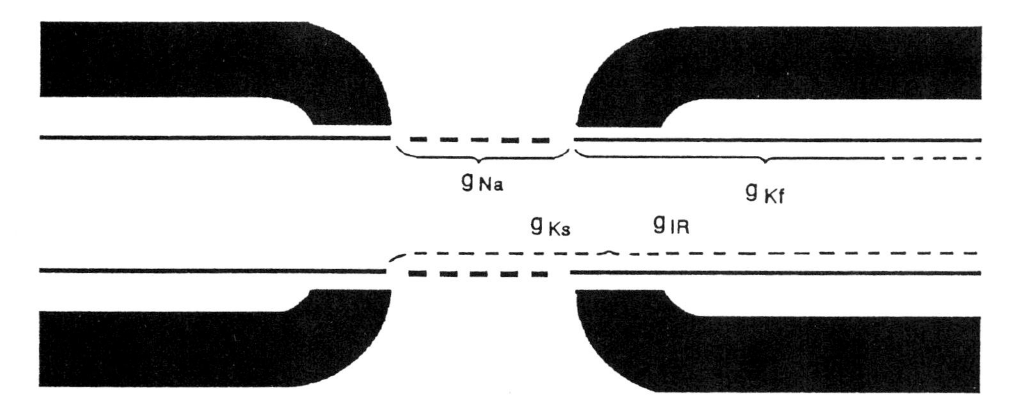

Fig. 1. Model showing putative localization of voltage-sensitive ion channels in the mammalian myelinated fiber. Abbreviations: g_{Na}, Na^+ channels; g_{Kf}, fast K^+ channels; g_{Ks}, slow K^+ channels; g_{IR}, inward rectifier. Sodium channels are clustered in high density in the axon membrane at the node of Ranvier and are present in much lower densities in the internode. Fast K^+ channels, present in the axon membrane under the myelin, are normally masked by myelin but are unmasked by demyelination. Slow Na^+ channels may be co-localized with rapidly-inactivating Na^+ channels, although a differential distribution has not been excluded.

tinctly heterogeneous, and includes a non-uniform distribution of Na^+ and K^+ channels. Figure 1 shows the current model of ion channel organization of the mammalian myelinated fiber. Na^+ channels are clustered in high density ($\sim 1000/\mu m^2$) in the axon membrane at the node of Ranvier, which is the site of action potential generation in normal myelinated axons (Ritchie and Rogart, 1977; Waxman, 1977). Within the internodal axon membrane (i.e. the axon membrane under the myelin sheath), the Na^+ channel density is much lower ($<25/\mu m^2$). At least three types of K^+ channels are present in the axon membrane in mammalian myelinated fibers: a "fast" K^+ channel, a "slow" K^+ channel, and an inward rectifier (for reviews, see Kocsis et al., 1993; Waxman and Ritchie, 1993). Fast K^+ channels display a distribution that is complementary to that of the Na^+ channels that are clustered in the nodal axon membrane and are expressed in highest density in the axon membrane under the myelin, but only in low densities in the axon membrane at the node (Chiu and Ritchie, 1981; Foster et al., 1982; Kocsis et al., 1982; Ritchie, 1982). Voltage-clamp studies indicate that the density of fast K^+ channels is maximal in the paranodal axon membrane, decreasing to 1/6 of the paranodal value in the node and internode (Röper and Schwarz, 1989).

Fast K^+ channels can be blocked by 4-aminopyridine (4-AP) when it is applied externally if it is given access to these channels (Kocsis et al., 1986; Baker et al., 1987). In normal non-myelinated fibers, application of 4-AP leads to delayed repolarization and prolongation of the action potential in both the PNS (Sherratt et al., 1980; Bostock et al., 1981) and CNS (Malenka et al., 1981; Preston et al., 1983); these results indicate that fast K^+ channels contribute to action potential repolarization in these fibers.

In contrast, in myelinated axons within the adult mammalian PNS and CNS, fast K^+ channels do not appear to play a significant role in action potential repolarization, and as shown in Fig. 2*B* (Kocsis et al., 1982), the action potential is not prolonged by exposure to 4-AP (Bostock et al., 1981; Kocsis and Waxman, 1980). This does not imply, however, that fast K^+ channels are not present in mature myelinated fibers. These channels are expressed in the internodal axon membrane, and are masked by the overlying myelin. This can be seen in developmental studies, which show a significantly larger effect of 4-AP on action potential duration in premyelinated axons, compared to mature myelinated fibers after the maturation of myelin (Fig. 2*A*) (Eng et al., 1988; Foster et al., 1982; Kocsis et al., 1982; Ritchie, 1982). The fast K^+ chan-

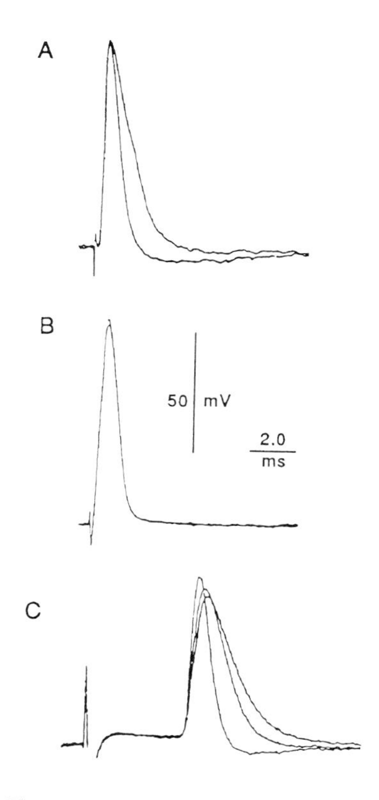

Fig. 2. Intra-axonal recordings showing the effect of 4-aminoyridine (0.5 mM) on action potential configuration in (*A*) preyelinated axon in regenerating sciatic nerve after nerve crush, (*B*) myelinated axon in adult rat sciatic nerve, and (*C*) demyelinated (lysophosphatidylcholine) ventral root axon. Note the conduction slowing after demyelination as evidenced by the long spike onset latency of the demyelinated ventral root axon. Extracellular application of 4-AP results in substantial prolongation of the action potential in premyelinated and demyelinated axons but not in mature myelinated axons where fast K^+ channels are masked. (*A*, *B* reproduced with permission from Kocsis et al. (1982); *C* reproduced with permission from Bowe et al. (1987)).

nels can also be seen following acute demyelination (Targ and Kocsis, 1986); when 4-AP is applied to demyelinated fibers where it has access to the exposed (formerly internodal) axon membrane, it produces a significant delay in repolarization of the action potential (Fig. 2*C*).

Since safety factor is decreased in demyelinated axons, it might be expected that maneuvers that prolong the action potential, thereby increasing the time integral of inward current, might improve conduction (Schauf and Davis, 1974). Bostock and co-workers (Sherratt et al., 1980; Bostock et al., 1981) demonstrated that 4-AP increases the temperature at which conduction failure occurs in demyelinated ventral root axons, in some cases reversing conduction block at physiologic temperatures. Experiments at the single fiber level have also demonstrated reversal of conduction block, with restoration of secure impulse conduction, following treatment of experimentally demyelinated sciatic nerve axons with 4-AP (Targ and Kocsis, 1985).

Figure 3 shows an example of reversal of conduction block in an experimentally demyelinated (lysophosphatidyl choline-treated) axon following application of 4-AP (Targ and Kocsis, 1985). Stimulating electrodes were positioned on both ends of the nerve, and recordings from single axons were obtained with

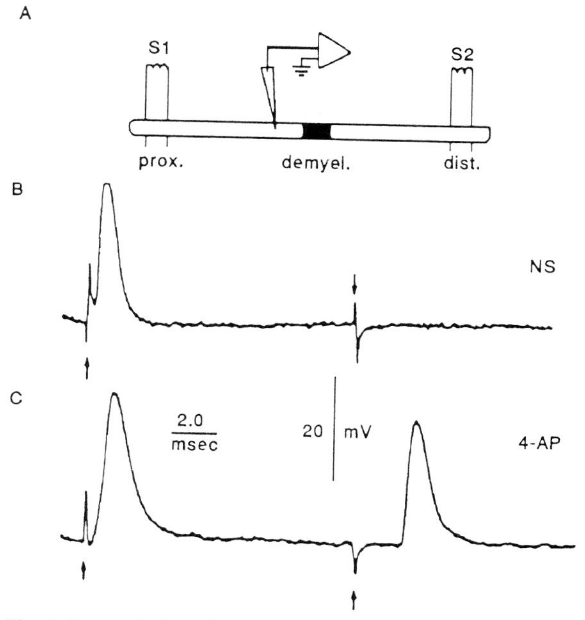

Fig. 3. Reversal of conduction block in demyelinated sciated nerve fiber with 4-AP (1.0 mM). The fiber was focally demyelinated by injection of lysophosphatidyl choline. (*A*) Experimental design permits examination of conduction through normal (stimulation at S_1) or demyelinated (stimulation at S_2) regions. (*B*) Stimulation at S_1 leads to an action potential, but conduction block occurs following S_2 stimulation, when the demyelinated zone is interposed between stimulation and recording site. (*C*) Following application of 4-AP, conduction block is overcome and the action potential propagates, with increased latency, through the demyelinated zone. Note the increased action potential duration, due to blockade of fast K^+ channels which tend to repolarize the fiber in demyelinated regions. NS = normal solution (modified from Targ and Kocsis, 1985).

microelectrodes positioned intra-axonally on one side of the lesion (Fig. 3*A*). Following stimulation proximal to the lesion (so that the conduction block did not include the demyelinated region), a propagated action potential could be recorded in normal Ringer solution. In contrast, following stimulation on the contralateral side of the lesion (so that conduction had to cross the zone of demyelination), an action potential could not be recorded since conduction block had occurred (Fig. 3*B*). The nerve was then exposed to 4-AP and, as shown in Fig. 3*C*, action potential duration was increased and conduction through the lesion was restored.

These observations have now been extended to the clinical arena. A number of clinical studies have been carried out with 4-AP, and with the related drug 3,4-diaminopyridine (3,4-DAP), to examine the effects of these agents on neurological status in patients with MS (see e.g. Jones et al., 1983; Stefoski et al., 1987; Davis et al., 1990). These studies have demonstrated improvements in motor function, improved brainstem function (e.g. improvement in extraocular movements), reduction in the size of scotomata, and improved critical flicker fusion in patients with MS. Similar studies have been carried out in the Lambert-Eaton myasthenic syndrome (on the basis of the rationale that prolongation of the action potential by blockade of fast K^+ channels in the preterminal axon will lead to an increase in acetylcholine release) and have shown improved motor function (Lundh et al., 1984; Murray and Newsom-Davis 1981).

Recently, Hansebout et al. (1993) reported a preliminary clinical study, in which 4-AP was administered to eight patients with chronic spinal cord injury. Beneficial effects were not detected in the two patients with complete paraplegia who were studied. However, in five of the six patients with incomplete spinal cord injury, there appeared to be significant transient neurologic improvement. This included improvement in sensory scores, as well as reduction in spasticity and in chronic pain and dysesthesias in the lower extremities. There was a tendency towards improvement in motor scores, although this did not reach a statistically significant level in this initial study of a small number of cases.

Further, systematic studies on larger numbers of patients with spinal cord injury are clearly needed. Hansebout et al. (1993) noted a decrease in vibratory sensation in some spinal cord injured patients following treatment with 4-AP, and this may provide some clues about the mechanism of this drug's action in spinal cord injury; in this regard, it is now well-established that there are differences in ion channel organization of mammalian sensory versus motor axons (Bowe et al., 1985) and between different types of sensory fibers (Honmou et al., 1994). Recent studies suggest that the repertoire of ion channels expressed by mammalian axons may be more complex than previously suspected (Kampe et al., 1992; Scholz et al., 1992; Kocsis et al., 1993; Stys et al., 1993). If axons in different CNS tracts display differences in ion channel expression, it may be possible to develop targeted pharmacologic interventions that will specifically alter conduction in a given pathway; this strategy would allow selective treatment of symptoms such as pain, spasticity and sensory loss.

Transplantation of myelin-forming cells

Theoretical studies demonstrate that remyelination with even thin, or short, myelin segments can support the conduction of action potentials through previously demyelinated fibers, if the remyelinated nodes of Ranvier develop membrane properties similar to those in normal fibers (Koles and Rasminsky, 1972; Waxman and Brill, 1978). The development of relatively normal Na^+ channel densities at remyelinated nodes of Ranvier is, in fact, suggested by cytochemical studies in the spinal cord, which show that newly formed nodes along remyelinated axons develop normal properties (Weiner et al., 1980), and by saxitoxin-binding experiments in sciatic nerve, which show an increase in the number of Na channels that is proportional to the increase in nodal membrane area imposed by the shorter spacing between remyelinated nodes (Ritchie, 1982).

In peripheral nerves demyelinated with lysophosphatidyl choline, increased conduction velocity and restoration of the ability to conduct high-frequency impulse trains are observed in association with remyelination (Smith and Hall, 1980). Electrophysiological studies also provide evidence for recovery of conduction following remyelination in CNS axons in the dorsal columns following injection of lysophosphatidyl choline, with refractory period for transmission returning to normal levels following remyelination even with abnormally thin and short myelin segments (Smith et al., 1983).

Recent experiments suggest that clinical recovery in experimental allergic encephalomyelitis is correlated with remyelination in the PNS and CNS (by Schwann cells and oligodendrocytes). It is notable, in this regard, that remyelination of CNS axons can be mediated by either oligodendrocytes or Schwann cells, and action potential conduction can be facilitated by both oligodendrocyte- and Schwann cell-mediated remyelination. Remyelination of dorsal column axons by Schwann cells, as well as oligodendrocytes, is effective in restoring secure action potential conduction (Blight and Young, 1989; Felts and Smith, 1991).

An important new strategy for enhancing conduction in demyelinated axons is based on the idea that transplantation of myelin-forming cells or their precursors may lead to the production of functional myelin associated with enhanced action potential conduction. Morphological studies have demonstrated that transplanted Schwann cells and oligodendrocytes can form myelin around demyelinated axons within the spinal cord (Duncan et al., 1988; Gout et al., 1988; Rosenbluth et al., 1990). The formation of myelin by transplanted glial cells does not, per se, insure secure impulse conduction, since action potential electrogenesis depends not only on myelin formation, but also on the deployment of adequate numbers and types of ion channels at the newly formed nodes of Ranvier. Mature paranodal axo-glial junctions must be formed following myelin formation by transplanted cells; otherwise there will be a shunt under the myelin which may interfere with conduction (Hirano and Dembitzer, 1978). Moreover, patchy or incomplete remyelination can lead to impedance mismatch at the junction between myelinated and demyelinated axon zones, thereby causing conduction failure (Sears et al., 1978; Waxman, 1978).

To determine whether conduction of action potentials in myelin-deficient axons can be improved by the replacement of cellular elements via transplantation, Utzschneider et al. (1993) recently used electrophysiological methods to study the amyelinated axons in the spinal cord of the myelin-deficient (md) rat following the transplantation of myelin-forming glial cells (Duncan et al., 1992) from normal litter mates. The absence of host myelin in this system (which is due to an absence of proteolipid protein) permits definitive confirmation that functional changes are due to myelination by transplanted cells, and are not due to background host myelination (Duncan et al., 1988). Electrophysiological studies in the md rat spinal cord have demonstrated that while the amyelinated fibers of the md rat are capable of secure impulse conduction, they exhibit conduction velocities that are approximately 1/4 of normal (Utzschneider et al., 1992).

To assess the conduction properties of axons in the md spinal cord after transplantation, we recorded from axons both within the transplant region, where histological examination confirmed the presence of morphologically normal myelin, and upstream and downstream along the same beams of axons, where there was no myelin (Utzschneider et al., 1993). In these experiments, affected md rats received a transplant, on postnatal day 5, from unaffected female litter mates. The transplant recipients were studied 15–17 days after transplantation (postnatal days 20, 21, 22). As shown in Fig. 4, field potential recordings obtained along axon trajectories both within the transplant region, and in the non-transplant region, demonstrated that conduction velocity of the most rapidly conducting fibers was about three times faster within the transplant region (3.2 ± 0.2 m/s ($\pm$ SEM); compared to 0.9 ± 0.03 m/s outside the transplant region). The field potential could be observed to propagate either into or out of the transplant area in all six of the rats that were studied. In those cases where a conduction track encompassed both myelinated and non-myelinated regions, there was an approximate doubling of conduction velocity as the action potential propagated into the myelinated region. Axons within the transplant region displayed frequency-response characteristics that were similar to those of fibers outside the trans

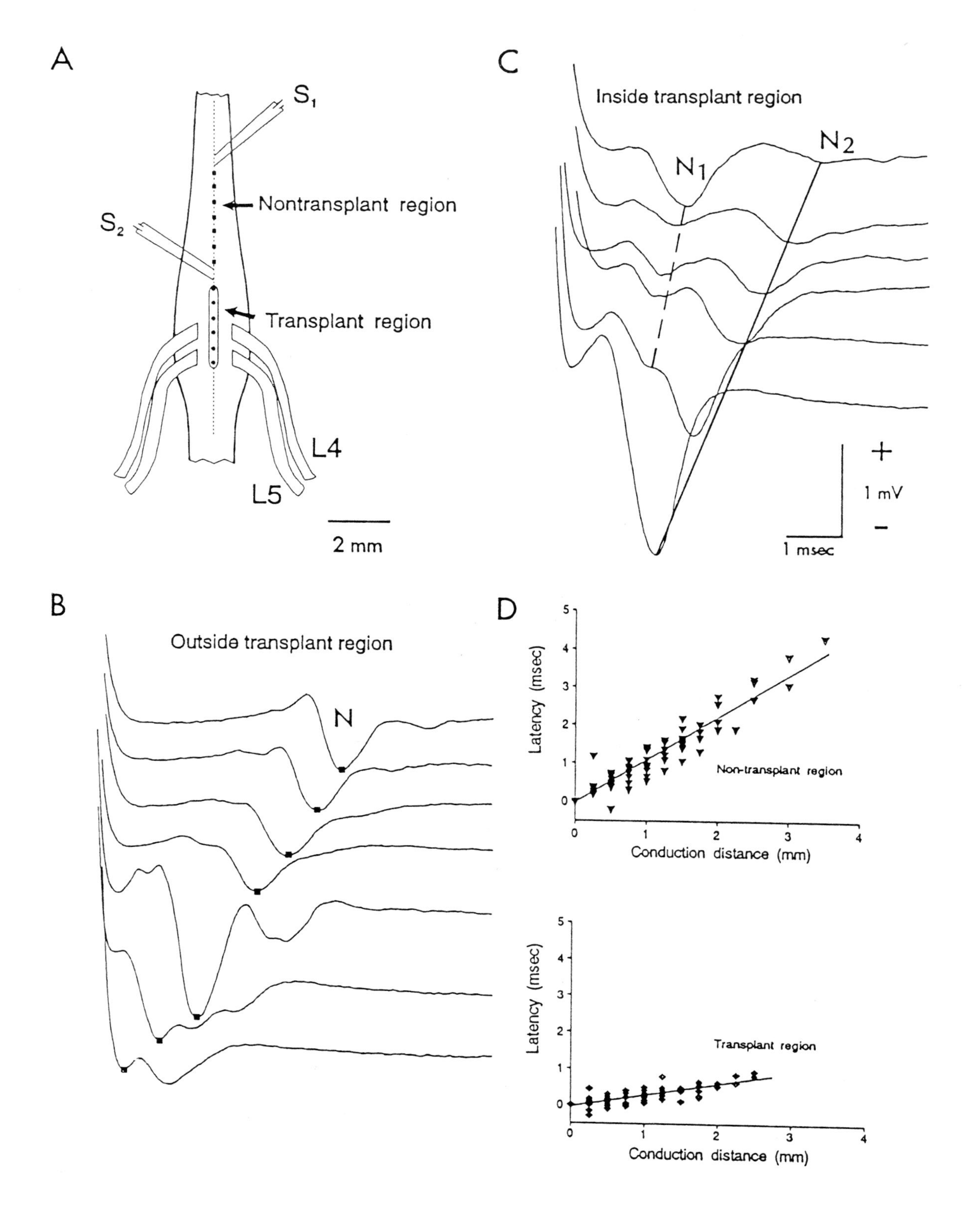
A
S1
S2
Nontransplant region
Transplant region
L4
L5
2 mm
B
Outside transplant region
N
C
Inside transplant region
N1
N2
+
1 mV
−
1 msec
D
Latency (msec)
Non-transplant region
Conduction distance (mm)
Latency (msec)
Transplant region
Conduction distance (mm)

plant region (Fig. 5*C,D*), and fibers within the transplant region could follow tetanic stimuli at frequencies up to 100 Hz, similar to fibers outside the transplant zone.

To further assess the physiological consequences of cell transplantation, intracellular recordings were obtained from single dorsal root ganglion (DRG) neurons, following antidromic stimulation of dorsal columns. These experiments demonstrated a significant increase (~3-fold faster) in conduction velocity within the transplant region. Conduction of the action potential from the transplant region, into non-transplanted areas, indicated that action potentials were securely conducted through the zone of potential impedance mismatch from the myelinated to non-myelinated parts of the host nervous system (Fig. 5*A*).

These results demonstrate that myelination of CNS axons, by exogenous CNS glial cells, is associated with significantly increased conduction velocities. Axons myelinated by the transplanted cells have refractory periods and frequency-following characteristics similar to those of axons in non-transplanted regions. Moreover, action potentials can be initiated outside the transplant region, propagate into the transplant area, and continue beyond the transplanted zone. Transplantation of exogenous glial cells into the amyelinated spinal cord thus can result in myelination that is associated with an enhancement of conduction properties.

While detailed biophysical data are not yet available, it is likely that the increased conduction velocity is due to myelination by the transplanted cells. The conduction velocity of axons in the transplanted zone increased approximately threefold relative to the non-transplant region, and is significantly greater than the conduction velocity of normal CNS non-myelinated axons (Waxman and Bennett, 1972). A very large increase (~9-fold) in axon diameter would be required to account for this increased conduction velocity in the absence of myelination. Saltatory conduction in normal myelinated axons requires a nodal Na^+ channel density (~1000/μm^2) that is much higher than in normal non-myelinated axons (Ritchie and Rogart, 1977). While Na^+ channel densities at the newly formed nodes along transplanted axons have not yet been measured, increased Na^+ channel densities have been demonstrated in chronically demyelinated spinal cord axons following the injection of ethidium bromide and X-irradiation (Black et al., 1991), and at the newly formed nodes along remyelinated axons following viral-induced demyelination in the mouse spinal cord (Weiner et al., 1980). The increased conduction velocity and normal frequency-following properties of axons in the transplanted zone suggest that relatively normal nodes of Ranvier are formed in association with myelination by exogenous glial cells following transplantation.

Since there is a fourfold reduction in conduction velocity in md spinal cord axons (Utzschneider et al., 1992), the threefold increase in conduction velocity demonstrates a significant return toward normal function following transplantation of myelin-forming glial cells in the amyelinated spinal cord. This provides a demonstration that functional properties, as well as morphological characteristics, of pathological white matter tracts can be favorably altered by transplantation of glial cells.

In these initial studies, the electrophysiological

Fig. 4. Increased conduction velocity in dorsal column axons of myelin-deficient (md) rat 16 days following transplantation of myelin-forming cells. Field potentials from transplant and non-transplant regions of the dorsal columns are shown. (*A*) A schematic showing the longitudinal extent of the transplant region (~3 mm). Two stimulation sites (S_1 and S_2) provide recording tracks within the transplant region and more rostrally outside the transplant region. The recording interval is 0.5 mm for both tracks. (*B*) Field potentials outside the transplant region usually show a single main negativity with occasional early or late components (fifth trace from top). (*C*) Field potentials from transplant region of the same animal show two separate negativities (N_1 and N_2) with increasingly distinct latencies as the recording electrode is moved further away from the stimulus site. Conduction velocity for the N_1 component is increased. The stimulus site is outside the transplant region, indicating propagation of the impulse across the amyelinated-myelinated junction. (*D*) Aggregate conduction latencies in non-transplant (upper graph) and transplant regions (lower graph) in the md dorsal columns. The upper graph shows the latency of the main N negativity (from 100 recording sites from 17 recording tracks) outside the transplant region. The slope of the linear regression indicates an average conduction velocity of 0.9 ± 0.03 m/s. The lower graph shows a significantly smaller increase in latency with increasing conduction distance in the transplant region, with an average conduction velocity of 3.2 ± 0.23 m/s (modified from Utzschneider et al., 1993).

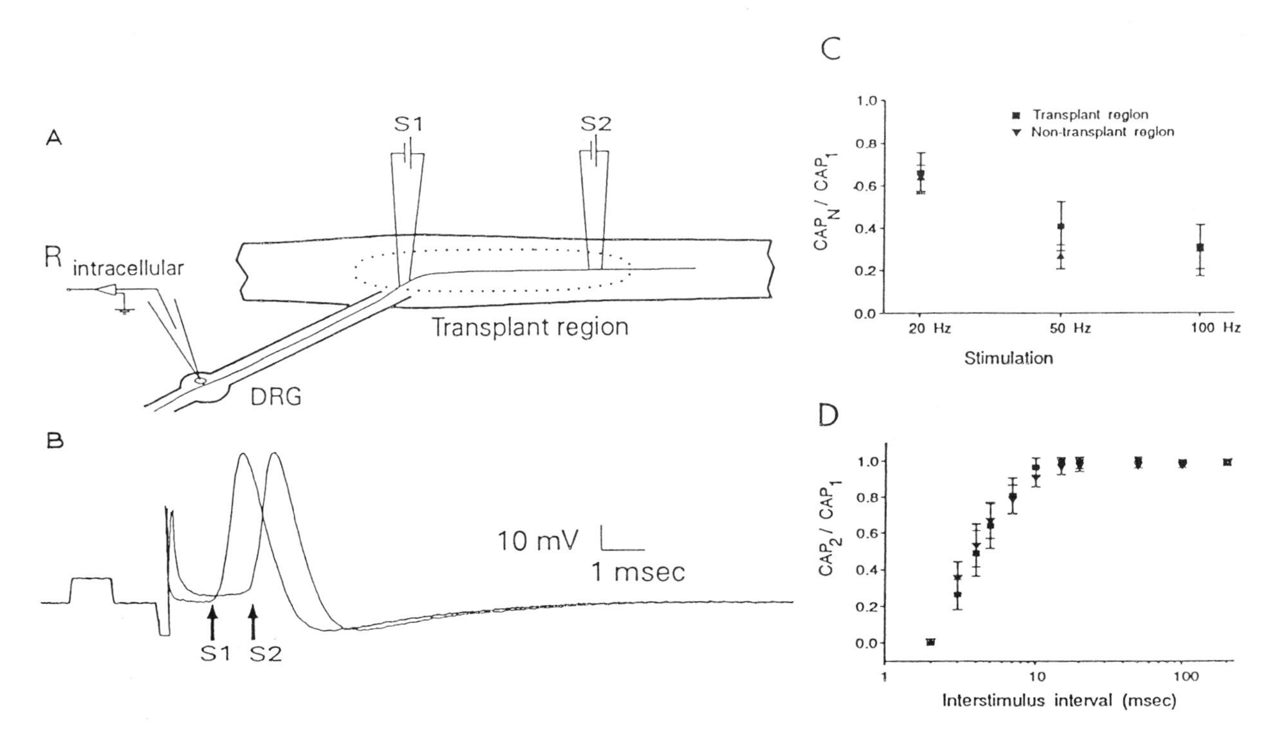

Fig. 5. (*A*) Single-cell recording of action potential conduction through the transplant region. The schematic shows the placement of two stimulating electrodes in the transplant zone and an intracellular recording electrode in a dorsal root ganglion cell. (*B*) Action potentials recorded from dorsal root ganglion cell following stimulation at two sites (S_1, S_2) in the transplant region. Propagation of the action potential from stimulating electrode S_2 to the dorsal root ganglion cell demonstrates that conduction occurred through the zone of potential impedance mismatch, from the transplant zone to non-transplanted parts of the host nervous system. From the latency shift and interstimulus distance, a conduction velocity of 2.6 m/s within the transplant zone can be calculated. (*C*) The ability of axons to follow tetanic stimuli is similar inside and outside the transplant region. The graph shows the ratio of the amplitudes of the first and last compound action potentials (CAPs) for repetitive stimuli of 20 Hz (10 s), 50 Hz (10 s), and 100 Hz (2 s). (*D*) Double-shock experiments showing the ratio of test CAP to control CAP for interstimulus intervals of 2–200 ms. The time-course of recovery for impulse conduction is similar inside and outside the transplant region.

consequences of transplantation of myelin-forming glial cells into immature (5-day-old) rats was studied within 3 weeks of transplantation; this strategy focused on examining the effects of glial cell transplantation during the normal period of myelination. Future studies will examine the long-term effects of glial cell transplantation and the effects of transplantation into adult recipients.

It is also possible that the transplantation strategy may be applicable to the development of new immunotherapies for inflammatory demyelinating diseases such as MS. A relatively new approach to controlling the immune response in MS involves attempts to downregulate T-cell-mediated activity directed against myelin antigens by using protein and peptide-based immunotherapy. Antigen-specific tolerization could result from a downregulation of self-reactive T-cells or a vaccination-like approach that would upregulate the capacity of regulatory T-cells to control self-reactive T-cells (Adorini, 1993; Hafler et al., 1993). Weiner et al. (1993) recently suggested, from an initial study in MS patients, that oral tolerization with bovine myelin may result in a significant reduction in myelin-basic-protein (MBP)-reactive T-cells associated with clinical improvement; the tolerization regimen with the bovine myelin did not result in apparent toxicity or side-effects. On the basis of our observations of appropriate function by transplanted glial cells, we are led to speculate that glial cell transplantation may provide an effective means of providing antigen-specific immunotherapy. The ability to genetically modify the MBP and proteolipid protein (PLP) epitopes expressed by the transplanted myelin-forming glial cells and to specify the localization of the newly expressed myelin may represent a uniquely effective method for presenting antigen to self-reactive T-cells and inducing tolerance. Thus, myelin-forming transplants may have a role in antigen-specific, peptide-based *immunotherapy* as well as in the *physiological restoration* of conduction.

Concluding remarks

As described above, several experimental strategies have demonstrated, in experimental models, that it is possible to improve conduction of action potentials along previously demyelinated axons. In the case of K^+ channel blockade with 4-AP and 3,4-DAP, transient beneficial effects have been well documented in some patients with MS. Initial studies in a small number of patients with spinal cord injury have provided provocative results, and further studies are underway. The observation of relatively long-lasting (48 h or more) changes in sensory symptoms in spinal cord injury patients following treatment with 4-AP (Hansebout et al., 1993) raises important questions about the mechanism of action of this drug. The multiplicity of K^+ channel subtypes that are expressed in different types of axons (Honmou et al., 1994; Vogel and Schwarz, 1994) suggests that it may be possible to develop ion channel blocking agents that selectively alter the conduction properties of demyelinated axons in specific pathways or tracts.

Much work remains to be done on transplantation of myelin-forming cells. Recently, it has been demonstrated that transplanted O2-A glial progenitor cells can differentiate and form myelin around myelin-deprived axons in the CNS (Groves et al., 1993), and it is possible that engineered cells may also be developed, as a source for transplantation. The presently available data are promising in that they indicate, in genetically amyelinated axons, that myelination by exogenous glial cells after transplantation can enhance axonal conduction. Nevertheless, considerable work will be needed before transplantation can be considered in the clinical domain, particularly for disorders such as MS, where demyelination is multi-focal and progressive. The occurrence of demyelination in spinal cord injury (where the pathology is presumably static rather than progressive) may present a situation that is especially appropriate for further study, in terms of restoration of action potential conduction via repair of demyelinated axons by transplantation.

Acknowledgments

Work in the authors' laboratories has been supported in part by grants from the National Multiple Sclerosis Society and the NINCDS, and by the Medical Research Service, US Department of Veterans Affairs. DAU was supported in part by the Medical Scientist Training Program, and by an EPVA Multiple Sclerosis Fellowship.

References

Adorini, L. (1993) Selective inhibition of T cell responses by protein and peptide-based immunotherapy. *Clin. Exp. Rheumatol.*, 11(Suppl. 8): S41–S44.

Baker, M., Bostock, P., Grafe, P. and Martins, P. (1987) Function and distribution of three types of rectifying channel in rat spinal root myelinated axons. *J. Physiol. (London)*, 383: 45–67.

Black, J.A., Felts, P., Smith, K.J., Kocsis, J.D. and Waxman, S.G. (1991) Distribution of sodium channels in chronically demyelinated spinal cord axons: immuno-ultrastructural localization and electrophysiological observations. *Brain Res.*, 544: 59–70.

Blight, A.R. (1985) Delayed demyelination and macrophage invasion: a candidate for secondary cell damage in spinal cord injury. *CNS Trauma*, 2: 299–314.

Blight, A.R. and Young, W. (1989) Central axons in injured cat spinal cord recover electrophysiological function following remyelination by Schwann cells. *J. Neurol. Sci.*, 91: 15–34.

Bostock, H., Sears, T.A. and Sherratt, R.M. (1981) The effects of 4-aminopyridine and tetraethylammonium ions on normal and demyelinated mammalian nerve fibers. *J. Physiol. (London)*, 313: 301–315.

Bowe, C.M., Kocsis, J.D. and Waxman, S.G. (1985) Differences between mammalian ventral and dorsal spinal roots in response to blockade of potassium channels during maturation. *Proc. R. Soc. London, Ser. B*, 224: 355–366.

Bowe, C.M., Kocsis, J.D., Targ, E.F. and Waxman, S.G. (1987) Physiological effects of 4-aminopyridine on demyelinated mammalian motor and sensory fibers. *Ann. Neurol.*, 22: 264–268.

Bunge, R.P., Puckett, W.R., Becerra, J.L., Marcillo, A. and

Quencer, R.M. (1993) Observations on the pathology of human spinal cord injury. A review and classification of 22 new cases with details from a case of chronic cord compression with extensive focal demyelination. In: F.J. Seil (Ed.), *Advances in Neurology,* Vol. 59. *Neural Injury and Regeneration*, Raven Press, New York, pp. 75–89.

Byrne, T.N. and Waxman, S.G. (1990) *Spinal Cord Compression*, F.A. Davis, Philadelphia.

Chiu, S.Y. and Ritchie, J.M. (1981) Evidence for the presence of potassium channels in the paranodal region of acutely demyelinated mammalian nerve fibres. *J. Physiol. (London)*, 313: 415–437.

Davis, F.A., Stefoski, D. and Rush, J. (1990) Orally administered 4-aminopyridine improves clinical signs in multiple sclerosis. *Ann. Neurol.*, 27: 186–192.

Duncan, I.D., Hammang, J.P., Jackson, K.F., Wood, P.M., Bunge, R.P. and Langford, L. (1988) Transplantation of oligodendrocytes and Schwann cells into the spinal cord of the myelin-deficient rat. *J. Neurocytol.*, 17: 351–360.

Duncan, I.D., Archer, D.R. and Wood, P.M. (1992) Functional capacities of transplanted cell-sorted adult oligodendrocytes. *Dev. Neurosci.*, 14: 114–122.

Eng, D.L., Gordon, T.R., Kocsis, J.D. and Waxman, S.G. (1988) Development of 4-AP and TEA sensitivities in mammalian myelinated nerve fibers. *J. Neurophysiol.*, 60: 2168–2179.

Felts, P.A. and Smith, K.J. (1991) Conduction properties of central nerve fibers remyelinated by Schwann cells. *Brain Res.*, 574: 178–192.

Foster, R.E., Connors, B.W. and Waxman, S.G. (1982) Rat optic nerve: electrophysiological, pharmacological, and anatomical studies during development. *Dev. Brain Res.*, 3: 361–376.

Gledhill, R.F., Harrison, B.M. and McDonald, W.I. (1973) Demyelination and remyelination after acute spinal cord compression. *Exp. Neurol.*, 38: 472–487.

Gout, O., Gansmuller, A., Baumann, N. and Gumpel, M. (1988) Remyelination by transplanted oligodendrocytes of a demyelinated lesion in the spinal cord of the adult shiverer mouse. *Neurosci. Lett.*, 87: 195–199.

Griffiths, I.R. and McCulloch, M.C. (1983) Nerve fibers in spinal cord impact injuries. 1. Changes in the myelin sheath during the initial five weeks. *J. Neurol. Sci.*, 58: 335–345.

Groves, A.K., Barnett, S.C., Franklin, R.J.M., Crang, A.J., Mayer, M., Blakemore, W.F. and Noble, M. (1993) Repair of demyelinated lesions by transplantation of purified O-2A progenitor cells. *Nature*, 362: 453–456.

Hafler, D.A., Zharg, J.W., LaSalle, J., Donnelly, C., Webster, H.L. and Wucherpfeffnig, K. (1993) The development of antigen-specific therapies for autoimmune diseases; investigations in multiple sclerosis as a paradigm for rheumatoid arthritis. *Clin. Exp. Rheumatol.*, 11(Suppl. 8): S39–S40.

Hansebout, R.R., Blight, A.R., Fawcett, S. and Reddy, K. (1993) 4-Aminopyridine in chronic spinal cord injury: a controlled, double-blind, crossover study in eight patients. *J. Neurotraum.*, 10: 1–18.

Harrison, B.M. and McDonald, W.I. (1977) Remyelination after transient experimental compression of the spinal cord. *Ann. Neurol.*, 1: 542–551.

Hirano, A. and Dembitzer, H.M. (1978) Morphology of normal central myelinated axons. In: S.G. Waxman (Ed.), *Physiology and Pathobiology of Axons*, Raven Press, New York, pp. 68–82.

Honmou, O., Utzschneider, D.A., Rizzo, M.A., Bowe, C.M., Waxman, S.G. and Kocsis, J.D. (1994) Delayed depolarization and slow sodium currents in cutaneous afferents. *J. Neurophysiol*, in press.

Jones, R.E., Heron, J.R., Foster, D.H., Snelgar, R.S. and Mason, R.J. (1983) Effects of 4-aminopyridine in patients with multiple sclerosis. *J. Neurol. Sci.*, 60: 353–362.

Kampe, K., Safronov, B. and Vogel, W. (1992) A Ca-activated and three voltage-dependent K channels identified in mammalian peripheral nerve. *Pflügers Arch., Eur. J. Physiol.*, 420(Suppl. 1): R28.

Kocsis, J.D. and Waxman, S.G. (1980) Absence of potassium conductance in central myelinated axons. *Nature*, 287: 348–349.

Kocsis, J.D., Waxman, S.G., Hildebrand, C. and Ruiz, J.A. (1982) Regenerating mammalian nerve fibres: changes in action potential waveform and firing characteristics following blockage of potassium conductance. *Proc. R. Soc. London Ser. B*, 217: 277–287.

Kocsis, J.D., Gordon, T.R. and Waxman, S.G. (1986) Mammalian optic nerve fibers display two pharmacologically distinct potassium channels. *Brain Res.*, 393: 357–361.

Kocsis, J.D., Black, J.A. and Waxman, S.G. (1993) Pharmacological modification of axon membrane molecules and cell transplantation as approaches to the restoration of conduction in demyelinated axons. In: S.G. Waxman (Ed.), *Molecular and Cellular Approaches to the Treatment of Neurological Disease*, Raven Press, New York, pp. 265–292.

Koles, Z.J. and Rasminsky, M. (1972) A computer simulation of conduction in demyelinated nerve fibres. *J. Physiol. (London)*, 227: 351–364.

Lundh, H., Nilsson, O. and Rosen, I. (1984) Treatment of Lambert-Eaton syndrome: 3,4-di-aminopyridine and pyridostigmine. *Neurology*, 34: 1324–1330.

Malenka, R.C., Kocsis, J.D., Ransom, B.R. and Waxman, S.G. (1981) Modulation of parallel fiber excitability by postsynaptically mediated changes in extracellular potassium. *Science*, 214: 339–341.

Murray, N.M. and Newsom-Davis, J. (1981) Treatment with oral 4-aminopyridine in disorders of neuromuscular transmission. *Neurology*, 31: 265–271.

Preston, R.J., Waxman, S.G. and Kocsis, J.D. (1983) Effects of 4-aminopyridine on rapidly and slowly conducting axons of rat corpus callosum. *Exp. Neurol.*, 79: 808–820.

Ritchie, J.M. (1982) Sodium and potassium channels in regenerating and developing mammalian myelinated nerves. *Proc. R. Soc. London Ser. B*, 215: 273–287.

Ritchie, J.M. and Rogart, R.B. (1977) The density of sodium channels in mammalian myelinated nerve fibers and the nature of the axonal membrane under the myelin sheath. *Proc. Natl. Acad. Sci. USA*, 74: 211–215.

Röper, J. and Schwarz, J.R. (1989) Heterogeneous distribution of fast and slow potassium channels in myelinated rat nerve fibers. *J. Physiol. (London)*, 416: 93–110.

Rosenbluth, J., Hasegawa, M., Shirasaki, N., Rosen, C.L. and Liu, Z. (1990) Myelin formation following transplantation of normal fetal glia into myelin-deficient rat spinal cord. *J. Neurocytol.*, 19: 718–730.

Schauf, C.L. and Davis, F.A. (1974) Impulse conduction in multiple sclerosis: a theoretical basis for modification by temperature and pharmacological agents. *J. Neurol. Neurosurg. Psychiatry*, 37: 152–161.

Scholz, A., Reid, G., Bostock, H. and Vogel, W. (1992) Na and K channels in human axons. *Pflügers Arch., Eur. J. Physiol.*, 420 (Suppl. 1): R28.

Sears, T.A., Bostock, H. and Sherratt, M. (1978) The pathophysiology of demyelination and its implications for the symptomatic treatment of multiple sclerosis. *Neurology*, 28: 21–26.

Sherratt, R.M., Bostock, H. and Sears, T.A. (1980) Effects of 4-aminopyridine on normal and demyelinated mammalian nerve fibers. *Nature*, 283: 570–572.

Sherwood, A.M., Dimitrijevic, M.R. and McKay, W.B. (1992) Evidence of subclinical brain influence in clinically complete spinal cord injury: discomplete SCI. *J. Neurol Sci.*, 110: 90–98.

Smith, K.J. and Hall, S.M. (1980) Nerve conduction during peripheral demyelination and remyelination. *J. Neurol. Sci.*, 48: 201–219.

Smith, K.J., Blakemore, W.F. and McDonald, W.I. (1983) Central remyelination restores secure conduction. *Nature*, 280: 395–396.

Stefoski, D., Davis, F.A., Faut, M. and Schauf, C.L. (1987) 4-Aminopyridine improves clinical signs in multiple sclerosis. *Ann. Neurol.*, 21: 71–77.

Stys, P.K., Sontheimer, H., Ransom, B.R. and Waxman, S.G. (1993) Non-inactivating, TTX-sensitive Na^+ conductance in rat optic nerve axons. *Proc. Natl. Acad. Sci. USA*, 90: 6976–6980.

Targ, E.F. and Kocsis, J.D. (1985) 4-Aminopyridine leads to restoration of conduction in demyelinated rat sciatic nerve. *Brain Res.*, 328: 358–361.

Targ, E.F. and Kocsis, J.D. (1986) Action potential characteristics of demyelinated rat sciatic nerve following application of 4-aminopyridine. *Brain Res.*, 363: 1–9.

Utzschneider, D., Black, J.A. and Kocsis, J.D. (1992) Conduction properties of spinal cord axons in the myelin-deficient rat mutant. *Neuroscience*, 49: 221–228.

Utzschneider, D.A., Archer, D.R., Kocsis, J.D., Waxman, S.G. and Duncan, I.D. (1994) Transplantation of glial cells enhances action potential conduction of amyelinated spinal cord axons in the myelin-deficient rat. *Proc. Natl. Acad. Sci. USA*, 91: 53–57.

Vogel, W. and Schwarz, J.R. (1994) Voltage-clamp studies in frog, rat, and human axons: macroscopic and single channel currents. In: S.G. Waxman, J.D. Kocsis and P.K. Stys (Eds.), *The Axon*, Oxford University Press, New York, in press.

Waxman, S.G. (1977) Conduction in myelinated, unmyelinated, and demyelinated fibers. *Arch. Neurol.*, 34: 585–590.

Waxman, S.G. (1978) Prerequisites for conduction in demyelinated fibers. *Neurology*, 28: 27–34.

Waxman, S.G. (1988) Clinical course and electrophysiology of multiple sclerosis. In: S.G. Waxman (Ed.), *Functional Recovery in Neurological Disease*, Raven Press, New York, pp. 157–184.

Waxman, S.G. and Bennett, M.V.L. (1972) Relative conduction velocities of small myelinated and non-myelinated fibers in the central nervous system. *Nat., New Biol.*, 238: 217–219.

Waxman, S.G. and Brill, M.H. (1978) Conduction through demyelinated plaques in multiple sclerosis: computer simulations of facilitation by short internodes. *J. Neurol. Neurosurg. Psychiatry*, 41: 408–417.

Waxman, S.G. and Ritchie, J.M. (1993) Molecular dissection of the myelinated axon. *Ann. Neurol.*, 33: 121–136.

Weiner, L.P., Waxman, S.G., Stohlman, S.A. and Kwan, A. (1980) Remyelination following viral-induced demyelination: ferric ion-ferrocyanide staining of nodes of Ranvier within the CNS. *Ann. Neurol.*, 8: 580–583.

Weiner, H.L., Mackin, G.A., Matsui, M., Orav, E.J., Khoury, S.J., Dawson, D.M. and Hafler, D.A. (1993) Double-blind pilot trial of oral tolerization with myelin antigens in multiple sclerosis. *Science*, 259: 1321–1324.

Young, W. (1989) Recovery mechanisms in spinal cord injury: implications for regenerative therapy. In: F.J. Seil (Ed.), *Neural Regeneration and Transplantation*, Alan R. Liss, New York, pp. 157–169.

F. Bloom (Editor)
Progress in Brain Research, Vol. 100

CHAPTER 30

Functional integrity of neural systems related to memory in Alzheimer's disease

Nancy A. Simonian, G. William Rebeck and Bradley T. Hyman

Neurology Service, Massachusetts General Hospital and Harvard Medical School, Boston, MA 02114, USA

Memory impairment is a hallmark of Alzheimer's disease

The initial symptom of the disease is most frequently loss of memory function, as exemplified by losses on short-term delayed recall tasks. Memory impairment, in one form or another, remains at the core of the clinical syndrome and dominates the illness as it progresses for the next 5–10 years, until death. The underlying pathophysiology of this progressive and relentless memory impairment remains unknown. When we began our studies of the causes of memory impairment in Alzheimer's disease 10 years ago, we were interested in knowing the functional status of the neural systems that subserve memory in the human brain. As a first approximation we began a systematic study of structural changes that occur in memory-related brain areas.

Anatomic basis for memory impairment in Alzheimer's disease

Structures whose integrity is crucial for normal memory function have been defined on the basis of lesion experiments in animals and as a result of surgical or pathological lesions in the human. A review of this literature is beyond the scope of this chapter, but certainly a neural system whose components include the hippocampus, entorhinal cortex (anterior parahippocampal gyrus), the cholinergic basal forebrain, and likely contributions from the amygdala, midline and anterior thalamic nuclei, mammillary bodies, and proisocortical areas surrounding the medial temporal lobe are implicated. The pathological changes of Alzheimer's disease seem to specifically affect this memory-related neural system, along with the neocortical association cortices that are crucial for information processing.

Histopathologically, the Alzheimer disease brain shows intraneuronal inclusions of cytoskeletal elements (neurofibrillary tangles) and extraneuronal deposits of β/A4 amyloid protein as senile plaques. Our initial studies (Hyman et al., 1984) showed that a specific set of neurons within the hippocampal formation consistently developed neurofibrillary tangles, while other anatomic fields were consistently spared. Layer II of the entorhinal cortex, layer IV of the entorhinal cortex, and the CA1/subicular field of the hippocampus were the most vulnerable regions for neuronal loss.

We interpreted these data in the context of known neuroanatomical connections as derived from studies in the non-human primate (for review, sees Rosene and Van Hoesen, 1987). Afferents from limbic areas, and unimodal, and multimodal association cortices converge on the entorhinal cortex rather than projecting directly to the hippocampus. The stellate neurons of layer II of entorhinal cortex give rise to the perforant pathway, which "perforates" across the subiculum and across the hippocampal fissure to terminate on the distal dendrites of pyramidal cells throughout the hippocampus and, most strongly, in the outer portion of the molecular layer of the dentate gyrus (Fig. 1). A series of intrinsic intrahippocampal projections lead from the dentate gyrus to CA3 (mossy fibers), from CA3 to CA1 (Schaeffer collaterals), and from CA1 to the subiculum (ammonic-subicular pathway). Cortically directed hippocampal output arises in great part from the pyramidal neurons of the CA1/subicular field. One of the major projections of these neurons is a reciprocal projection back to layer IV of entorhinal cortex, which in turn gives rise to widespread cortical projections. Thus neuronal lesions in the neurons of layer II of entorhinal cortex would disrupt the flow of information from the cortex, via the perforant pathway, to the hippocampus, and neuronal lesions in CA1/subiculum and layer IV of entorhinal cortex would disrupt the flow of information from the hippocampal formation back towards the cortex. We postulated that these lesions would isolate the hippocampus from the cortex, and, together with the already known loss of cholinergic projections to the hippocampus, contribute to the memory impairment of Alzheimer's disease.

These observations, based in part on the studies of Ball (1978) and Kemper (1978), have since been confirmed and expanded. We later demonstrated that senile plaques often occur in the terminal zones of neurons that contain neurofibrillary tangles (Hyman et al., 1986, 1990). A detailed study of the amygdala showed that it, too, was severely affected in Alzheimer's disease in a fashion somewhat analogous to the way the hippocampus was affected. Specific nuclei (e.g. accessory basal nucleus) that had strong projections with the hippocampus tended to accumulate neurofibrillary tangles and senile plaques, whereas other nuclei were consistently spared (Hyman et al., 1990; Kromer Vogt et al., 1990). A survey of degree of pathological change in each of 49 cytoarchitectural fields in 17 hemispheres of individuals with Alzheimer's disease revealed a striking, consistent hierarchical pattern of involvement of various brain areas, with neurofibrillary tangles most severe in the entorhinal cortex, hippocampus, amygdala, and adjacent anatomically closely related perirhinal cortex, temporal pole, and posterior parahippocampal gyrus (Arnold et al., 1991). We confirmed the observation that high order association areas were more severely affected than unimodal association cortices, which in turn were more affected than the primary sensory and motor areas (Brun and Gustafson, 1976; Arnold et al., 1991; Braak and Braak., 1991). Moreover, within association cortex, the neurons of layers V and III were preferentially affected. These neurons are large pyramidal neurons that give rise to many cortico-cortical projections, and their loss was interpreted as leading to a widespread disconnection syndrome (Hof and Morrison, 1990; Hyman et al., 1990; Lewis et al., 1987; Arnold et al., 1991).

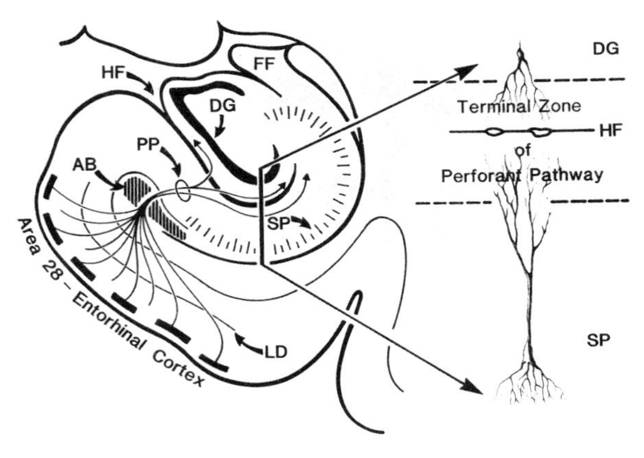

Fig. 1. Line drawing of the perforant pathway and its disruption in Alzheimer's disease. Reprinted from Hyman et al., 1986.

Our recent studies have shown that the hierarchical pattern of vulnerability seems to reflect the temporal order of various brain areas developing neurofibrillary tangles or senile plaques. We examined a cohort of Alzheimer patients who had been followed clinically and with neuropsychometric testing for 1–16 years. All these individuals had marked changes in the entorhinal cortex and CA1/subicular hippocampal fields. With increasing duration or severity of illness, the degree of pathological changes in neocortical and subcortical ascending neurotransmitter specific areas increased, and this increase, especially in the high order association cortices, was the factor that was best correlated with clinical parameters (Arriagada et al., 1992a). We have also examined a cohort of 25 presumed normal, non-demented individuals and found that many of them had small numbers of neurofibrillary tangles and senile plaques. When these were present, they were invariably in the cytoarchitectural fields and lamina predicted by the hierarchical vulnerability scheme we had defined in Alzheimer's disease itself, again with entorhinal cortex layer II being most vulnerable (Arriagada et al., 1992b; see also Hof et al., 1992; Price et al., 1991). Finally, an ongoing study of individuals with Down's syndrome of various ages also demonstrates this same pattern of vulnerability of projection neurons in these cytoarchitectural fields, and again emphasizes entorhinal lesions as an early signature of the disease (Hyman and Mann, 1991).

The perforant pathway in Alzheimer's disease as a model of deafferentation

As noted above, layer II of the entorhinal cortex provides nearly all the cortical afferents to the outer two-thirds of the molecular layer of the dentate gyrus via the perforant pathway. In Alzheimer's disease, these neurons invariably develop neurofibrillary tangles, and senile plaques frequently occur in the terminal zone (Fig. 2). Synaptic loss and re-innervation have also been demonstrated, presumably as a result of neurofibrillary tangles in the cells of origin, or senile plaques in the terminal zone, of this major projection. Evidence of dentate gyrus deafferentation in Alzheimer's disease has been established in several ways: there is a plasticity response analogous to that seen after entorhinal lesions in experimental animals in kainate receptors and acetylcholinesterase staining (Geddes et al., 1985; Hyman et al., 1987a); there is loss of the putative neurotransmitter glutamate in the terminal zone (Hyman et al., 1987b), and there is loss of immunoreactive synaptic markers in the terminal zone (Hamos et al., 1989; Masliah et al. 1991; Cabalka et al., 1992). Moreover, senile plaques have been implicated in the pathogenesis of altered synaptic transmission. Reduced axonal numbers and disrupted morphology have been found both within and around plaques and synaptic loss as determined by synaptophysin staining has been demonstrated in "mature" plaques (Masliah et al., 1990). These studies suggest that SPs through local pathogenetic mechanisms related to plaque formation, may disrupt axons en passage and contribute to deafferentation.

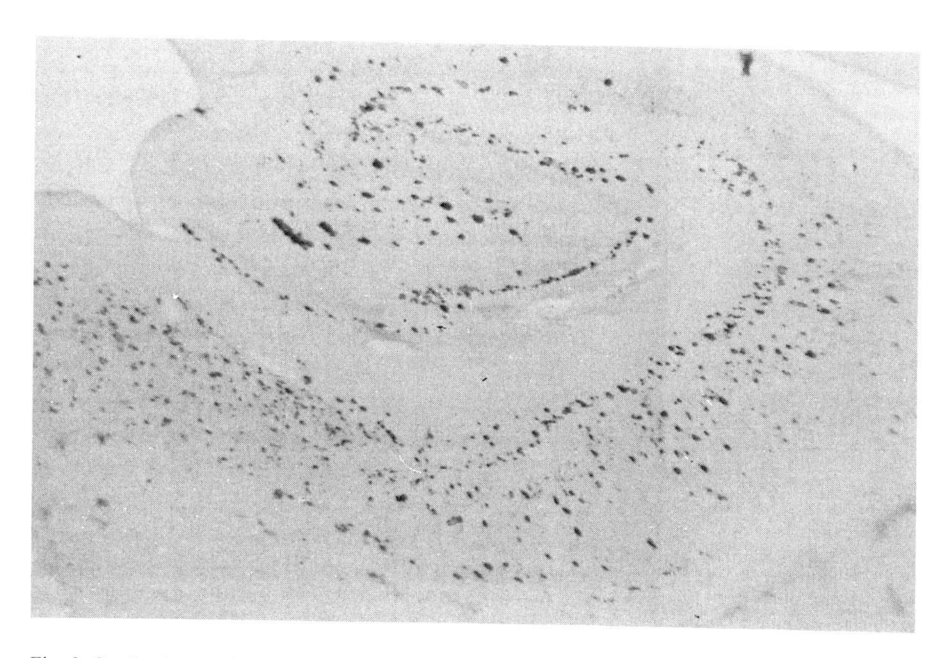

Fig. 2. Senile plaques frequently occupy the middle and outer portions of the molecular layer of the dentate gyrus, the terminal zone of the perforant pathway.

Markers of diminished afferent activity in experimental systems

Intact neuronal projections clearly play a crucial role in the early development of the nervous system and in the establishment of neural systems. Although it is widely believed that the adult central nervous system has a much more limited repertoire of responses to changes in the level of afferent activity, recent studies have shown that a remarkable degree of plasticity is still present in the adult. Deafferentation of the visual system, the somatosensory system, and the hippocampal formation (Geddes et al., 1985) in the adult animal result in alterations in the expression of a variety of gene products. In several instances, the effects are seen not only in neurons that are directly deafferented by a lesion but also trans-synaptically, sometimes several synapses away. For example, after monocular deprivation by eyelid suture or injection of tetrodotoxin, effects are seen most prominently in layer IV (which receives the major thalamic relay input) but also extend to other layers within the vertical organization of the ocular dominance columns (Jones, 1990).

Cytochrome oxidase decreases as a functional marker of deafferentation

Cytochrome oxidase, or complex IV, is the terminal enzyme of the electron transport chain and is one of the energy-generating enzymes most strongly correlated with neuronal functional activity (Wong-Riley, 1989). Cytochrome oxidase activity is highest in dendrites and cell bodies and can easily be detected in tissue using diaminobenzidine histochemistry (Kageyama and Wong-Riley, 1982). This method provides precise localization of reaction product at the regional, laminar, cellular and subcellular levels and relative changes in enzymatic activity can be quantitated by optical densitometry. The intensity of reaction product detected by optical density in rat brain is closely correlated with cytochrome oxidase measured spectrophotometrically in punch biopsies of brain (Darriet et al., 1986).

Cytochrome oxidase activity in the outer two-thirds of the molecular layer of the dentate gyrus is significantly reduced after disruption of the perforant pathway by entorhinal cortex lesions in rats (Borowsky and Collins, 1989). Similarly, in monkey, cytochrome oxidase activity is reduced within 24 h in neurons in ocular dominance columns in area 17 deprived of visual input by lid suture or enucleation (Horton and Hubel, 1981).

Cytochrome oxidase abnormalities in Alzheimer's disease

While evidence is emerging that alterations in energy metabolism exist in Alzheimer's disease (Beal, 1992), the pathophysiologic basis for these changes is not known. Hypometabolism of higher order association areas has been demonstrated by positron emission tomography (Duara et al., 1986) and reduced cytochrome oxidase activity has been reported in brain homogenates of frontal cortex (Kish et al., 1992) in Alzheimer's disease. Cytochrome oxidase histochemistry and in situ hybridization provide a tool to measure changes in energy metabolism in individual neurons in specific cytoarchitectural areas and to address whether these changes are primary or secondary. Based on our deafferentation model, we predicted that cytochrome oxidase activity would be decreased in the terminal zone of the perforant pathway, and polysynaptically in neurons downstream in the circuit. In support of this, Chandrasekaran et al. (1992) recently reported significantly decreased mRNA for two mitochondrial encoded cytochrome oxidase subunits in the dentate gyrus, CA3 and CA1 of Alzheimer's disease patients.

We recently studied the distribution and intensity of cytochrome oxidase activity in the human hippocampal formation using the Wong–Riley histochemical technique (Simonian and Hyman, 1993). Overall, in control individuals, the hippocampal formation is fairly intensely stained compared to other cortical regions. By contrast, however, in Alzheimer's disease there is a marked diminution in staining. This loss is specific for certain areas. For example, the loss of staining is greater in the outer than in the inner molecular layer of the dentate, likely because of the loss of perforant path afferents to the outer portion of the

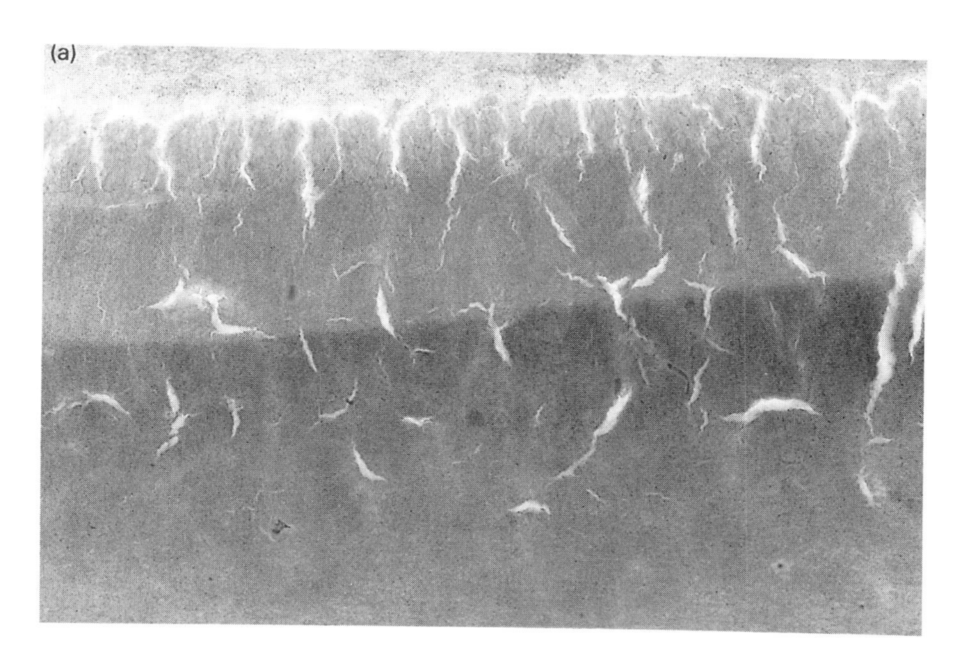

Fig. 3. Cytochrome oxidase histochemical staining on a hippocampal section from a non-demented control patient (a) and a patient with Alzheimer's disease (b). In the control brain, a dark band of cytochrome oxidase staining is seen in the outer two-thirds of the molecular layer of the dentate gyrus (DG) (arrow marks border). In the Alzheimer brain, there is a decrease in staining in the dentate molecular layer and loss of demarcation between the outer and inner layers (b). Magnification bar = 100 μm.

TABLE I

Cytochrome oxidase activity is reduced in the hippocampal formation, but not visual cortex, in Alzheimer's disease (from Simonian and Hyman, 1993)

	Diffuse density ($X \pm$ SEM)	
	Control	Alzheimer's disease
Dentate gyrus: outer 2/3	0.88 ± 0.10	0.60 ± 0.05*
Dentate gyrus: inner 1/3	0.78 ± 0.07	0.62 ± 0.05*
CA4	0.83 ± 0.08	0.62 ± 0.05*
CA3	0.86 ± 0.07	0.65 ± 0.05*
CA1	0.85 ± 0.08	0.65 ± 0.05*
Area 17	0.69 ± 0.10	0.71 ± 0.05

*$p < 0.05$.

molecular layer (Fig. 3). In addition, CA3, CA1 and subiculum also contain diminished activity. This may be due to a transynaptic effect, with diminished excitatory input to the dentate gyrus reflected by diminished activity throughout the hippocampal projection fields. By contrast, examination of the primary visual cortex in Alzheimer's disease showed no difference in cytochrome oxidase staining compared to controls, consistent with the clinical and anatomical observations suggesting that primary visual input is not affected in Alzheimer's disease (Table I).

Glut 3 glucose transporter isoform is the primary neuronal glucose transport protein: loss in the perforant pathway terminal zone in Alzheimer's disease may reflect diminished energy metabolism

At least five glucose transport proteins have been identified that provide the molecular basis for facilitated glucose transport across cell membranes. Of these, two appear to be predominant in brain: Glut 1 in brain capillaries, responsible for transporting glucose across the tight junctions of the blood–brain barrier, and Glut 3, present on neurons, presumably responsible for regulating glucose uptake to neurons (Mantych et al., 1992). Neurons are dependent on glucose for most of their energy needs, and functional activity in creases glucose utilization. It is of course this principle that underlies the use of 2-deoxyglucose for PET studies.

Mantych et al., (1992) showed that the Glut 3 transporter molecule is located principally in neurons in the human brain. We have taken advantage of this to examine whether or not alterations in the pattern or level of expression occur in Alzheimer's disease. We hypothesized that deafferented areas would down-regulate the amount of Glut 3 that was expressed. Our preliminary analyses suggest that there is a loss of Glut 3 immunoreactivity (East Acres Biologicals, Southbridge, MA, diluted 1:500) in the Alzheimer's disease hippocampal formation that in many ways parallels the type of changes we have seen for cytochrome oxidase. In particular, the outer portion of the molecular layer of the dentate gyrus, i.e. the perforant pathway terminal zone, shows a dramatic loss of Glut 3 immunoreactivity in each of the eight Alzheimer brains we have examined to date.

NADPH diaphorase (nitric oxide synthase) activity is diminished in the perforant pathway terminal zone in Alzheimer's disease

We have recently examined the population of nitric oxide synthase containing neurons in Alzheimer's disease to determine whether this population of neurons, which are resistant to a variety of metabolic insults, NMDA excitotoxicity, and to degeneration in the striatum in Huntington's disease, was also spared from degeneration in Alzheimer's disease. Indeed, in the hippocampal formation, the number of nitric oxide synthase immunoreactive neurons was unchanged in Alzheimer's disease (Hyman et al., 1992).

However, we noticed that the processes of these neurons appeared to be distorted and somewhat atrophic. We therefore examined the distribution and intensity of staining of NADPH diaphorase (a histochemical stain that identifies nitric oxide synthase). Again, there was an alteration in the pattern of staining, with the staining in the perforant pathway terminal zone depleted in Alzheimer's disease (Rebeck et al., 1993).

Alteration of substance P in the perforant pathway terminal zone in Alzheimer's disease

Substance P levels are known to inversely reflect levels of afferent activity in the visual cortex in models of acute deprivation (Jones, 1990). We postulated that the same may be true in the hippocampal formation in Alzheimer's disease. There have been relatively few prior studies of substance P immunoreactivity in Alzheimer's disease. Bouras et al. (1990) describe "significantly reduced substance P like immunoreactivity in the neocortical areas and in the hippocampus". Quigley and Kowall (1991) also reported that substance P neurons are depleted in Alzheimer's disease and noted a band of terminal like staining in the molecular layer of the dentate gyrus.

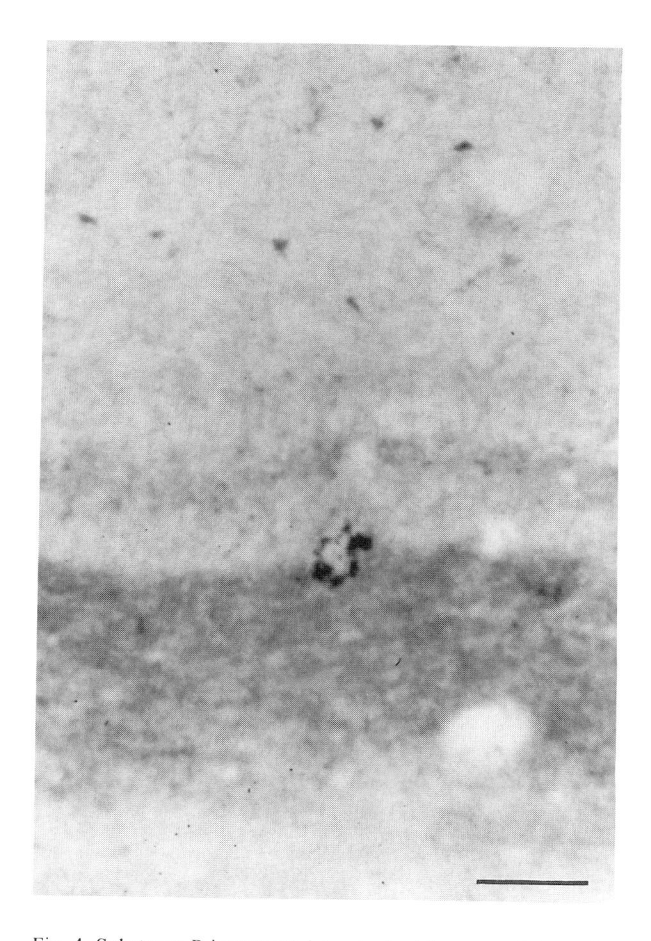

Fig. 4. Substance P immunostaining in the hippocampal formation reveals three bands of terminal-like staining in the molecular layer of the dentate gyrus: an intense supragranular band, a somewhat lighter band in the inner one-third of the molecular layer, and a heavier band in the outer two-thirds of the molecular layer. Magnification bar = 100 μm.

We used substance P immunohistochemistry (Accurate Immunochemicals, 1:5000) to assess the pattern of substance P immunostaining in the hippocampal formation. We also found a distinct band of immunoreactivity in the dentate gyrus molecular layer in both control and Alzheimer individuals. There is an intense supragranular band, a light stain in the inner one-third of the dentate molecular layer, and a more intense band in the outer two-thirds (Fig. 4). We used a Bioquant image analysis system to quantitate the amount of immunostaining in these bands. Surprisingly, in contrast to Bouras' description, our quantitative analysis of 10 Alzheimer and 10 control individuals shows an increase in staining intensity in all three bands in Alzheimer's disease (Fig. 5). This may reflect reinnervation of the deafferented areas by remaining intact afferents, such as the substance P positive neurons of the hilus.

This type of remodelling of afferents has also been noted for acetylcholinesterase activity (Geddes et al., 1985; Hyman et al., 1987a) and for some glutamate receptor binding (Geddes et al., 1985). However, we have recently demonstrated by immunohistochemistry that at least some types of glutamate receptors (GluR1, GluR2,3 and GluR4) are unaltered in the dentate gyrus molecular layer in Alzheimer's disease (Hyman et al., 1994). Thus, neurotransmitter system remodelling occurs in several, but not all inputs to the deafferented hippocampus in Alzheimer's disease.

Conclusion

Anatomical evidence suggests loss of projection neurons and widespread disconnection of the hippocampal formation and association cortices in Alzheimer's disease. This is perhaps most pronounced in the perforant pathway, the projection from the entorhinal cortex to the dentate gyrus which is uniformly destroyed in Alzheimer's disease. We have recently developed new methodologies to examine the functional

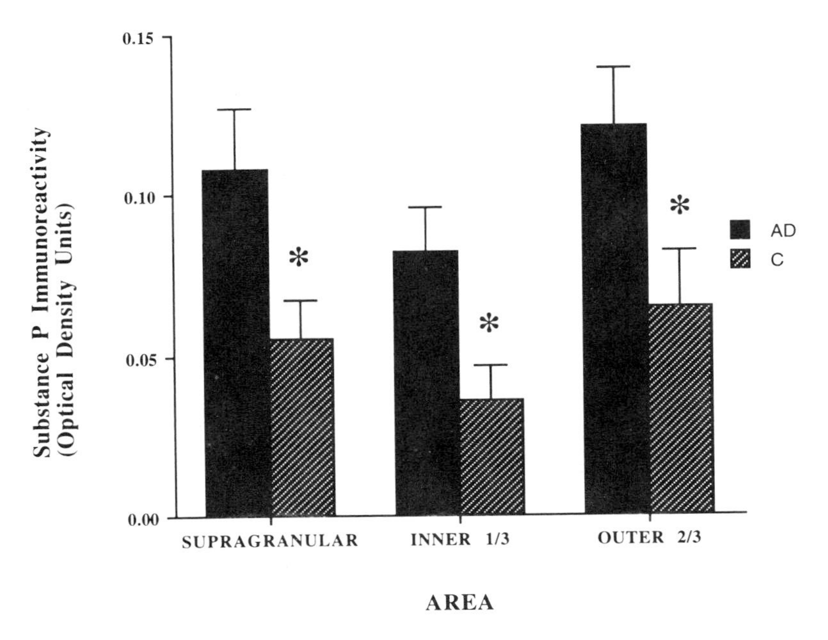

Fig. 5. Substance P immunoreactivity increases in the dentate gyrus in Alzheimer's disease. Optical density measurements of substance P immunoreactivity was measured using a Bioquant image analysis system in the supragranular, inner one-third, and outer two-thirds of the molecular layer of the dentate gyrus. Ten Alzheimer and 10 control individuals were studied. There is a statistically significant increase in staining in each area in Alzheimer's disease ($p < 0.05$).

integrity of hippocampal projections in the post mortem human brain, and have found that there is impairment both of neural elements that are directly affected by the disease process (i.e. by neurofibrillary tangles) as well as "downstream" of the pathological changes. These results highlight the effect of the disease process in disrupting neural systems, no doubt causing the impairment of memory function that is so prominent in patients with Alzheimer's disease.

Acknowledgments

We thank Ted Elvhage and Steven Harr for excellent technical assistance. Supported by NIH AG08487, and grants from the Brookdale Foundation and the Alzheimer Association. We thank the Massachusetts Alzheimer Disease Research Center Brain Bank (Dr. E.T. Hedley-Whyte, Director) for tissue used in these studies, S. Melanson for assistance with the manuscript and H. West for photographic expertise.

References

Arnold, S.E., Hyman, B.T., Flory, J., Damasio, A.R. and Van Hoesen, G.W. (1991) The topographical and neuroanatomical distribution of neurofibrillary tangles and neuritic plaques in the cerebral cortex of patients in Alzheimer's disease. *Cerebral Cortex*, 1: 103–116.

Arriagada, P.V., Growdon, J.H., Hedley-White, E.T. and Hyman, B.T. (1992a) Neurofibrillary tangles but not senile plaques parallel duration and severity of Alzheimer's disease. *Neurology*, 42: 631–639.

Arriagada, P.V., Marzloff, K.M. and Hyman, B.T. (1992b) The distribution of Alzheimer type pathological changes in nondemented elderly individuals matches the pattern in Alzheimer disease. *Neurology*, 42: 1681–1688.

Ball, M.J. (1978) Topographic distribution of neurofibrillary tangles and granulovacuolar degeneration in hippocampal cortex of aging and demented patients. *Acta Neuropathol.*, 42: 73–80.

Beal, M.F. (1992) Does impairment of energy metabolism result in excitotoxic neuronal death in neurodegenerative illnesses? *Ann. Neurol.*, 31: 119–130.

Borowsky, W. and Collins, R.C. (1989) Histochemical changes in enzymes of energy metabolism in the dentate gyrus accompany deafferentation and synaptic reorganization. *Neuroscience*, 33: 253–262.

Bouras, C., Vallet, P.G., Hof, P.R., Charnay, Y., Golaz, J. and Constantinidis, J. (1990) Substance P immunoreactivity in Alzheimer disease: a study in cases presenting symmetric or asymmetric cortical atrophy. *Alzheimer Dis. Assoc. Disord.*, 4: 24–34.

Braak, H. and Braak, E. (1991) Neuropathological staging of Alzheimer-related changes. *Acta Neuropathol.*, 82: 239–259.

Brun, A. and Gustafson, L. (1976) Distribution of cerebral degeneration in Alzheimer's disease. *Arch. Psychiatr. Nervenk*, 223: 15–33.

Cabalka, L.M., Hyman, B.T., Goodlett, C.R., Ritchie, T.C. and Van Hoesen, G.W. (1992) Alteration in the pattern of nerve terminal protein immunoreactivity in the perforant pathway in Alzheimer's disease and in rats after entorhinal lesions. *Neurobiol. Aging*, 13: 283–291.

Chandrasekaran, K., Stoll, J., Brady, D.R., and Rapoport, S.I. (1992a) Distribution of cytochrome oxidase(COX) activity and mRNA in monkey and human brain: COX mRNA distribution correlates with neurons vulnerable to Alzheimer pathology. *Neuroscience*, 557 (Abstr).

Darriet, D., Der, T. and Collins, R.C. (1986) Distribution of cytochrome oxidase in rat brain: studies with diaminobenzidine histochemistry in vitro and [^{14}C] cyanide tissue labeling in vivo. *J. Cereb. Blood Flow. Metab.*, 6: 8–14.

Duara, R., Grady, C., Haxby, J., Sundaram, M., Cutler, N.R., Heston, L., Moore, A., Schlageter, N., Larson, S. and Rapoport, S.I. (1986) Positron emission tomography in Alzheimer's disease. *Neurology*, 36: 879–887.

Geddes, J.W., Monaghan, D.T. and Cotman, C.W. (1985) Plasticity of the hippocampal circuitry in Alzheimer's disease . *Science*, 230: 1179–1181.

Hamos, J.E., DeGennaro, L.J. and Drachman, D.A . (1989) Synaptic loss in Alzheimer's disease and other dementias. *Neurology*, 39: 355–361.

Hof, P.R. and Morrison, J.H. (1990) Quantitative analysis of a vulnerable subset of pyramidal neurons in Alzheimer's disease: II. Primary and secondary visual cortex. *J. Comp. Neurol.*, 301: 55–64.

Hof, P.R., Bierer, L.M., Perl, D.P., Delacourte, A., Buee, L., Bouras, C. and Morrison, J.H. (1992) Evidence of early vulnerability of the medial and inferior aspects of the temporal lobe in an 82 year old patient with preclinical signs of dementia. *Arch. Neurol.*, 49: 946–953.

Horton, J.C. and Hubel, D.H. (1981) Regular patchy distribution of cytochrome oxidase staining in primary visual cortex of macque monkey. *Nature*, 292: 762–764.

Hyman, B.T. and Mann, D.M.A. (1991) Alzheimer type pathological changes in Down's syndrome individuals of various ages. In: K. Iqbal, D.R.C. McLachlan, B. Winblad and H.M. Wisniewski (Eds.), *Alzheimer's Disease: Basic Mechanisms, Diagnosis, and Therapeutic Strategies*, Wiley New York, pp. 105–113.

Hyman, B.T., Damasio, A.R., Van Hoesen, G.W. and Barnes, C.L. (1984) Alzheimer's disease: cell specific pathology isolates the hippocampal formation. *Science*, 298: 83–95.

Hyman, B.T., Van Hoesen, G.W., Kromer, L.J. and Damasio, A.R. (1986) Perforant pathway changes and the memory impairment of Alzheimer's disease. *Ann. Neurol.*, 20: 472–481.

Hyman, B.T., Kromer, L.J. and Van Hoesen, G.W. (1987a) Reinnervation of the hippocampal perforant pathway zone in Alzheimer's disease. *Ann. Neurol.* 21: 259–267.

Hyman, B.T., Van Hoesen, G.W. and Damasio, A.R. (1987b) Alzheimer's disease: glutamate depletion in perforant pathway terminals. *Ann. Neurol.*, 22: 37–40.

Hyman, B.T., Van Hoesen, G.W., Kromer, L.J. and Damasio, A.R. (1990) Memory-related neural systems in Alzheimer's disease: an anatomic study. *Neurology*, 40: 1721–1730.

Hyman, B.T., Marzloff, K.M., Wenniger, J.J., Dawson, T.M., Bredt, D.S. and Snyder, S.H. (1992) Relative sparing of nitric oxide synthase containing neurons in the hippocampal formation in Alzheimer's disease. *Ann. Neurol.*, 32: 818–821.

Hyman, B.T., Penney, J.B., Blackstone, C.D. and Young, A.B. (1994) Localization of non-*N*-methyl-D-aspartate glutamate receptors in normal and Alzheimer hippocampal formation. *Ann. Neurol.*, 35: 31–37.

Jones, E.G. (1990) The role of afferent activity in the maintenance of primate neocortical function. *J. Exp. Biol.*, 153: 155–176.

Kageyama, G.H. and Wong-Riley, M.T.T. (1982) Histochemical localization of cytochrome oxidase in the hippocampus: correlation with specific neuronal types and afferent pathways. *Neuroscience*, 7: 2337–2361.

Kemper, T.L. (1978) Senile dementia: a focal disease in the temporal lobe. In: K. Nandy (Ed.), *Senile Dementia: A Biomedical Approach*, Elsevier, Amsterdam, pp. 105–113.

Kish, S.J., Bergeron, C., Rajput, A., Dozie, S., Mastrogiacomo, F., Chang, L.J., Wilson, J.M., DiStefano, L.M. and Nobrega, J.N. (1992) Brain cytochrome oxidase in Alzheimer's disease. *J. Neurochem.*, 59: 776–779.

Kromer Vogt, L.J., Hyman, B.T., Van Hoesen, G.W. and Damasio, A.R. (1990) Pathological alterations in the amygdala in Alzheimer's disease. *Neuroscience*, 37: 377–385.

Lewis, D.A., Campbell, J.M., Terry, R.D. and Morrison, J.H. (1987) Laminar and regional distributions of neurofibrillary tangles and neuritic plaques in Alzheimer's disease: a quantitative study of visual and auditory cortices. *J. Neurosci.*, 7: 1799–1808.

Mantych, G.J., James, D.E., Chung, H.D. and Devaskar, S.U. (1992) Cellular localization and characterization of Glut 3 glucose transporter isoform in human brain. *Endocrinology*, 131: 1270–1278.

Masliah, E., Terry, R.D., Mallory, B.S., Alford, M. and Hansen, L. (1990) Diffuse plaques do not accentuate synaptic loss in Alzheimer's disease. *Am. J. Pathol.*, 137: 1293–1297.

Masliah, E., Terry, R.D., Alford, M., DeTeresa, R. and Hansen, L.A. (1991) Cortical and subcortical patterns of synaptophysin immunoreactivity in Alzheimer's disease. *Am. J. Pathol.*, 138: 235–246.

Price, J.L., David, P.B., Morris, J.C. and White, D.L. (1991) The distribution of tangles, plaques and related immunohistochemical markers in healthy aging and Alzheimer's disease. *Neurobiol. Aging*, 12: 295–312.

Quigley, B.J., Jr. and Kowall, N.W. (1991) Substance P-like immunoreactive neurons are depleted in Alzheimer's disease cerebral cortex. *Neuroscience*, 41: 41–60.

Rebeck, G.W., Marzloff, K.M. and Hyman, B.T. (1993) The pattern of NADPH-diaphorase staining, a marker of nitric oxide synthase activity, is altered in the perforant pathway terminal zone in Alzheimer's disease. *Neurosci. Lett.*, 152: 165–168.

Rosene, D.L. and Van Hoesen, G.W. (1987) The hippocampal formation of the primate brain. In: E.G. Jones and A. Peters (Eds.), *The Cerebral Cortex*, Vol. 6, Plenum, New York, pp. 345–456.

Simonian, N.A. and Hyman, B.T. (1993) Functional alterations in Alzheimer's disease: diminution of cytochrome oxidase in the hippocampal formation. *J. Neuropathol. Exp. Neurol.*, 52: 580–585.

Wong-Riley, M.T.T. (1989) Cytochrome oxidase: an endogenous metabolic marker of neuronal activity. *Trends Neurosci.*, 12: 94–101.

F. Bloom (Editor)
Progress in Brain Research, Vol. 100

CHAPTER 31

The search for a manic depressive gene: from classical to molecular genetics

J. Mendlewicz

Department of Psychiatry, Free University Clinics of Brussels, Erasme Hospital, route de Lennik 808, 1070 Brussels, Belgium

A growing number of researchers in the last decade have addressed the issue of genetic factors in affective illness and its various subtypes (for review see Mendlewicz et al., 1993). The twin method allows comparison of concordance rates for a trait between sets of monozygotic (MZ) and dizygotic (DZ) twins in bipolar manic depression. The concordance rates in MZ twins vary between 50 and 90% (mean 70%) as compared to 0–39% in DZ twins (mean 20%). These results strongly support the presence of a genetic factor in the etiology of bipolar disorder (BP). Among pairs of identical twins who had been reared apart since early childhood and who were characterized by at least one of the twins being diagnosed as affectively ill, eight out of 12 pairs were concordant for the disease, an observation suggesting that the predisposition to BP will usually express itself regardless of the early environment.

In adoption studies, depressive disorders in adulthood are significantly more frequent in adopted away offspring of affectively ill biological parents compared to adoptees whose biological parents were well or had other psychiatric conditions (Cadoret, 1978). Similarly, psychopathology of the affective spectrum is found more frequently in biological parents of bipolar adoptees than in their adoptive parents (Mendlewicz and Rainer, 1977).

Most of the early studies on BP have shown that this illness tends to be familial. The lifetime risk for the disease in relatives of bipolar probands is significantly higher than the risk in the general population (about ten times higher).

Bipolar patients show a greater genetic loading for affective disorders with more hypomanic temperaments in relatives. Moreover, bipolar and unipolar illnesses are present in the relatives of bipolar patients whereas only unipolar illnesses were present in the relatives of unipolar patients. After reviewing all family studies, the risk for manic depressive illness in the relatives of bipolar patients can be estimated at somewhere between 15 and 35%. There is, however, a large proportion of relatives of bipolar probands who exhibit unipolar illness only.

Linkage analysis is a promising method to study the genetics of manic depressive illness (BPI). It explores a major single genetic transmission, and evaluates the degree of co-segregation between genetic markers, including deoxyribonucleic acid (DNA) polymorphisms and illness traits in informative pedigrees. This method tests the hypothesis of a potential linkage relationship between a known genetic marker and a trait known to be genetically determined, but not yet mapped on the chromosome. DNA polymorphisms in various regions of the human genome are being explored using the DNA recombinant method and, more recently, the polymerase chain reaction (PCR) for gene amplification is also being used.

Unfortunately, several factors limit the results of linkage analysis. BP is a complex disorder lacking clear-cut mendelian patterns of inheritance. Although the true mode of inheritance may involve the interaction of alleles at more than one locus, the major contributing loci may still be detected by assuming a single mendelian locus model in the linkage analysis.

Assumptions are also to be made on numerous parameters such as gene frequency, penetrance, genetic heterogeneity, variable age of onset and diagnostic uncertainties. Since the underlying genetic model is not known, penetrance and allele frequency may be misspecified and may reduce the linkage results. Lack of replication between studies is often attributed to genetic heterogeneity. The latter occurs when one disease phenotype is caused by different mutant alleles at different loci.

The vulnerability to affective illness could be linked to more that one gene, and for such common disorders as affective illness, phenocopies (or false positive) may also be present in large pedigrees. Because of variable age of onset, relatives of probands may be diagnosed as unaffected at the time of study, and may become affectively ill in follow-up studies, resulting in a significant change in linkage scores. Moreover, co-morbidity of other psychiatric disorders with depressive illness may modify the expression of the affective disorder which may result in misclassification. Other factors such as assortative mating, the change in the rate of mental illness over time (cohort effect) and laboratory errors may also bias the results.

Linkage analysis results may be improved by defining age-specific and cohort-specific penetrances. Because of assortative mating, spouses and their relatives should be evaluated systematically, and families with evidence of illness on both paternal and maternal sides should be excluded from linkage analysis or should be analyzed separately. Notwithstanding these limitations, linkage with DNA markers in manic depression has been studied in three distinct chromosomal regions: the subterminal region of the long arm of the X chromosome (Xq26–28), and regions of the short (11p15) and long (11q21–23) arms of chromosome 11. So far, two main hypotheses of genetic transmission for affective illness have been tested: an X-linked and an autosomal dominant transmission.

Besides X-linked transmission, a major autosomal dominant gene with reduced penetrance for bipolar illness has been postulated. Indeed, a preponderance of affected females, as compared to males in first-degree relatives has not been found in some studies, and a male to male transmission of the disease is present in some families (Mendlewicz et al., 1993). Although it is nevertheless a rare event in the kindreds of bipolar probands, it has been observed in about 10% of most samples. The hypothesis of an autosomal transmission has been investigated in association studies with the O blood group located on chromosome 9, as well as linkage studies on chromosome 6 with the human leucocyte antigen (HLA) haplotypes and on chromosome 11 with DNA markers for the following genes: D2 dopamine receptor, tyrosinase, C-Harvey-Ras-A (HRAS) oncogene, insulin (ins), and tyrosine hydroxylase (TH).

The O blood group has been found to be more frequent in BP patients in some studies. Although poorly understood, the association between a blood group factor and a major psychosis indicates that the ABO genotype located on 9q34 may play a role in the predisposition to BP. Although a linkage to HLA genes located on the short arm of chromosome 6 has been proposed for affective illness, it has not been confirmed. On the long arm of chromosome 11, a balanced translocation from 11q23.3 to chromosome 9p22 was described in some bipolar patients and in some others, a translocation from region 11q21–22 to region q43 of chromosome 1 was reported suggesting, a linkage between psychiatric illness and genes at the site of the translocation. The human D2 dopamine receptor gene located on 11q22–23, and the tyrosinase gene also located on the long arm of chromosome 11 may be close to the translocation point observed. Consequently, linkage analysis between these markers and BP was performed. However, no evidence of linkage has been found so far. Concerning the short arm of chromosome 11, a positive linkage between BP and the HRAS oncogene as well as the INS marker on the short arm of chromosome 11 (11p15) was reported in studying a large pedigree of the old order Amish Community (Egeland et al., 1987). However, linkage analysis in American bipolar pedigrees of non-Amish origin, in other European pedigrees of bipolar disorders, and in pedigrees of unipolar disorder could not confirm these results. Additionally, the probability of linkage of affective illness to the 11p15 region of chromosome 11 was almost excluded by a re-analysis of the original Amish pedigree with two lateral extensions (Kelsoe et al., 1989). Because of a close link between the genes coding for TH, INS and HRAS loci on chromosome 11, linkage between BPI and the TH locus has also been investigated in BP, but with negative results so far (Mendlewicz et al., 1991b). In association studies, positive results between the TH gene and affective illness have been reported but not yet confirmed.

Rosanoff et al. (1934) first postulated a chromosome X transmission for bipolar illness, which was also suggested by studies reporting a sex ratio of two females to one male in the distribution of bipolar illness and an observed excess of females over males in the relatives of bipolar probands. Colour blindness (CB) and glucose-6-phosphate dehydrogenase (G6PD) deficiency are two loci known to be located in the region Xq28 on the long arm of the X chromosome.

Previous studies have provided evidence of a linkage between these loci and a dominant gene involved in the transmission of manic depressive illness in some families (Mendlewicz et al., 1992). Additionally (Mendlewicz et al., 1987, 1991a), in 11 informative pedigrees, Lucotte et al. (1992) in one French pedigree, reported DNA results suggestive of a linkage between MDI and blood coagulation factor 9 (F9) in the Xq27 region. In addition, Gill et al. (1992) and Craddock and Owen (1992) reported segregation of affective disorders with Christmas disease and new data from Berretini (personal communication) indicated the presence of an X-linked gene in some MDI families proximal of the F9 gene. However, the logarithm of the odds ratio (LOD scores) for F9 and MDI were not very robust in our study (Mendlewicz et al., 1991a). A subsequent study using additional pedigrees could not confirm linkage with factors 9 as was also the case in the study of Bredbacka et al. (1993). Nevertheless, the results suggested the presence of a MDI dominant gene located in the region Xq27–Xq28 (Mendlewicz et al., 1987). However, X-linked transmission has not been observed in all families studied with classical markers (Gershon et al., 1979) or DNA polymorphisms (Berretini et al., 1990; Gejman et al., 1990; Baron et al., 1993; Bredbacka et al., 1993). Moreover, a possible linkage between the fragile X syndrome (Fra-x) and affective illness has been observed (Mendlewicz and Hirsch, 1991).

These contradictory findings are usually attributed to genetic heterogeneity. Accordingly, only a subgroup of bipolar pedigrees will show close linkage to the X chromosome and thus carry the X-linked gene.

Another hypothesis about these inconsistent findings is the presence of diagnostic uncertainties, and the possibility of spurious linkage has also been discussed (Mendlewicz et al., 1991c).

In order to address the issue of diagnostic uncertainties, previous X-linkage data of our group were re-

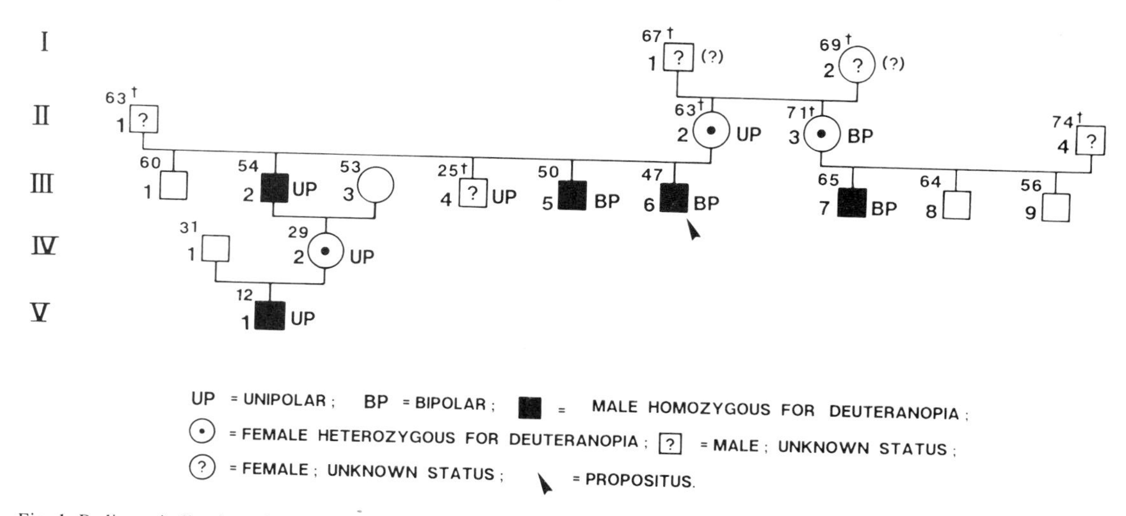

Fig. 1. Pedigree indicative of X-linkage of manic depressive illness.

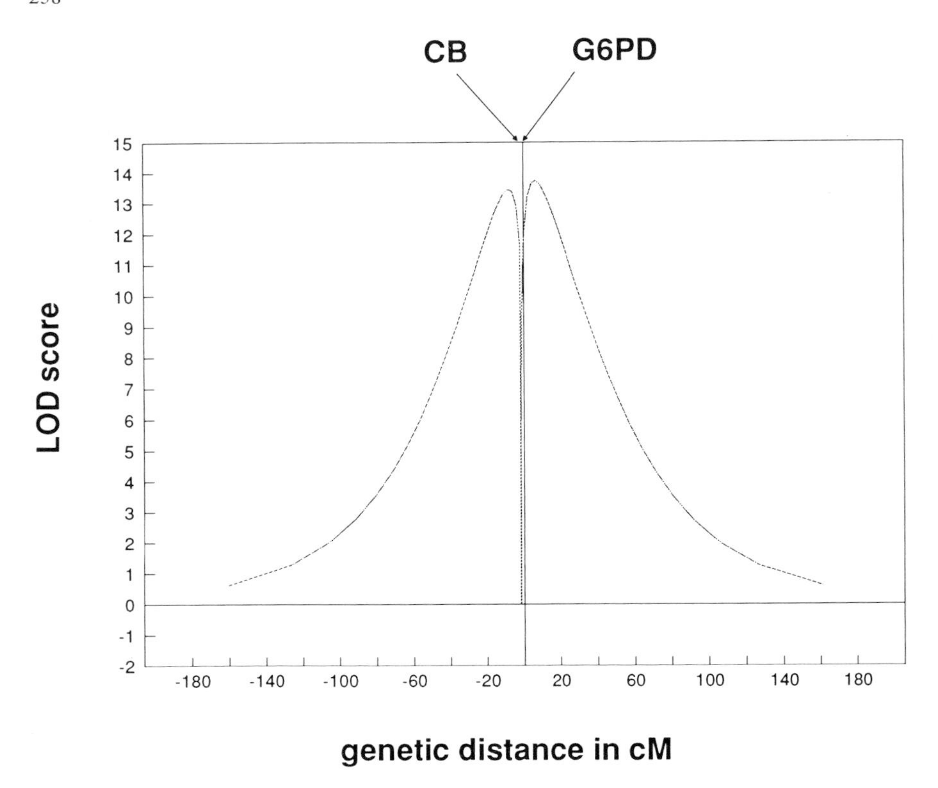

Fig. 2. Linkage of MDI to the X-chromosome: model BP + UP.

analysed by using narrow and broad definitions of MDI.

Thirty-five families of bipolar probands informative for X-linked transmission who participated in previous studies were included in the linkage re-analysis: 23 pedigrees had been analyzed for color blindness, 1 pedigree for glucose-6-phosphate dehydrogenase and 11 pedigrees for F9 linkage. A section of a pedigree illustrating the co-segregation of deuteranopia and bipolar-unipolar illness in successive generations of a family informative for linkage analysis is provided in Fig. 1.

Analyses were performed using two models (M1 and M2). M1 assumes that unipolar (UP) and bipolar (BP) individuals are affected. M2 is more conservative than M1 and assumes that only BP individuals are affected.

The sum of the two-point LOD scores with CB and G6PD genes under classification M1 (BP + UP) gives a maximum LOD score of 13.8 at 6.2 centiMorgan (cM) to the left of the CB gene and a maximum LOD score of 13.8 at 6.2 cM to the right of the G6PD gene (Fig. 2).

Two-point linkage analyses results thus confirm our previous finding of a conclusive linkage between the MDI and the CB-G6PD genes.

The present re-analysis confirms the X-linked hypothesis of a MDI genetic transmission, but this susceptibility gene may probably account for only a fraction of bipolar patients, and may represent one of the major genes involved in the genetic vulnerability to bipolar disorder. In a recent study, an X-linked susceptibility gene in the transmission of affective disorders has also been postulated by Vailland et al. (1992) who demonstrated an association between affective disorders in males and early mortality in their maternal grandfathers.

Although non-X-linked forms of the illness are most likely to be present, a major susceptibility gene for BP on chromosome 11 is at present not confirmed.

Nevertheless, results of linkage studies are limited by several factors such as uncertainty in mode of inheritance, genetic heterogeneity, methodological and statistical problems originating from the use of several disease definitions and genetic models, ascertainment bias, number of marker loci, assortative mating, non-genetic environmental factors and cohort effect. Hence, the linkage results should be interpreted with caution to avoid making premature claims of linkage because of the possibility of a spurious linkage especially when dealing with selective ascertainment of frequent disorders and common genetic traits, as seems to be the case in the area of psychiatric disorders. Because of the complex nature and the heterogeneity of MDI, replication of results and re-analysis of existing data will have to be emphasized in future studies through collaborative efforts in such projects as carried out by the European Science Foundation (ESF) or the National Institute for Mental Health (NIMH) in the United States. Hopefully, a major susceptibility locus for BP can be detected through a systematic screen covering the whole human genome together with non-parametric tests (sib pair method, affected pedigree member method), and population based association studies.

Acknowledgments

The support of the Association for Mental Health Research is acknowledged.

References

Baron, M., Freimer, N.F., Risch, N., Lerer, B., Alexander, J.R., Straub, R.E., Asokan, S., Das, K., Peterson, A., Amos, J., Endicott, J. and Gilliam, C. (1993) Diminished support for linkage between manic depressive illness and X-chromosome markers in three Israeli pedigrees. *Nature Genet,*, 3: 49–55.

Berretini, W.H., Goldin, L.R., Gelernter, J., Gejman, P.V., Gershon, E.S. and Detera-Wadleigh, S. (1990) X-Chromosome markers and manic depressive illness: rejection of linkage to Xq28 in nine bipolar pedigrees. *Arch. Gen. Psychiatry*, 47: 366–373.

Bredbacka, P.E., Pekkarinen, P., Peltonen, L. and Lönnqvist, J. (1993) Bipolar disorder in an extended pedigree with a segregation pattern compatible with X-linked transmission: exclusion of the previously reported linkage to F9. *Psychiatric Genet.*, 3: 79–87.

Cadoret, R.J. (1978) Evidence for genetic inheritance of primary affective disorders in adoptees. *Am. J. Psychiatry*, 134: 463–466.

Craddock, N. and Owen, M. (1992) Christmas disease and major affective disorder. *Br. J. Psychiatry*, 160: 715.

Egeland, J.A., Gerhard, D.S., Paul, D.C., Sussex, J.N., Kidd, K.K., Allen, C.R., Hostetter, A.M. and Housman, D.E. (1987) Bipolar affective disorder linked to DNA markers on chromosome 11. *Nature*, 325: 783–787.

Gejman, P.V., Detera-Wadleigh, S., Martinez, M.M., Berretini, W.H., Goldin, L.R., Gelernter, J., Hsieh, W.-T. and Gershon, E.S. (1990) Manic depressive illness not linked to factor IX region in an independent series of pedigrees. *Genomics*, 8: 648–655.

Gershon, E.S., Targum, S.D., Matthysse, S. and Bunney, W.E. (1979) Color blindness not closely linked to bipolar illness. *Arch. Gen. Psychiatry*, 36: 1423–1430.

Gill, M., Castle, D. and Duggan, C. (1992) Cosegregation of Christmas disease and major affective disorder in a pedigree. *Br. J. Psychiatry*, 160: 112–114.

Kelsoe, J.R., Ginns, E.I., Egeland, J.A., Gerhard, D.S., Gostein, A.M., Bale, S.J., Pauls, D.L., Long, R.J., Kidd, K.K., Conte, G., Housman, D.E. and Paul, S.M. (1989) Re-evaluation of the linkage relationship between chromosome 11p loci and the gene for bipolar affective disorder in the Old Order Amish. *Nature*, 342: 238–243.

Lucotte, G., Landoulsi, A., Berriche, S., David, F. and Babron, M.C. (1992) Manic depressive illness is linked to factor IX in a french pedigree. *Ann. Génét.*, 35: 93–95.

Mendlewicz, J. and Hirsch, D. (1991) Bipolar manic depressive illness and X-fragile syndrome. Biol. Psychiatry, 29: 295–308.

Mendlewicz, J. and Rainer, J.D. (1977). Adoption study supporting genetic transmission in manic depression illness. *Nature*, 268: 327–329.

Mendlewicz, J., Simon, P., Sevy, S., Charon, F., Brocas, H., Legros, S. and Vassart, G. (1987) A polymorphic DNA marker on X chromosome and manic depression. *Lancet*, i: 1230–1232.

Mendlewicz, J., Leboyer, M., De Bruyn, A., Malafosse, A., Sevy, S., Hirsch, D., Van Broeckhoven, C. and Mallet, J. (1991a) Absence of linkage between chromosome 11p15 markers and manic depressive illness in a Belgian pedigree. *Am. J. Psychiatry*, 148: 12.

Mendlewicz, J., Sevy, S., Charon, F. and Legros, S. (1991b) Manic depressive illness and X chromosome. *Lancet*, 338: 1213.

Mendlewicz, J., Sandkuyl, L.A., De Bruyn, A., Van Broeckhoven, C. (1991c) X-linkage in bipolar illness (letter). Biol. Psychiatry, 29: 730–734.

Mendlewicz, J., Sevy, S. and Mendelbaum, K. (1993) Molecular genetics in affective illness. *Life Sci.*, 52: 231–242.

Rosanoff, A.H., Handy, L.M. and Rosanoff-Plesset, I.B.A. (1934) The etiology of manic depressive syndromes with special reference to their occurrence in twins. *Am. J. Psychiatry*, 91: 725–762.

Vaillant, G.E., Roston, D. and McHugo, G.J. (1992) An intriguing association between ancestral mortality and male affective disorder. *Arch. Gen. Psychiatry*, 49: 709–715.

F. Bloom (Editor)
Progress in Brain Research, Vol. 100

CHAPTER 32

Age, sex and light: variability in the human suprachiasmatic nucleus in relation to its functions

D.F. Swaab and M.A. Hofman

Graduate School of Neurosciences Amsterdam, Netherlands Institute for Brain Research, Meibergdreef 33, 1105 AZ Amsterdam, The Netherlands

Introduction

In the last decade, when dealing with the subject of the human brain, the major research effort has undoubtedly been the clinically highly relevant comparison of material from neurological patients with that of controls. For such studies, controls are generally matched with the pathological cases for a few factors, e.g. postmortem interval, age and sex. However, the possibility of extracting fundamental information from the controls has generally attracted little or no attention. On the contrary, studies on human brain structures in controls are generally discouraged in view of the large variability often obtained in these investigations. Our 10 years of experience in studying human hypothalamic nuclei using a combination of immunocytochemistry and morphometrics have confirmed the presence of a considerable variability in functional-anatomical parameters. However, it has also been shown that this variability is not necessarily of a disturbing nature. Quite the reverse, it may bear important functional information. In this chapter, the idea is presented that variability of data on the human brain may become an increasingly useful tool in research on the relationship between structure and function.

As an example, we present data on the suprachiasmatic nucleus (SCN) of the human hypothalamus, a small structure (0.25 mm^3) on top of the optic chiasm. It is one of the few brain structures for which a main function is known. The SCN generates and coordinates circadian rhythms (Moore, 1992), but this is certainly not its only function (see below). It is not possible to reliably localize the SCN in conventionally stained human brain sections (Swaab et al., 1990), but it shows up clearly following staining of its peptidergic neurotransmitters, e.g. vasopressin (VP) (Swaab et al., 1985) or vasoactive intestinal polypeptide (VIP) (Moore, 1992). These peptides are stable in postmortem material. The number of SCN neurons expressing them varies widely in relation to age, sex and stage of rhythms providing information on the functional involvement of the SCN in these processes.

Age and SCN structure-function

Circadian rhythms are already present in the fetus, i.e. in rest-activity, breathing movements, hormone levels and heart rate variability. In addition, a circadian rhythm is found in the pattern of birth in humans, with a peak at approximately 0300–0400 h and a nadir at 1700–1800 h. These rhythms are probably mainly driven by the mother and not by the fetus (Honnebier et al., 1989). Yet, circadian rhythms in body temperature are found in 50% of low risk "healthy" premature infants with a gestational age of 29–35 weeks (Mirmiran and Kok, 1991). However, in contrast to adult rhythms, these preterm rhythms are more variable and not synchronized to the time of day, possibly through a lack of functional SCN afferents and efferents.

In order to assess the maturity of the human SCN at the moment of birth, the number of neurons expressing

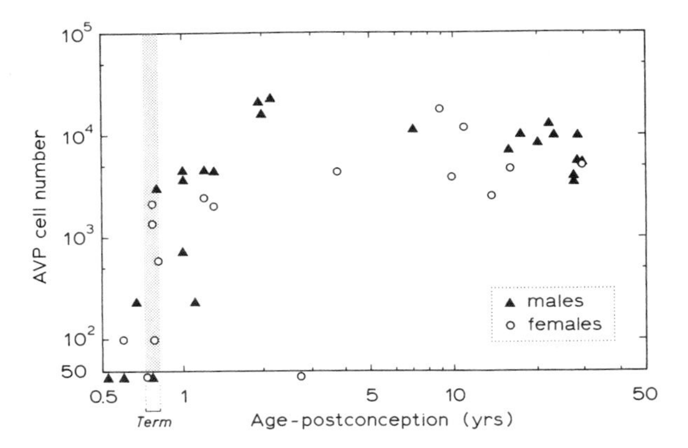

Fig. 1. Development of vasopressin (VP) cell number in the human suprachiasmatic nucleus (SCN) of the hypothalamus. Log–log scale. The period at term (38–42 weeks of gestation) is indicated by the vertical bar (from Swaab et al. (1990) with permission).

VP were determined (Fig. 1). VP staining in the SCN was present from 31 weeks onwards. However, most cells only start to express VP after birth. At term (38–42 weeks of gestation), only 13% of the adult numbers of VP-expressing neurons was present. VP cell numbers rose rapidly in the first neonatal months and at the same time overt sleep-wakefulness, temperature and *N*-acetyltransferase rhythms developed (Swaab et al., 1990) (Fig. 1). The presence of an immature SCN in premature children and the fact that normally the mother guides the circadian rhythms of the fetus has practical consequences. Exposure of premature children in the neonatal care unit to a light-dark environment improves their development (Mann et al., 1986; Fajardo et al., 1990).

In addition, at the other end of our lifespan, the SCN shows clear changes that can be related to, e.g. a fragmentation of sleep/wakefulness patterns in senescence. The neurological basis for these sleep changes may be found in the SCN and its input. A marked decrease in the number of SCN cells expressing VP was found in subjects older than 80 years, and even more so in Alzheimer patients (Fig. 2) (Swaab et al., 1985). Of practical importance in this respect may be the observation of Witting et al. (1993) that similar circadian disturbances in the aged rat can be countered by increasing the SCN input, i.e. by increasing the environmental light intensity. This way the circadian amplitude of sleep-wakefulness in old rats reached the level of young rats. Current research shows that circadian disturbances in Alzheimer patients can also be improved by light therapy (Okawa et al., 1991; Satlin et al., 1992).

Sex and SCN structure-function

Although the exact role of the SCN in sexual behaviour and reproduction has not yet been crystallized, data on various species strongly indicate the existence of such a role. Neuronal activity in the SCN increases around puberty, whereas circadian functions mature much earlier. The ovarian cycle of the rat is controlled by the SCN, and in the female rabbit, postcoital ultrastructural changes have been observed in the SCN (for references, see Swaab et al., 1994). In addition, neonatal castration of gerbils results in a 62% decrease in SCN volume (Holman and Hutchison, 1991).

With respect to the possible role of the SCN in sexual behaviour and reproduction, differences in the human SCN in relation to gender and sexual orientation are of interest. So far, two sex differences have been

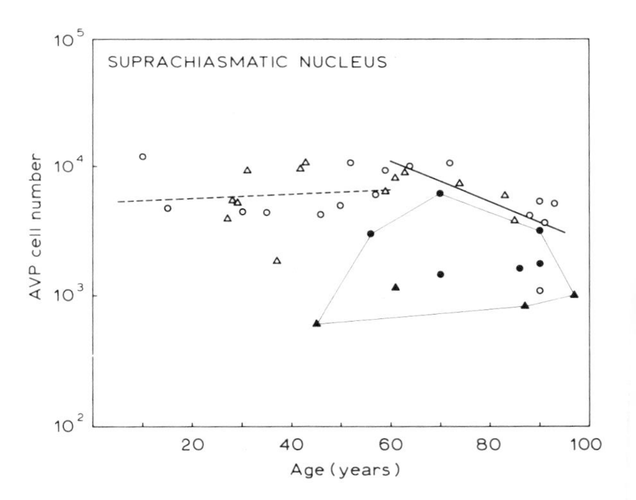

Fig. 2. Linear regression between VP cell numbers in the human SCN and age. A statistically significant decrease was observed in controls after 60 years of age ($P < 0.05$). Triangles represent males and circles represent females. Values of Alzheimer patients (closed symbols) are delineated by a minimum convex polygon and were reduced as compared to age-matched controls ($P < 0.01$). Redrawn using data from Swaab et al. (1985, 1987).

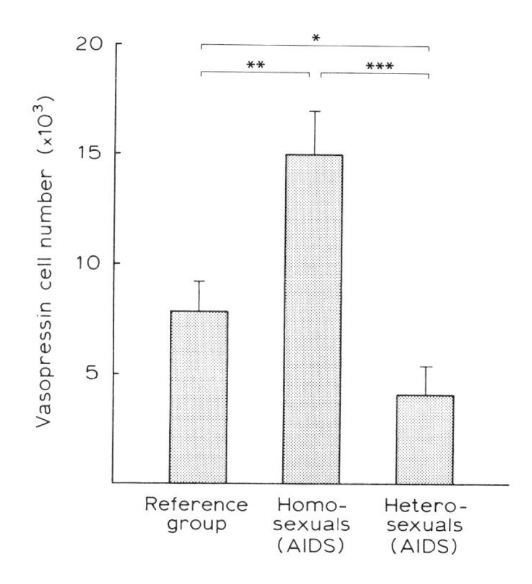

Fig. 3. The number of VP neurons in the human SCN in homosexuals contains 1.9 times more VP-producing neurons than the reference group and 3.6 times as many VP neurons as the SCN of heterosexual AIDS patients (from Swaab and Hofman (1990) with permission).

found in the human SCN. A sex difference in shape is present in the VP compartment of the SCN as the shape of the SCN is more spherical in men and more elongated in women (Swaab et al., 1985). Another, remarkable, recent finding is the sex difference in cell number in the VIP population of SCN neurons; between 10 and 30 years of age, twice as many VIP-expressing neurons were observed in males as in females (Swaab et al., 1994). Regarding sexual orientation, at least twice as many cells were found in the VP compartment of the SCN of homosexual as in heterosexual men (Swaab and Hofman, 1990) (Fig. 3).

What the exact functional meaning of the variability of the SCN is in relation to gender and sexual orientation is currently being studied. The observation of an enlarged SCN in homosexual men shows that male homosexuals do not have a "female hypothalamus", as was proposed by Dörner (1988).

Rhythms and SCN structure-function

Strong circannual and circadian fluctuations have been observed in the human SCN. The season in which the patients died appeared to be responsible for a considerable amount of variation in the SCN. The volume of the VP cell population of the SCN was 2.5 times larger in October–November than in May–June, and contained 2.7 times as many VP-immunoreactive neurons in the autumn period (Hofman and Swaab, 1992; Hofman et al., 1993) (Fig. 4). The annual cycle of the human SCN showed a non-sinusoidal pattern, reaching maximum values in early autumn, a lower plateau in winter and a deep trough in late spring and early summer. The VP neurons in the PVN did not show such changes over the year, which is an indication of the specificity of the SCN rhythm. The annual SCN rhythms appeared to depend on the photoperiod cycle rather than on the annual temperature cycle, and these data, therefore, indicate that human beings are much more influenced by photoperiodic changes than is generally assumed. The annual variations in VP immunoreactivity in the human SCN coincide with (1) variations in plasma testosterone levels which are high during late summer and early autumn and low during spring, (2) the amount of sleep per 24-h period, which is the lowest in May–June and the highest in September–October, and (3) annual reproductive differences (see Hofman et al., 1993).

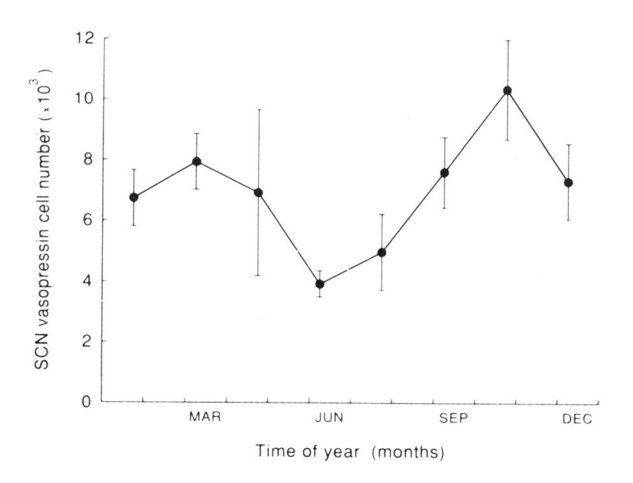

Fig. 4. VP cell number of the human SCN as a function of time of year. The values represent the mean ± SEM The seasonal variation in the cell number of the SCN is statistically significant (Kruskal–Wallis multiple comparisons test, $P = 0.05$). The human SCN contains, on average, 2.7 times as many cells in October–November as in May–June (from Hofman et al. (1993) with permission).

Since the environmental light-dark cycle is the main Zeitgeber for the biological clock, it will probably not come as too much of a surprise that the human SCN also shows clear fluctuations over the 24-h period, i.e. in relation to the hour of the day when the patient died. The volume of the VP cell population was, on average, 1.4 times larger during the day than during the night, and contained 1.8 times as many VP-immunoreactive neurons in subjects between 6 and 47 years of age (Hofman and Swaab, 1993).

Such changes are, of course, consistent with the circadian clock function of the SCN. A lesion in the suprachiasmatic region of the anterior hypothalamus, e.g. as the result of a tumour, indeed results in disturbed circadian rhythms in humans (Schwartz et al., 1986; Cohen and Albers, 1991). Totally blind people may show free-running temperature, cortisol and melatonin rhythms. In addition, they may suffer from sleep disturbances (Sack et al., 1992). The circadian variability in the SCN emphasizes the importance of the light-dark cycle for synchronization and of the SCN for the generation and coordination of circadian rhythms in humans.

The data discussed show that the variation in number of hypothalamic neurons expressing a certain neuropeptide may reveal a wealth of data on processes in which such neurons are involved. A practical consequence of the finding that structural characteristics of hypothalamic nuclei are to a large extent determined by age, gender, sexual orientation, season, hour of the day, and other factors that have not even been discussed, such as agonal state, lateralization and postmortem delay (Ravid et al., 1992), is that the marginal matching procedure generally used in neuropathology have to be greatly extended. Careful and systematic documentation of premortem and postmortem factors that might influence the later outcome of brain research are absolute prerequisites, which, in turn, demand a well-organized brain bank (Ravid et al., 1992).

As the factors influencing the morphometry of our brain are gradually becoming known, the unexplained variation remaining is comparable to the variation that occurs in similar structures in the hypothalamus of rats that have been inbred for generations and kept under well-standardized conditions. The unexplained variation remaining in the human SCN (age, season and hour of the day having been accounted for) is comparable to that of studies in rat.

Of course, observations on the human brain are primarily of a correlative nature, showing a relationship, for instance, between the number of SCN cells expressing a certain peptide and age, sex, season or hour of death. Animal experiments are, in principle, best suited for the study of the causality of such relationships, although human pathological conditions affecting a particular system in the human brain and interference studies may also provide such causal relationships. However, this last possibility usually demands more time and effort. Examples of such observations in the human hypothalamus are: (1) the degenerative changes in the SCN in Alzheimer's disease (Swaab et al., 1985) in relation to the functional circadian changes (Witting et al., 1990); (2) lesions in the SCN area, caused by tumours, that result in circadian rhythm disturbances (Schwartz et al., 1986; Cohen and Albers, 1991); (3) the free-running system in blind people (Sack et al., 1992), and (4) the disappearance of circadian behavioral disturbances in Alzheimer patients by exposure to bright light (Okawa et al., 1991; Satlin et al., 1992).

In conclusion, the large variability generally found in data on human brain structures in controls is not necessarily a disturbing factor. It may provide useful functional information on such structures in relation to, e.g. development, reproduction and aging. Such information will become more and more useful in the coming years now that powerful modern neurobiological techniques have been developed that can be applied to human postmortem brain tissue.

Acknowledgements

The authors would like to express their thanks to Ms. W.T.P. Verweij and Ms. O. Pach for their secretarial help. Brain material was obtained from the Netherlands Brain Bank (coordinator Dr. R. Ravid).

References

Cohen, R.A. and Albers, H.E. (1991) Disruption of human circadian and cognitive regulation following a discrete hypothalamic lesion: a case study. *Neurology*, 41: 726–729.

Dörner, G. (1988) Neuroendocrine response to estrogen and brain differentiation in heterosexuals, homosexuals, and transsexuals. *Arch. Sexual Behav.*, 17: 57–75.

Fajardo, B., Browning, M., Fisher, D. and Paton, J. (1990) Effect of nursery environment on state regulation in very-low-birth-weight premature infants. *Inf. Behav. Dev.*, 13: 287–303.

Hofman M.A. and Swaab, D.F. (1992) Seasonal changes in the suprachiasmatic nucleus of man. *Neurosci. Lett.*, 139: 257–260.

Hofman M.A. and Swaab, D.F. (1993) Diurnal and seasonal rhythms of neuronal activity in the suprachiasmatic nucleus of humans. *J. Biol. Rhythms*, 8: 283–295.

Hofman, M.A., Purba, J.S. and Swaab, D.F. (1993) Annual variations in the vasopressin neuron population of the human suprachiasmatic nucleus. *Neuroscience*, 53: 1103–1112.

Holman, S.D. and Hutchison, J.B. (1991) Differential effects of neonatal castration on the development of sexually dimorphic brain areas in the gerbil. *Dev. Brain Res.*, 61: 147–150.

Honnebier, M.B.O.M., Swaab, D.F. and Mirmiran, M. (1989) Diurnal rhythmicity during early human development. In: S.M. Reppert (Ed.), *Development of Circadian Rhythmicity and Photoperiodism in Mammals*, Perinatology Press, Ithaca, NY, pp. 83–103.

Mann, N.P., Haddow, R., Stokes, L., Goodley, S. and Rutter, N. (1986) Effect of night and day on pre-term infants in a newborn nursery: randomised trial. *Br. Med. J.*, 293: 1265–1267.

Mirmiran, M. and Kok, J.H. (1991) Circadian rhythms in early human development. *Early Hum. Dev.*, 26: 121–128.

Moore, R.Y. (1992) The organization of the human circadian timing system. In: D.F. Swaab, M.A. Hofman, M. Mirmiran, R. Ravid and F.W. Van Leeuwen (Eds.), *The Human Hypothalamus in Health and Disease, Progress in Brain Research,* Vol. 93, Elsevier, Amsterdam, pp. 99–117.

Okawa, M., Hishikawa, Y., Hozumi, S. and Hori, H. (1991) Sleep-wake rhythm disorder and phototherapy in elderly patients with dementia. In: G. Racagni et al. (Eds.), *Biological Psychiatry,* Vol. 1, Elsevier, Amsterdam, pp. 837–840.

Ravid, R., Van Zwieten, E.J. and Swaab, D.F. (1992) Brain banking and the human hypothalamus - factors to match for, pitfalls and potentials. In: D.F. Swaab, M.A. Hofman, M. Mirmiran, R. Ravid and F.W. Van Leeuwen (Eds.), *The Human Hypothalamus in Health and Disease, Progress in Brain Research,* Vol. 93, Elsevier, Amsterdam, pp. 83–95.

Sack, R.L., Lewy, A.J., Blood, M.L., Keith, L.D. and Nakagawa, H. (1992) Circadian rhythm abnormalities in totally blind people: incidence and clinical significance. *J. Clin. Endocrinol. Metab.*, 75: 127–134.

Satlin, A., Volicer, L., Ross, V., Herz, L. and Campbell, S. (1992) Bright light treatment of behavioral and sleep disturbances in patients with Alzheimer's disease. *Am. J. Psychiatry*, 149: 1028–1032.

Schwartz, W.J., Bosis, N.A. and Hedley-Whyte, E.T. (1986) A discrete lesion of ventral hypothalamus and optic chiasm that disturbed the daily temperature rhythm. *J. Neurol.*, 233: 1–4.

Swaab, D.F. and Hofman, M.A. (1990) An enlarged suprachiasmatic nucleus in homosexual men. *Brain Res.*, 537: 141–148.

Swaab, D.F., Fliers, E. and Partiman, T.S. (1985) The suprachiasmatic nucleus of the human brain in relation to sex, age and senile dementia. *Brain Res.*, 342: 37–44.

Swaab, D.F., Roozendaal, B., Ravid, R., Velis, D.N., Gooren, L. and Williams, R.S. (1987) Suprachiasmatic nucleus in aging, Alzheimer's disease, transsexuality and Prader-Willy syndrome. In: R. de Kloet, V.M. Wiegany and D. de Wied (Eds.), *Neuropeptides and Brain Function. Progress in Brain Research,* Vol. 72, Elsevier, Amsterdam, pp. 301–310.

Swaab, D.F., Hofman. M.A. and Honnebier, M.B.O.M. (1990) Development of vasopressin neurons in the human suprachiasmatic nucleus in relation to birth. *Dev. Brain Res*, 52: 289–293.

Swaab, D.F., Zhou, J.N., Ehlhart, T. and Hofman. M.A. (1994) Development of vasoactive intestinal polypeptide (VIP) neurons in the human suprachiasmatic nucleus (SCN) in relation to birth and sex. *Dev. Brain Res.*, in press.

Witting, W., Kwa, I.H., Eikelenboom, P., Mirmiran, M. and Swaab, D.F. (1990) Alterations in the circadian rest-activity rhythm in aging and Alzheimer's disease. *Biol. Psychiatry*, 27: 563–572.

Witting, W., Mirmiran, M., Bos, N.P.A. and Swaab, D.F. (1993) Effect of light intensity on diurnal sleep-wake distribution in young and old rats. *Brain Res. Bull.*, 30: 157–162.

F. Bloom (Editor)
Progress in Brain Research, Vol. 100

CHAPTER 33

Schizophrenia: neurobiological perspectives

C.N. Stefanis

Department of Psychiatry, Athens University Medical School, Eginition Hospital, 72–74 Vas Sophias Ave., 115 28 Athens, Greece

Introduction

Schizophrenia despite recent advances, undoubtedly still remains the most enigmatic and elusive of all psychopathological conditions. The present revSwaab, D.F., piew cannot be extensive enough to include all the available information, derived from clinical, neuropsycological, psychosocial, neurogenetic and neuroscience research areas, on its etiology and pathogenesis. It is rather limited to surveying the field from a combined clinical and neurobiological perspective and more specifically, recent findings are presented supporting the author's view that schizophrenia, as an aberrant emotional, cognitive and behavioral state rather than a homogeneous distinct entity, is closely related to (determined by) a dysfunction of the brain's information processing system involving the associative cortical areas, limbic system, prefrontal cortex loop and primarily arising from a defective signal gating and signal modulating mechanism in limbic structures (mainly in hippocampus and parahippocampal cortex). The proposition lies in a three-dimensional conceptual framework comprising neurodevelopmental, activity dependent neuroplastic and functional connectivity processes interacting in an integrative fashion.

The profile of schizophrenia

A unified neurobiological model of schizophrenia should be capable of accounting, partly or in full, for the following features of the illness (Kales et al., 1990). It is a condition recognized only by a variety of affective, cognitive and behavioral manifestations defined only on phenomenological grounds by operational-clinical criteria.

The diversity of the symptoms is enormous and a distinction is currently made between positive (hallucinations, delusions etc) and negative (apathy, poor thought and speech, inattention, etc.) symptoms often co-occurring and overlapping with other psychiatric disorders.

The course is variable. The first episode usually presents with positive symptoms accompanied by agitation, while in long-standing cases the positive symptoms tend to be attenuated and the negative prevail. The peak of onset is 18–30 years. In a percentage of cases, some signs of cognitive dysfunction and social maladjustment are retrospectively detected to precede the onset of the first episode. Current treatments, be they pharmacological or psychosocial are symptom-oriented.

Although, almost immediate and full occupancy of DA receptors is accomplished by low dose neuroleptics treatment, duration exceeding 3 weeks is usually required for the core clinical symptoms to be modified.

Neuroanatomical and neurophysiological correlations

Among the limbic structures, endorhinal cortex (EC) is considered to play a crucial role as a relay and integrative station of the sensory information converging on the hippocampal formation. EC receives inputs from widespread parts of the somatosensory, auditory, visual, olfactory, gustatory and nociceptive cortical areas, either directly or by way of the paralimbic and perirhinal cortex.

Apart from the strong projections from the subiculum and the EC to the nucleus accumbens, all limbic areas have reciprocal connections with widespread parts of the telencephalic isocortex. The projections from primary sensory areas to the EC are very weak compared with those of the multimodal association areas. The EC may thus receive highly processed information that is committed to several modalities.

Because the most massive afferent projection to the hippocampus originates in the endorhinal cortex, it is clear that lesions of the parahippocampal cortex deprive the hippocampus from its main input. It is, therefore, to be expected that the behavioral deficits found after such lesions are essentially similar to those obtained after damage to the hippocampus.

According to a proposed model (Teyler and DiScenna, 1984), the hippocampus is considered to represent a coordinate system capable of reciprocally addressing neocortical loci in space and time.

Although there is no indication, as yet, that neuronal responses in the limbic cortical areas are elicited by specific features of sensory stimuli, responses in these areas are invariably dependent on the behavioral meaning of the stimulus, namely when the applied stimulus is used as a conditioned one in a discrimination task. Experimental findings suggest that the changes in firing rate of hippocampal cells in response to CS signal the formation of hippocampus memory traces and that the hippocampal neuronal populations process combinations of stimulus features that animals use in memory tasks (Lopes Da Silva et al., 1990).

It may thus be concluded that the limbic cortical areas (a) receive multimodal sensory information and form associations between stimuli of different kinds and (b) are involved in the formation of the substrate for temporary storage of information effected by the strong plastic properties of the synapses in the circuitry, namely initiating LTP-like processes.

The role of the reticular slow activity (RSA) seems to be pertinent. In all limbic cortical areas, RSA is present and appears to modulate the fast synaptic transmission and the long-term changes in synaptic strength (LTP). It largely depends on the brain stem inputs that are related to arousal and motivation.

The hippocampal circuit (perforant path, dentate gyrus, CA_3, CA_1, subiculum) subserves the function of a selector. Each synapse of the hippocampal trisynaptic circuit represents a gate, the state of which depends on the behavior of the organism and the RSA.

Structural changes in schizophrenia

Post mortem studies on brains of schizophrenics have shown: (a) a substantial increase in the total volume of the ventricular system mainly of the temporal horn (Brown et al., 1986); (b) a considerable volume reduction of temporal lobe structures (hippocampal formation, amygdala, parahippocampal gyrus) (Bogerts et al., 1985); and (c) a reduction of neuronal cells in the hippocampal formation, pyramidal cell bands, hippocampal segments and endorhinal cortex. In all cases, the number of neuronal cells were reduced, but there was no evidence of gliosis (Falkai and Bogerts, 1986).

Morphological changes in the temporal lobe of schizophrenics included alterations in the sulcogyral pattern of the cortex and cell loss from the cortical lamina of the ventral insula (claustro-cortex) and parahippocampal gyrus. The cytoarchitecture of the remaining temporal cortex was normal. The abnormalities were more conspicuous in the left hemisphere and were considered to indicate abnormal ontogenetic development of a small part of the endorhinal cortex.

Thus, alterations to the cellular organization of the parahippocampal gyrus and hippocampus appear to be a primary and central feature of the cellular biology of schizophrenia.

Neuroimaging findings

Both computed tomography (CT) and magnetic resonance (MR) studies yielded results comparable to those obtained from post mortem studies.

Two neuroimaging techniques to visualize the functional state of the brain, single photon emission computed tomography (SPECT) and positron. emission tomography (PET) have extensively been used recently to assess possible physiological abnormality in schizophrenia. Metabolic activity measured at the resting state as well as during a cognitive challenge suggested "hypofrontality" frequently associated with negative symptoms of schizophrenia (Andreassen et al., 1992). Conversely, "hyperfrontality" during the resting state was observed in drug-naive, recent onset schizophrenic patients by a Canadian group of investigators (Gleghorn et al., 1990). The use of PET for the investigation of DA activity in the brain yielded conflicting results by two groups of investigators and failed to either definitely refute or support the dopamine hypothesis.

Synaptic transmitters and modulators in schizophrenia

In the hippocampal formation, a large number of neurotransmitters and modulators of synaptic transmission are involved. Cardinal among them are the amino acids, glutamate and aspartate, that are present not only in all intrinsic pathways (mossy fibers, schaffer collaterals), but also in the major input pathway, the perforant path. Their direct effects are mainly excitatory through membrane depolarization, although the different types of receptors are activated in different ways and have a specific regional distribution. An important feature of these receptors (mainly the NMDA type) is that their activation depends critically on the state of neuronal membrane potential, i.e. on the immediate past history of the neuronal circuit. In this way, they play a crucial role in synaptic plasticity through the regulation of the influx of Ca^{2+}. In the hippocampus, NMDA receptors are not involved directly in normal excitatory transmission but appear to mediate synaptic potentiation such as long-term potentiation (LTP).

The excitatory amino acids (EAA) have recently been implicated in the pathogenesis of schizophrenia either as excitotoxic agents accounting for the observed structural abnormalities or as hyper or hypo-activators of an aberrant neurotransmission in schizophrenia. The latter has to be viewed in relation to the long standing dopaminergic hypothesis in schizophrenia. The validity of a dopaminergic overactivity hypothesis rests on the induction of some of the symptoms by DA receptor agonists and, conversely, on their amelioration by DA receptor blockers. However, both of these pharmacological findings do not necessarily imply that the apparent DA overactivity constitutes the primary chemical abnormality. It may very well reflect a state of imbalance brought about by a primary abnormality in another neurotransmission system, much like the Ach/DA imbalance of Parkinson's disease for which an Ach relative hyperactivity was postulated in the past. The glutamatergic system deficiency, more specifically in glutamatergic projections from the prefrontal cortex and hippocampal regions to the nucleus accumbens with their modifying and reciprocal effect on DA activity is the likely candidate of such an imbalance.

Other aminergic systems, and particularly, the 5-HT system have also to be considered as evidenced by the established efficacy of the new atypical neuroleptics with antiserotonergic action. Moreover, the observation that D_2 and D_1 receptor blockers, if at all effective, modify the negative symptoms for the worse and it takes weeks before their clinical effects are manifested, as well as the observation that patients with tardive dyskinesia with their D_2 receptors presumably fully occupied still present with psychotic symptoms, argues against the primary and sole pathogenetic factor of DA hyperactivity in schizophrenic symptomatology.

The reciprocal function of the glutamatergic and DAminergic system at the neurotransmission and at the circuit level has been adequately documented in the past few years. It has been shown that: glutamate regulates DA release from DAminergic neuron terminals in the frontal cortex, NMDA receptors modulate striatal cortex DA–D_2 transmission and neuroleptics, typical and atypical in addition to their antidopaminergic activity also exert a direct effect on the NMDA receptor function.

Plasticity of limbic neuronal networks

Limbic cortical neuronal networks are characterized by their readiness to undergo plastic changes dependent on the past experience of the organism. This explains why they are involved in learning and memory processes and also why they are prone to epileptic seizures. A form of synaptic plasticity that received considerable attention in recent years is the phenomenon of long-term potentiation (LTP), defined as an increase in the efficacy of synaptic transmission, lasting >15 min, even hours or days, following a short-lasting high-frequency stimulation. LTP has been induced in hippocampal and several other limbic structures. Polysynaptic LTP (pp-dentate gyrus, CA_3, CA1) has also been produced, while the induction of LTP in the output pathway from the hippocampus to the prefrontal cortex, may provide a useful model for a functional analysis of hippocampo-neocortical communication in learning and behavior.

In the induction stage of LTP, glutamate by binding to the AMPA receptor depolarizes the neuron and activates the *N*-methyl-D-aspartate (NMDA) receptor, to allow Na^+ and Ca^{++} to pass through its ionic channel. Fast transmission is mediated by the AMPA receptor but during intense synaptic activity, NMDA receptor is activated and triggers long-term changes. Drugs blocking the NMDA receptor also abolish synaptic plasticity as well as all forms of LTP.

LTP was regarded at first as a unitary process, but accumulating data indicate that at least two stages can be differentiated: induction and maintenance (Ben-Ari et al., 1992). The induction phase, during which a cascade of events triggers synaptic potentiation includes a NMDA component. The maintenance phase results in long-lasting and selective enhancement of the synaptic responses mediated by AMPA receptors. In the presence of D-2-amino-5-phosphonovalerate (APV) the specific antagonist of the NMDA receptor or Ca^{2+} chelators (PCP and MK-801), a tetanic stimulation fails to enhance the synaptic response. In the presence of protein kinase C inhibitors, a tetanic stimulation enhances the synaptic response, but it is accompanied by a smaller slope of the rising phase and faster early decay of the potentiation to the control level. Other than glutamate neurotransmitters active in the region and known for their modulatory functions may contribute to modifying the neuronal mechanisms underlying the late and long-lasting maintenance phase of LTP as the NE input to the dentate.

Dopamine receptor-mediated signals may also contribute to the production of the late, maintenance LTP since DA receptor blockers during tetanization prevent the occurrence of the late LTP maintenance. Moreover, in addition to enhancing protein synthesis a dopamine (D_1) mediated increase in glycoprotein fucosylation was shown to be necessary for the maintenance of the late stage of LTP.

An interesting aspect of the hippocampal LTP is its relationship with the typical rhythmic slow activity (RSA) (Larson and Lynch, 1986). In the dentate gyrus and the CA1 field, LTP induction is optimal when the time interval between stimuli is approximately 200 ms, corresponding to the frequency band of the spontaneously occurring RSA in the hippocampus. It appears that brief high-frequency stimulation elicits a weak NMDA receptor response that is amplified when the bursts are delivered in a pattern within the frequency range of the RSA.

In a recent study, non-linear mathematical analysis of hippocampal EEG epochs recorded during the maintenance phase of hippocampal LTP, have shown a relative reduction in the correlation dimension compared to the values estimated prior to the LTP induction. This suggests that the physiological mechanisms underlying the plastic neuronal changes occurring in the process of the LTP, behave as a non-linear deterministic system with lower degrees of freedom (Koutsoukos et al., 1993).

Evidence for involvement of glutamatergic system in schizophrenia

Interest in the search for more direct evidence implicating glutamatergic system involvement in schizophrenia was mainly initiated by the observation that phencyclidine (PCP), as well as ketamine, both non-competitive NMDA receptor antagonists evoke in normals and exacerbate in schizophrenics, psychotomimetic symptoms most characteristic of schizophrenia. Consistent with these clinical observations are the results obtained in recent years from the administrations of the MK-801, the highly selective non-competitive antagonist of the PCP binding site of the NMDA-ion channel to experimental animals. Overall, the PCP receptor blockers in animals exhibit amphetamine-like behavioral effects on a variety of tasks usually employed in psychosis-simulation experiments.

A number of recent post mortem studies on brains of schizophrenics have demonstrated changes in EAA and NMDA receptor complex such as presynaptic and postsynaptic upregulation of glutamate receptors in frontal and temporal cortex (Deakin et al., 1989), elevated binding to NMDA receptor PCP sites in the hippocampus and endorhinal cortex (Kornhuber et al., 1989), a significant reduction in the mRNA that encodes the KA/AMPA-R non-NMDA glutamate receptor within the CA_3 hippocampal field (Harrison et al., 1991) and significantly higher concentrations of serine glycine in the medial temporal lobe (Waziri et al., 1992).

Concluding remarks

At the present state of knowledge, no single brain mechanism can be invoked to account for the multiformity and highly complicated clinical and biological profile of schizophrenia. It would suffice to state that several brain mechanisms subserved by multiregional, multineuronal, multisynaptic, multitransmitional, multiansportational, multitranductional and multieffectorial systems are involved. To reduce this complexity to a primary and single cause initiating the cascade of phenomena observed in the clinical setting and in the experimental laboratory seems at present an impossible task. However, the data presented deriving from seemingly disparate observations seem to converge in a fashion that may be used as inferential evidence to formulate empirically testable hypotheses. It is in this frame that our three-dimensional proposition involving neurodevelopmental, neuroplastic and functional neuroconnectivity processes has been articulated. The available morphological findings are consistent with the view advanced recently by several authors, that the observed structural abnormalities exist in schizophrenia as residues of a neurodevelopmental deficit either of genetic or environmental origin (during the perinatal period of development) mainly affecting the temporolimbic cortical areas and taxing them with an increased functional vulnerability to subsequent (in adulthood) adverse stimuli. The strategic position within the brain's neuronal network of the hippocampus and of the parahippocampal complex with their vast incoming and outgoing connections makes them the most likely candidates for the distributor's and amplifier's role of any disturbance in the patterning of the neuronal signals that may occur. That in schizophrenia such a disturbance may originate or be amplified in the hippocampal-endorhinal complex is inferred from this structure's neuroplastic properties displayed by its marked capacity to gate, process, encode and temporarily store information through LTP-like NMDA receptor mediated cascade of synaptic and intercellular events. Such an inference derives empirical support from the psychotomimetic behavioral effects of PCP receptor blocking agents.

Consequent to the above, an altered state of functional connectivity with other brain areas, such as the prefrontal lobe in which information fine tuning and storage occur, follows a sequential logic. Considering further the observation already mentioned that LTP or LTP-like processes subserving learning and mneumonic functions behave as non-linear deterministic systems with lower degrees of freedom, we may speculate that the altered functional connectivity of hippocampo-cortical networks may develop irregularities in the neurocognitive substrate either by the efficacious reproduction of the same pattern of information, as in the case of LTD, or with erroneous association of the information, results in disturbances in the plasticity process that may affect the integration process of information in schizophrenia.

Acknowledgements

I am indebted to my co-workers E. Angelopoulos and E. Koutsoukos for their helpful advice and comments.

References

Andreassen, N.C., Rezai, K., Alliger, R., Swayze, V.W., Falum, M., Kirchener, P., Cohen, G. and O'Leary, D.S. (1992) Hypofrontality in neuroleptic-naive patients and in patients with chronic schizophrenia. Assessment with xenon 133 single-photon emission computed tomography and the Tower of London. *Arch. Gen. Psychiatry*, 49: 943–958.

Ben-Ari, Y., Aniksztejn, L. and Bregestovski, P. (1992) Protein kinase C modulation of NMDA currents: an important link for LTP induction. *Trends Neurosci.*, 15: 333–339.

Bogerts, B., Meertz, E. and Schoenfeldt-Bausch, R. (1985) Basal gaglia and limbic system pathology in schizophrenia. A morphometric study of brain volume and shrinkage. *Arch. Gen. Psychiatry*, 42: 784–791.

Brown, R., Colter, N., Corsellis, J.A.N. et al. (1986) Postmortem evidence of structural brain changes in schizophrenia. Differences in brain weight, temporal horn area, and parahippocampal gyrus compared with affective dis*order. Arch. Gen. Psychiatry*, 43: 36–42.

Deakin, J.F.W., Slater, P., Simpson, M.D.C., Gilchrist, A.C., Skan, W.J., Royston, M.C., Reynolds, G.P. and Cross, A.J. (1989) Frontal cortical and left temporal glutamatergic dysfunction in schizophrenia. *J. Neurochem.*, 52: 1781–1786.

Falkai, P. and Bogerts, B. (1986) Cell loss in the hippocampus of schizophrenics. *Eur. Arch. Psychol. Neurol. Sci.*, 236: 154–161.

Gleghorn, J.M., Garnett, E.S., Nahmias, C., Brown, G.M., Kaplan, R.D., Szetchman, H., Szechtman, B., Franco, S., Dermer, S.W. and Cook, P. (1990) Regional brain metabolism during auditory hallucinations in chronic schizophrenia. *Br. J. Psychiatry*, 157: 562–570.

Harrison, P.J., McLaughin, D. and Kerwin, R.W. (1991) Decreased hippocampal expression of a glutamate receptor gene in schizophrenia. *Lancet*, 337: 450–452.

Kales, A., Stefanis, C. and Talbot J. (1990) Recent advances in schizophrenia. In: A. Kales and C. Stefanis (Eds.), *Int. Perspective Series: Psychiatry, Psychology and Neurosciences*, ??Publisher???

Kornhuber, J., Mack-Burkhardt, F., Riederer, P., Hebenstreit, G.F., Reynolds, G.P., Andrews, H.B. and Beckmann, H. (1989) [^{3}H]MK-801 binding sites in postmortem brain regions of schizophrenic patients. *J. Neural Transmission*, 77: 231–236.

Koutsoukos, E., Angelopoulos, E., Maillis, A. and C. Stefanis (1993) Does learning mean more order in the CNS? *9th World Congress of Psychiatry Abstract Book*, 1893, p. 482.

Larson, J. and Lynch, G. (1986) Induction of synaptic potentiation in hippocampus by patterned stimulation involves two events. *Science*, 232: 986–988.

Lopes da Silva, F.H., Witter, M.P., Boeijninga, P.H. and Lohman, A.H.M. (1990) Anatomical organization and physiology of the limbic cortex. *Physiol. Rev.*, 70: 453–511.

Teyler, T.J. and Discenna, R. (1984) The topological anatomy of the hippocampus: a clue to its function. *Brain Res. Bull.*, 12: 711–719.

Waziri, R., Baruah, S. and Sherman, A.D. (1992) Abnormal serine-glycine metabolism in the brains of schizophrenics. *Schizophrenia Res.*, 8: 233–243.

SECTION VI

Informatics and Progress in Brain Research

F. Bloom (Editor)
Progress in Brain Research, Vol. 100

CHAPTER 34

New solutions for neuroscience communications are still needed

Floyd E. Bloom and Warren G. Young

Department of Neuropharmacology, The Scripps Research Institute, La Jolla, CA, USA

Introduction

More than 15 years ago, Bloom and Melnechuk (1978) asserted that "even the most active neuroscientist spends more working hours in reading, reviewing and writing scientific reports than on direct experimental effort". The dilemma continues: to do or to read? In 1994, it seems nearly impossible to maintain an active ongoing comprehension of the scientific literature in any special corner of neuroscience research, let alone a broad awareness of new discoveries, or an in-depth awareness of any but the most narrowly defined field . The explosions of data about the brain, its cells and their molecules that startled us in the late 1970s and 1980s have grown unwaveringly throughout this first half of the Decade of the Brain. Yet the fact remains that there have been only a few changes in the habits of scientific information gathering, sharing and analysing that have been the traditional standards of neuroscientists: namely reading research journals and travelling to scientific meetings. This despite the obvious recognition by all participants in the profession of neuroscience that the data they are reading or hearing are months to years behind the actual state of experimental progress at the bench tops of our field.

For this milestone volume in the *Progress of Brain Research*, we note here some of the steps we are taking to begin to harness the flow of scientific data and to develop within the community of neuroscientists some means of information handling that rival the sophistication of the instruments and methods by which we acquire our data.

A database of the brain

For the past 5 years, we have devoted considerable effort towards the development of a complex hierarchical, relational, object oriented database of published neuroscience information, built around the orientation of a brain atlas template (see Bloom, 1990) for additional background on neuroscience databases). We have taken this path as our initial approach to an eventual superhighway of neuroscience information traffic control. We describe here how we plan to use such an information handling system to establish the normative parameters of molecular, cellular and behavioral data (in terms composed of genetic, metabolic, physiologic and structural details of chemistry, circuitry and physiology), which could then be used for the detection and definition of pathological variations. It is our perspective that a comprehensive neuronal circuitry database for each of the major vertebrate central nervous systems is an essential tool for understanding the known molecular, cellular and macroscopic features of vertebrate brains and their interspecies relationships.

We also take the pragmatic position that such a data management system is necessary to illuminate essential missing elements of information. As a consequence, our efforts have been dedicated to the development of a technology to combine informatics (the science of data collection, organization and interpretation) with neuroscience with the goal of improved management and distribution of neuroscience information. A program to encourage such research efforts has recently been promulgated by several of the NIH institutes with ultimate international applications in mind (see Huerta et al., 1993).

We should note in passing that neuroscience is not alone among active areas of biomedical science in recognizing the need for such information handling tools. For example, those several hundred scientists involved in the Human Genome Project (see Pearson and Söll 1991; Cuticchia et al., 1993) have acknowledged comparable problems with the acquisition, analysis, and sharing of information on an essentially linear, but extremely long two-dimensional dataset consisting of alternations in four nucleic acid bases, and a far cry from the sort of complexities to be faced in developing a brain database (see below). Nevertheless, despite the differences in the complexity of the information sets being studied, those involved in the Human Genome Project have been explicit in advocating for informatics investments, and have stated that "the success of the genome project will depend in large part on the ease with which biologists can gain access to and use the information produced". Therefore, increased emphasis on data handling, its organization and its distribution remain major elements of the second 5 years of planning for the Genome Project (Collins and Galas, 1993).

The growth of neuroscience information

Scientific interest in the neurosciences has grown enormously over the past several years, as witnessed by the growth in membership of the national societies of neurosciences throughout the world, by the proliferation of scientific journals and magazines focused on the neurosciences, and by the programmatic interests of a wide range of governmental and non-governmental agencies. The sheer volume of accumulated published original reviewed articles in the neurosciences over the past 5 years probably rivals that over the entire previous history of neuroscience research. This increased level of activity has in part been fueled by the development of large series of scientific technologies (many of which are described in other chapters of this volume) for the rapid acquisition of rigorous data that were in the past elusive and capricious. At present, one can expect to reveal in rich detail far more reliable information on the detailed connections and mechanisms of interaction of neuronal circuitry at the cellular and molecular levels of understanding.

Although there can be little doubt that high quality data have expanded explosively in the neurosciences, there are several implicit barriers to the optimal utilization of this information, and in particular to the practical convergence of these myriad observations into testable hypotheses of normal brain function or the pathogenesis of the major human brain disorders such as Alzheimer's, Huntington's and Parkinson's diseases (see Morrison et al., 1985; Rogers and Morrison, 1985; Love et al., 1989; Masliah et al., 1990a,b; Morrison et al., 1991; Ludwig, 1993) or those with more subtle forms of pathology (e.g. schizophrenia, Andreasen et al., 1986; Benes and Bird, 1987; Benes et al., 1987a,b; Benes 1988; Andreasen, 1989; Beckmann, 1992; Akbarian et al., 1993a,b) or HIV-associated neurocognitive disorders (e.g. Masliah et al., 1992a–c; Wiley et al., 1992).

Aside from these pathological considerations, there is a striking need for serious scholarly attempts to model human cognitive operations through incorporation of rigorous data from chemical neuroanatomy and neurophysiology into modeling algorithms. However, when one seeks to do so, there is an immediate awareness that we lack reliable quantitative information on most aspects of human and experimental neuroanatomy. As noted by Cherniak (1990), estimates published by highly regarded neuroscientists for such elemental factual considerations as the actual area or volume of the human cortex, the density of the neurons within this sheet, and the average number of synapses within the cortical neuropil differ by orders of magnitude. Some estimates are clearly incompatible with known quantitative details such as the volume of the skull and the actual assembly of the predicted numbers of synapses with their afferent axons and target dendrites onto the predicted numbers and volumes of neurons within the predicted cortical volume. While there have been quantitative estimates of some such features of the human brain published years ago (Blinkov and Glezer, 1968), these data are not well known, and clearly antedate the major advances in defining neurochemical markers of interest or the advances in non-invasive imaging.

When one considers further the differences in brain shape between individuals, the difficulties in applying rigorously cytoarchitectonic and cortical connectivity criteria to define specific cortical regions, and the inability now to apply to human brain the connectivity tracing tools of experimental neuroanatomy (see Crick and Jones, 1993), details of the structure and function of the human brain may appear unapproachable. However, one promising path through these technical complexities may be through the development of principles to relate human brain structure and function to non-human primates, as well as to be able to develop procedures by which variations in brain shapes available from non-invasive imaging of the human brain (see Pfefferbaum et al., 1988; Gazzaniga 1989; Oppenheim et al., 1989) can be employed to assist in such inferential linkages. However, even with animal brains, there have been few attempts at objective identification of nuclear or laminar boundaries, or the areal boundaries of cortical regions; see Fleischhauer et al. (1980), Rehkamper et al. (1984), Rehkamper et al. (1985) and Ahrens et al. (1990) for examples of attempts to do so, but which lack actual volumetric parameters for cells or layers using partially automated discrimination of Nissl-stained neuronal packing densities.

Quite apart from these pathological and quantitative correlates, there appear to be finite limits on the ability of any individual scientist to absorb, digest and interpret the existing studies and to monitor, evaluate and incorporate new data into one's appreciation for a given brain region, system, or question. The characteristic motif of "neuroscience", namely the interdisciplinary merging of data acquired by anatomists, chemists and physiologists working at their preferred levels of resolution from the molecular to the organismic constitutes its own a major barrier to substantive intellectual consolidation of the data.

Neuroscience tools for organized data gathering

Paralleling the data explosion in the neurosciences comes frustration: While it is generally possible to retrieve relevant reliable information on brain molecules one may be quickly stymied for information to understand the cell systems which express these genes and then to relate those genes and cells to the pertinent behaviors governed by these cells and cell systems. One hungers for a means to perform such vertical integrations of information (from the molecular to the behavioral) in a manner that would meet rigorous scientific standards and yet permit individual scholars the intellectual opportunity to conduct investigation of the accumulated data for their own specific relationships and for hypothesis generation. In our view, several sorts of information management tools are most needed and therefore motivate our efforts in this activity.

High on our list is an integrated software system for the quantitative acquisition, display and analysis of cellular and subcellular morphological information from the microscope, in a manner that can be integrated with a textual database management system (i.e. the literature awareness library of individual or working groups of scientists) and a structural icon of the places in the brain of the species of animals to which those pieces of textual information are connected (we use the generic term "templates of atlases" to symbolize the graphical counterpart to the designated structural linkage dataset). Although the application of such a tool to structural information may be readily visualized, comparable tools are also required for the data obtained by neurochemical and neurophysiological research strategies, similarly tied to the cells and regions in which those facts are acquired.

One eventual goal for such tools is to provide the neuroscience scholar, regardless of prior experience, with access to a computer or microscope system of their selection and the capability to move from the synaptic level, through cellular, multi-cellular (like layers of specific cortical areas or nuclei of defined subcortical locations), and regional microscopic levels up to the macroscopic framework of our atlases and databases within a quantitatively accurate, and platform (i.e. computer type) independent graphic display environment.

Concentrating for purposes of illustration strictly on structural information, we are developing a set of graphical software to merge and analyze existing whole brain macro-structural data sets (section data in experimental animals and MRIs in humans) to understand the volumetric and spatial variations in defined brain macro-structures according to age, gender and experimental or health status; such software would permit the development of statistically definable volumetric and stereotaxic properties for specific macroscopically defined brain regions of interest, and allow for within- and between-species comparisons.

Since experimental verification of detailed structural and functional information on human brains is unlikely to be obtainable, we look to the non-human primate brain as a likely experimental route through which human brain scholars could access the much richer database of neuronal circuitry, chemistry and cellular function in other species. However, this approach is not fully dependable, given the growing number of situations in which monkey and rodent (or other non-primate) differ substantially; see Lewis et al. (1986, 1987, 1988) and Campbell et al. (1987) for aminergic differences between primate and rodent.

A realistic brain database

We are also actively developing software to create and distribute an interactive Brain Object Database of the nervous system oriented within species on the templates representing the "pages" (electronically speaking) of a classical brain structural atlas. Comparisons of datasets across species are made with reference to the definable homologies between brain regional structures. Our working model, is composed of many of the standard classes of objects that are encountered in Neurosciences, namely, neuroanatomic structures at various levels of resolution (from the top down: areas, regions, groups, nuclei, cells, cellular organelles and macromolecules), neurochemical objects (from gene and mRNA sequences "upwards" (in size) to proteins, organelles and the regulatory molecular machinery for intracellular metabolic maintenance and intercellular transductive signaling) neurofunctional objects (cells, synapses, receptors, transductive mechanisms including ion channels, and their interactions on membrane properties) providing for cell–cell interactions in the sense of defined circuits.

Realistic classes of generic neurons are initially encoded with the actual known details of their generalized features, and enhanced by their exceptional properties, which may be defined when determined. Such biologically based neurons may then be collected into defined assemblies of neurons which represent any of the several defined functional systems. This collection can be applied to both normal and pathological states and carried from the molecular specifications up to the behavioral levels. We also envision such a data representational system to encode other neurodata objects (classical data renditions EEG, and event-related potentials) as well as imaging modalities (MRI, PET, CAT, MEG). This Brain Object Database will be constantly extendible in the classes of objects (molecules, organelles, neurons, neuronal subtypes, etc.) and derived classes (circuits, circuit operations) based upon the latest research information. Because the classes of objects and their relationships will be linked as life-like metaphors to their biological structures, the system can be suitable not only for encoding and comparing data across levels of analysis and species, but should also be suitable for work at the theoretical level of cellular or systems simulations.

In this manner, we are working to establish a truly comprehensive database that can link graphic image sets as well as textually defined qualitative characteristics, based upon progressively accumulated and refined (and eventually quantitatively established) data. The data would range from the level of whole brains down to the level of DNA and protein sequence, along with archival reference lists to papers containing those and other data and would lend itself to other online commentary forums. Eventually one can envision an ongoing global neuroscience forum for informal and cooperative data analysis and concept formulation among those collecting the data and those hungry for data to interpret.

Attracting the users and data producers

Ultimately the users of this or some subsequent iteration of a whole brain database may decide to create an online intellectual community. Such a User Group would share their common interest in linking human brain data pertinent to neurological and psychiatric diseases with the collected wisdom deducible from experimental brain research. To build such a group of users, and more particularly to convince them that the overall database effort is sufficiently attractive to invest their own time in getting their data into a form that can be entered into the database will require a community-debated and harmoniously formulated database masterplan, and with community-accepted standard descriptors as well as standards of rigor for data inclusion.

Our experience with prior efforts to develop even primitive databases of this type for the rat (see Bloom et al., 1989) is that widespread user acceptance is essential to the effective participation of the community to get data analyzed in ways suitable for inclusion in a database, and that the only way to achieve effective and active user participation is through progressive iteration and modification with potential users.

Thus, a driving justification for establishing a Brain Object Database of realistic and modifiable objects representing real molecules in real linkages with other molecules, organelles, cells, circuits and functions is that it will be realistic enough to attract data producers and data analysts. It should take on the significance of a "deep" knowledge system in which the pieces are not just mechanically linked, but rather contain realistic intellectual connections based on defined properties. Furthermore, the Brain Object Database can lead to a shared conceptual view of the brain, and of specific brains as a databased suprastructure onto which new data, new linkages and new concepts can be superimposed/incorporated.

The necessity for user acceptance, motivation and participation also means to us that one cannot coerce users to any single form of computer platform. Thus, users should be free to use the device with which they are most comfortable, and in which they may already have substantial capital and intellectual investment. For that reason, we intend to use software that is portable across heterogeneous computer platforms, starting with the Macintosh and Windows based computer systems, and then developing it for UNIX/Motif systems. This developmental environment would make the entire database and its class structures executable on virtually any computer presently used by neuroscientists.

The database environment we envision should also provide means to integrate the efforts of many individual neuroscientists, regardless of where they work. There is no question that research institutions are already well connected via the Internet, and that having tasted this new communication, capacity has generated a boundless demand for higher speeds on new data superhighways. We envision the databases as being distributed among the neuroscientists such that database actions (such as searching for a specific combination of facts or diseases) will automatically be empowered to reach out into the worldwide network to retrieve the desired data with no more delay than if those data were already present within the network of the local research laboratory.

A more informed future?

Perhaps in this fashion, one may begin to change the information gathering habits of the neuroscientific community and allow the science to rise to even more powerful means to capture the information residing within today's data and to ask more heuristic questions tomorrow.

In addition to establishing a user-friendly "deep knowledge" database of brain information across levels of resolution (molecules to cells to systems to behaviors) and across grouped members of a species of brain, the Brain Object Database should also be operated in conjunction with an intelligent data filtering, indexing and database entry system such that the interested users can in fact be on top of current information that matches their user-defined profiles of interest and to which new interest elements can later be defined, dropped or re-assigned. Both parts of this neuroscientific communications solution are necessary to solve the current information problem. While we have only spoken here about the database tool, the development of data gathering tools cannot lag far behind.

Acknowledgment

This work is supported by HBP Grant MH52154.

References

Ahrens, P., Schleicher, A., Zilles, K. and Werner, L. (1990) Image analysis of Nissl-stained neuronal perikarya in the primary visual cortex of the rat: automatic detection and segmentation of neuronal profiles with nuclei and nucleoli. *J. Microsc.*, 157(Pt 3): 349–365.

Akbarian, S., Bunney Jr., W.E., Potkin, S.G., Wigal, S.B., Hagman, J.O., Sandman, C.A. and Jones, E.G. (1993a) Altered distribution of nicotinamide-adenine dinucleotide phosphate-diaphorase cells in frontal lobe of schizophrenics implies disturbances of cortical development. *Arch. Gen. Psychiatry*, 50: 169–177.

Akbarian, S., Vinuela, A., Kim, J.J., Potkin, S.G., Bunney Jr., W.E. and Jones, E.G. (1993b) Distorted distribution of nicotinamide-adenine dinucleotide phosphate-diaphorase neurons in temporal lobe of schizophrenics implies anomalous cortical development. *Arch. Gen. Psychiatry,* 50: 178–187.

Andreasen, N.C. (1989) Neural mechanisms of negative symptoms. *Br. J. Psychiatry*, 7(Suppl.): 93–98.

Andreasen, N., Nasrallah, H.A., Dunn, V., Olson, S.C., Grove, W.M., Ehrhardt, J.C., Coffman, J.A. and Crossett, J.H. (1986) Structural abnormalities in the frontal system in schizophrenia. A magnetic resonance imaging study. *Arch. Gen. Psychiatry*, 43: 136–144.

Beckmann, H. (1992) Temporal lobe cytoarchitectural neuropathology in schizophrenia. *Clin. Neuropharmacol.*, 15: 493A–494A.

Benes, F.M. (1988) Post-mortem structural analyses of schizophrenic brain: study designs and the interpretation of data. *Psychiatr. Dev.*, 6: 213–226.

Benes, F.M. and Bird, E.D. (1987) An analysis of the arrangement of neurons in the cingulate cortex of schizophrenic patients. *Arch. Gen. Psychiatry*, 44: 608–616.

Benes, F.M., Majocha, R., Bird, E.D. and Marotta, C.A. (1987a) Increased vertical axon numbers in cingulate cortex of schizophrenics. *Arch. Gen. Psychiatry*, 44: 1017–1021.

Benes, F.M., Matthysse, S.W., Davidson, J. and Bird, E.D. (1987b) The spatial distribution of neurons and glia in human cortex based on the poisson distribution. *Anal. Quant. Cytol. Histol.*, 9: 531–534.

Blinkov, S. and Glezer, I. (1968) *The Human Brain in Figures and Tables: A Quantitative Handbook.* Plenum, New York.

Bloom, F.E. (1990) *Databases of Brain Information. Three-Dimensional Neuroimaging.* Raven Press, New York, 273–306.

Bloom, F.E. and Melnechuk, T. (1978) New solutions for science communication problems needed now. *Trends Neurosci.*, 1: I–II.

Bloom, F.E., Young, W.G. and Kim, Y. (1989) *Brain Browser.* Academic Press, San Diego, CA.

Campbell, M.J., Lewis, D.A., Foote, S.L. and Morrison, J.H. (1987) Distribution of choline acetyltransferase-, serotonin-, dopamine-beta-hydroxylase-, tyrosine hydroxylase-immunoreactive fibers in monkey primary auditory cortex. *J. Comp. Neurol.*, 261: 209–20.

Cherniak, C. (1990) The bounded brain: toward quantitative neuroanatomy. *J. Cognitive Neurosci.*, 2: 58–68.

Collins, F. and Galas, D. (1993) A new five-year plan for the U.S. human genome project. *Science*, 262: 43–46.

Crick, F. and Jones, E.G. (1993) Backwardness of human neuroanatomy. *Nature*, 361: 109–110.

Cuticchia, A.J., Chipperfield, M.A., Porter, C.J., Kearns, W. and Pearson, P.L. (1993) Managing all those bytes: the human genome project. *Science*, 262: 47–48.

Fleischhauer, K., Zilles, K. and Schleicher, A. (1980) A revised cytoarchitectonic map of the neocortex of the rabbit (*Oryctolagus cuniculus*). *Anat. Embryol.*, 161: 121–143.

Gazzaniga, M.S. (1989) Organization of the human brain. *Science*, 245: 947–952.

Huerta, M.F., Koslow, S.H. and Leshner, A.I. (1993) The human brain project: an international resource. *Trends Neurosci.*, 16: 436–438.

Lewis, D.A., Campbell, M.J., Foote, S.L. and Morrison, J.H. (1986) The monoaminergic innervation of primate neocortex. *Hum. Neurobiol.*, 5: 181–188.

Lewis, D.A., Campbell, M.J., Foote, S.L., Goldstein, M. and Morrison, J.H. (1987) The distribution of tyrosine hydroxylase-immunoreactive fibers in primate neocortex is widespread but regionally specific. *J. Neurosci.*, 7: 279–290.

Lewis, D.A., Foote, S.L., Goldstein, M. and Morrison, J.H. (1988) The dopaminergic innervation of monkey prefrontal cortex: a tyrosine hydroxylase immunohistochemical study. *Brain Res.*, 449: 225–243.

Love, S., Burrola, P., Terry, R.D. and Wiley, C.A. (1989) Immunoelectron microscopy of Alzheimer and Pick brain tissue labelled with the monoclonal antibody Alz-50. *Neuropathol. Appl. Neurobiol.*, 15: 223–231.

Ludwig, F.C. (1993) Pathology in historical perspective. *The Pharos*, 56(2): 5–10.

Masliah, E., Iimoto, D.S., Saitoh, T., Hansen, L.A. and Terry, R.D. (1990a) Increased immunoreactivity of brain spectrin in Alzheimer disease: a marker for synapse loss? *Brain Res.*, 531: 36–44.

Masliah, E., Terry, R.D., Alford, M. and DeTeresa, R. (1990b) Quantitative immunohistochemistry of synaptophysin in human neocortex: an alternative method to estimate density of presynaptic terminals in paraffin sections. *J. Histochem. Cytochem.*, 38: 837–844.

Masliah, E., Achim, C.L., Ge, N., DeTeresa, R., Terry, R.D. and Wiley, C.A. (1992a) Spectrum of human immunodeficiency virus-associated neocortical damage. *Ann. Neurol.*, 32: 321–329.

Masliah, E., Ge, N., Achim, C.L., Hansen, L.A. and Wiley, C.A. (1992b) Selective neuronal vulnerability in HIV encephalitis. *J. Neuropathol. Exp. Neurol.*, 51: 585–593.

Masliah, E., Ge, N., Morey, M., DeTeresa, R., Terry, R.D. and Wiley, C.A. (1992c) Cortical dendritic pathology in human immunodeficiency virus encephalitis [see comments]. *Lab. Invest.*, 66: 285–291.

Morrison, J.H., Rogers, J., Scherr, S., Benoit, R. and Bloom, F.E.

(1985) Somatostatin immunoreactivity in neuritic plaques of Alzheimer's patients. *Nature*, 314: 90–92.

Morrison, J.H., Hof, P.R. and Bouras, C. (1991) An anatomic substrate for visual disconnection in Alzheimer's disease. *Ann. N. Y. Acad. Sci.*, 640: 36–43.

Oppenheim, J.S., Skerry, J.E., Tramo, M.J. and Gazzaniga, M.S. (1989) Magnetic resonance imaging morphology of the corpus callosum in monozygotic twins. *Ann. Neurol.*, 26: 100–104.

Pearson, M.L. and Söll, D. (1991) The human genome project: a paradigm for information management in the life sciences. *FASEB J.*, 5: 35–39.

Pfefferbaum, A., Zipursky, R.B., Lim, K.O., Zatz, L.M., Stahl, S.M. and Jernigan, T.L. (1988) Computed tomographic evidence for generalized sulcal and enlargement in schizophrenia. *Arch. Gen. Psychiatry*, 45: 633–640.

Rehkamper, G., Zilles, K. and Schleicher, A. (1984) A quantitative approach to cytoarchitectonics. IX. The areal pattern of the hyperstriatum ventrale in the domestic pigeon, *Columba livia* f.d. *Anat. Embryol.*, 169: 319–327.

Rehkamper, G., Zilles, K. and Schleicher, A. (1985) A quantitative approach to cytoarchitectonics. X. The areal pattern of the neostriatum in the domestic pigeon, *Columba livia* f.d. A cyto- and myeloarchitectonical study. *Anat. Embryol.*, 171: 345–355.

Rogers, J. and Morrison, J.H. (1985) Quantitative morphology and regional and laminar distributions of senile plaques in Alzheimer's disease. *J. Neurosci.*, 5: 2801–2808.

Wiley, C.A., Johnson, R.T. and Reingold, S.C. (1992) Neurological consequences of immune dysfunction: lessons from HIV infection and multiple sclerosis. *J. Neuroimmunol.*, 40: 115–119.

Subject Index